Benjamin Noack

State Estimation for Distributed Systems
with Stochastic and Set-membership Uncertainties

Karlsruhe Series on Intelligent Sensor-Actuator-Systems

Volume 14

ISAS | Karlsruhe Institute of Technology
Intelligent Sensor-Actuator-Systems Laboratory

Edited by Prof. Dr.-Ing. Uwe D. Hanebeck

State Estimation for Distributed Systems with Stochastic and Set-membership Uncertainties

by
Benjamin Noack

Dissertation, Karlsruher Institut für Technologie (KIT)
Fakultät für Informatik, 2013

Impressum

 Scientific
Publishing

Karlsruher Institut für Technologie (KIT)
KIT Scientific Publishing
Straße am Forum 2
D-76131 Karlsruhe

KIT Scientific Publishing is a registered trademark of Karlsruhe
Institute of Technology. Reprint using the book cover is not allowed.

www.ksp.kit.edu

Print on Demand 2014

ISSN ISSN 1867-3813
ISBN 978-3-7315-0124-4

State Estimation for Distributed Systems with Stochastic and Set-membership Uncertainties

zur Erlangung des akademischen Grades eines

Doktors der Ingenieurwissenschaften

von der Fakultät für Informatik
des Karlsruher Instituts für Technologie (KIT)

genehmigte

Dissertation

von

Benjamin Noack

aus Kappeln

Tag der mündlichen Prüfung: 21.01.2013

Erster Gutachter: Prof. Dr.-Ing. Uwe D. Hanebeck

Zweiter Gutachter: Dr. Simon J. Julier

Acknowledgements

Without the support, encouragement, and advice of many people, a project like this could not have been accomplished. It is due to them that the right questions have been asked, that seemingly insurmountable obstacles have become easy to overcome, and that this journey has been rich in experience and precious moments.

This thesis is the result of my research conducted at the Intelligent Sensor-Actuator-Systems lab at the Karlsruhe Institute of Technology, where I was a doctoral researcher in the Research Training Group 1194 "Self-organizing Sensor-Actuator-Networks" funded by the German Research Foundation. I wish to thank, in particular, my advisor *Uwe D. Hanebeck* for his guidance, continuous encouragement, and appreciation of my work. He gave me the opportunity to work in this highly interesting and fascinating field of research. I am especially grateful to my co-advisor *Simon J. Julier*, who made possible a fantastic research stay at the University College London. I particularly appreciate his kindness to share his office with me and his enlightening ideas. Actually, the stay in London remains an unforgettable experience.

My colleagues at the Intelligent Sensor-Actuator-Systems lab deserve special thanks and recognition for creating a pleasant and constructive atmosphere to work in. *Dietrich Brunn* and *Vesa Klumpp* inspired my enthusiasm for sets of probability densities and imprecise probabilities. I am always happy that Vesa shares my passion for liquid drum and bass. I wish to thank *Marcus Baum* and *Marc Reinhardt* for many fruitful discussions and collaborations. Many thanks go to my former and current colleagues *Evgeniya Ballmann, Frederik Beutler, Christof Chlebek, Florian Faion, Jörg Fischer, Igor Gilitschenski, Achim Hekler, Peter Krauthausen,*

Gerhard Kurz, Daniel Lyons, Ferdinand Packi, Antonia Pérez Arias, and *Patrick Ruoff* for the time we had together. The outside-lab activities with Vesa, Ferdi, Jörg, and Patrick and the Space Alert missions with Marc, Jörg, Patrick, and Gerhard remain a special memory. I also owe a debt of gratitude to the students *Michael Berg, Dominik Itte, Matthias Nagel, Nikolay Petkov*, and *Florian Pfaff*, who worked together with me on different projects. Furthermore, I would like to thank my colleagues from the research training group for all the interesting discussions and meetings.

This dissertation has benefited greatly from discussions and collaborations with many interesting people I met at conferences. I am deeply grateful to *Alessio Benavoli* for co-organizing special sessions on imprecise probabilities for many years now and sharing his inspiring ideas with me. Furthermore, I wish to sincerely thank *Joris Sijs* for his interest in my work, his invaluable suggestions and inspirations.

I would like to express my heartfelt thanks to all my friends outside academia, who have always reminded me that research is not all that life is about. In particular, I would like to name *Daniel*, the DSA hero party, and the CoC investigators. My mother *Ute*, my father *Ulli*, and my brother *Christian* receive my deepest gratitude for always believing in me and for their incredible support when the life has been tough. They always had just the right encouraging words that I needed to hear. I would not be the person I am today without their support and unconditional love. Finally, I wish to express my warmest thanks to my partner *Nicole*, who patiently put up with me while I worked many late nights and weekends. You believe in me, even when I do not, and always look at things in the right light. Your love, support, and patience allowed me to finish this thesis.

Thank you!

Karlsruhe, October 2013   Benjamin Noack

Contents

Notation

Conventions

x Scalar (lowercase)

$\underline{x}$ Vector (underlined, lowercase)

$\boldsymbol{x}, \underline{\boldsymbol{x}}$ Random variable/vector (bold, lowercase) that is precisely characterized by a single probability density function or imprecisely characterized by a set of densities

$\sim$ Random quantity $\underline{\boldsymbol{x}} \sim D$ has probability distribution D

$\hat{\underline{x}}$ Realization of $\underline{\boldsymbol{x}}$, estimate of $\underline{\boldsymbol{x}}$, or midpoint of ellipsoid

$\mathbf{A}$ Matrix (bold, uppercase)

$\mathbf{A}^{\mathrm{T}}$ Transposed matrix

$\mathbf{A}^{-1}$ Inverse matrix

$\mathbf{I}$ Identity matrix

$\mathcal{A}, \mathcal{B}, \ldots, \mathcal{Z}$ Sets (calligraphic, uppercase) in $\mathbb{R}^n$

$\mathcal{E}(\hat{\underline{x}}, \mathbf{X})$ Ellipsoidal set with midpoint $\hat{\underline{x}}$ and shape matrix $\mathbf{X}$

$\mathbb{N}$ Set of all natural numbers

$\mathbb{R}$ Set of all real numbers

$\mathbb{R}^n$ n-dimensional vector space over the field of the real numbers

$\boldsymbol{\mathcal{E}}$ Random set

$\mathcal{A} \oplus \mathcal{B}$ Minkowski sum of sets

$\mathcal{A} \cap \mathcal{B}$ Intersection of sets

$\mathcal{A} \cup \mathcal{B}$ Union of sets

$\underline{x}_k$	Quantity at time step k
$\underline{x}_{0:k}$	Sequence/set of quantities $\{\underline{x}_0, \dots, \underline{x}_k\}$
$\underline{x}_A$	Information provided by sensor node A
$\underline{x}_{A \cap B}$	Common information shared by two sensor nodes A and B
L^1	Space of integrable functions
L^2	Space of square-integrable functions
$\langle f_1, f_2 \rangle_{L^2}$	Inner product of two square-integrable functions f_1 and f_2
$\mathrm{supp}(f)$	Support of function f

Symbols Used in Estimation Algorithms

$f^{\mathrm{p}}, f^{\mathrm{e}}$	Estimated probability density of state after prediction/filtering step
f^w, f^v	Probability density of process/measurement noise
$\mathcal{X}^{\mathrm{p}}, \mathcal{X}^{\mathrm{e}}$	Predicted/filtered set of possible states
$\mathcal{X}^u, \mathcal{X}^z$	Bounding set for process/measurement errors
$\boldsymbol{\mathcal{F}}^{\mathrm{p}}, \boldsymbol{\mathcal{F}}^{\mathrm{e}}$	Set of predicted/filtered probability densities
$\boldsymbol{\mathcal{F}}^{\mathrm{T}}, \boldsymbol{\mathcal{F}}^{\mathrm{L}}$	Set of transition densities/likelihoods
$\hat{x}^{\mathrm{p}}, \hat{\underline{x}}^{\mathrm{p}}$	Prior or predicted state estimate
$\hat{x}^{\mathrm{e}}, \hat{\underline{x}}^{\mathrm{e}}$	State estimate after filtering step incorporating current measurements
$\mathbf{C}^{\mathrm{p}}, \mathbf{C}^{\mathrm{e}}$	Covariance matrix of stochastic error after prediction/filtering step
$\mathbf{X}^{\mathrm{p}}, \mathbf{X}^{\mathrm{e}}$	Shape matrix of ellipsoidal error bound after prediction/filtering step

$\hat{u}, \underline{\hat{u}}$ Scalar/vector-valued control input of process model

$\mathbf{C}^u, \mathbf{X}^u$ Covariance/shape matrix of stochastic/set-membership error affecting process model

$\hat{z}, \underline{\hat{z}}$ Scalar/vector-valued sensor measurement

$\mathbf{C}^z, \mathbf{X}^z$ Covariance/shape matrix of stochastic/set-membership error affecting measurement

Functions and Operators

$\mathcal{N}(\underline{\hat{x}}, \mathbf{C})$ Normal or Gaussian distribution with mean $\underline{\hat{x}}$ and covariance matrix $\mathbf{C}$

$\mathcal{N}(\,\cdot\,; \underline{\hat{x}}, \mathbf{C})$ Probability density of normal distribution $\mathcal{N}(\underline{\hat{x}}, \mathbf{C})$

δ Dirac delta distribution

$\mathrm{E}[\underline{x}]$ Expected value of $\underline{x}$

$\mathrm{Cov}(\underline{x})$ Variance or covariance matrix of $\underline{x}$

$\mathrm{trace}(\mathbf{A})$ Trace of matrix $\mathbf{A}$

$\det(\mathbf{A})$ Determinant of matrix $\mathbf{A}$

Glossary

MSE   Mean squared error

LMMSE   Linear minimum mean squared error

RMSE   Root mean squared error

EKF   Extended Kalman filter

CI   Covariance intersection

EMD   Exponential mixture density

w.l.o.g.   Without loss of generality

Zusammenfassung

Eine in vielen Anwendungen zu lösende Aufgabe ist die Bestimmung eines Schätzwertes bzw. –vektors für den unbekannten Zustand eines dynamischen Prozesses. Da in der Regel weder das Systemverhalten umfassend modelliert werden kann, noch äußere Störeinflüsse genau identifiziert werden können, sind sowohl die zeitliche Entwicklung des Systemzustands als auch der Zusammenhang zwischen Zustand und den zur Verfügung stehenden Messgrößen mit Unsicherheiten behaftet. Auch bei Verwendung hochpräziser Messsysteme können die auftretenden Unsicherheiten im Allgemeinen nicht vernachlässigt werden, sondern müssen systematisch in die Berechnung des Schätzwertes einbezogen werden. Die Forschungen im Bereich der Zustandsschätzung lassen sich im Wesentlichen in zwei Richtungen unterteilen: Am weitesten verbreitet ist die Betrachtung stochastischer Schätzverfahren, bei denen Fehlergrößen bekannten Wahrscheinlichkeitsverteilungen folgen. Daneben finden vor allem mengenbasierte Schätzverfahren Anwendung, welche dazu geeignet sind, unbekannte, aber amplitudenbegrenzte Fehler zu berücksichtigen.

Der Schwerpunkt dieser Arbeit liegt auf der umfassenden Untersuchung von Schätzverfahren in monolithischen, verteilten sowie dezentralen Systemen. Auftretende Unsicherheiten werden dabei systematisch einbezogen.

Simultane stochastische und mengenbasierte Zustandsschätzung:
Im Rahmen dieser Arbeit werden Schätzverfahren untersucht und hergeleitet, die eine simultane Berücksichtigung stochastischer und mengenbasierter Störgrößen ermöglichen und auf diese Weise die Vorteile beider Betrachtungsweisen vereinen können. Vor allem erlauben diese Schätzverfahren dem Anwender eine sehr flexible Modellierung verschiedenartiger Fehlerquellen.

Eine unsichere Größe, die sowohl von einem zufälligen als auch von einem unbekannten, aber amplitudenbegrenzten Fehler beeinflusst wird, kann durch eine Menge von Wahrscheinlichkeitsdichten charakterisiert werden. Im einfachsten Fall handelt es sich hierbei um die Dichte einer Normalverteilung mit einem unbekannten, aber durch eine Menge beschriebenen Mittelwert. Im Rahmen dieser Arbeit wird zunächst der Kalman-Filter-Algorithmus verallgemeinert, um beide Arten von Unsicherheiten simultan zu verarbeiten. Das entwickelte Schätzverfahren wird mit bekannten Ansätzen verglichen und auf lineare und nichtlineare Schätzprobleme angewendet. Darauf aufbauend werden weitere Schätzalgorithmen hergeleitet, die entweder die gesamte mittlere quadratische Abweichung minimieren oder es dem Anwender sogar erlauben, anhand eines kontinuierlichen Parameters die Minimierung der Varianz bzw. die der mengenbasierten Unsicherheit zu bevorzugen.

Verteilte und dezentrale Kalman-Filterung:
Neben den genannten Störeinflüssen, die direkt auf das System und die Sensoren einwirken, können sich auch aus der Informationsverarbeitung selbst zusätzliche Unsicherheiten ergeben. Insbesondere bei der Implementierung von Schätzalgorithmen in verteilten Rechnersystemen muss beachtet werden, dass Abhängigkeiten zwischen den auf verschiedenen Rechnereinheiten modellierten Fehlergrößen bestehen können. Ein zentrales Beispiel hierfür stellt ein Sensornetzwerk dar, dessen Knoten dasselbe Phänomen vermessen und jeweils lokal einen Schätzwert vorhalten. In diesem Fall sind die Schätzwerte dadurch miteinander korreliert, dass die zeitliche Entwicklung des Systemzustands in jedem Knoten durch dasselbe Prozessmodell beschrieben wird. Um die lokalen Schätzwerte zusammenzuführen, dabei jedoch ein verzerrtes Fusionsergebnis zu vermeiden, müssen die zugrundeliegenden Korrelationsmatrizen entweder fortlaufend mitberechnet oder konservativ abgeschätzt werden.

Der zentrale Ansatz besteht in der Herleitung eines mengenbasierten Informationsfilters, welches eine sehr effiziente Verarbeitung einer hohen Anzahl an Sensordaten erlaubt. Durch eine einfache Erweiterung der Fusionsmethode wird dieser Algorithmus zudem robust gegenüber unbekannten

Korrelationen. Ein interessantes Ergebnis ist, dass Abhängigkeiten zwischen mengenbasierten Fehlermodellen keinen Einfluss auf das Fusionsergebnis haben, aber dass das konservative Abschätzen der unbekannten stochastischen Abhängigkeiten in der Regel mit Hilfe mengenbasierter Methoden erfolgt.

Nichtlineare verteilte Filterung unter nichtlinearen Abhängigkeiten:
In vielen Anwendungen lassen sich System- und Messmodelle nur durch nichtlineare Funktionen beschreiben. Demzufolge können die Abhängigkeiten zwischen verteilt berechneten Schätzwerten dann nicht mehr linear durch Korrelationsmatrizen charakterisiert werden. Für die Behandlung und Charakterisierung nichtlinearer Abhängigkeiten lassen sich bisher nur wenige Ansätze benennen. Das in dieser Arbeit entwickelte Verfahren beruht daher auf der Idee, den Systemzustand in einen höherdimensionalen Raum zu transformieren, in dem die System- und Messzusammenhänge wieder linear beschrieben werden können. Nach einer solchen Transformation des Zustandsraums lassen sich die stochastischen Unsicherheiten als normalverteilt mit Hilfe sogenannter Pseudo-Gaußdichten charakterisieren und die Abhängigkeiten eindeutig durch Korrelationsmatrizen darstellen. Die Erkenntnisse aus der linearen Schätztheorie lassen sich so unmittelbar auf verteilte nichtlineare Schätzprobleme übertragen. Für weitere wesentliche Konzepte aus der linearen Schätztheorie können Verallgemeinerungen herleitet werden, die es in nichtlinearen Schätzproblemen erlauben, effizient eine große Anzahl an Sensordaten zu verarbeiten und Teilwissen über Abhängigkeiten auszunutzen.

Abstract

Many research directions and applications share the common objective of computing an estimate for the unknown state of a dynamic system. In general, the system's state can only be attributed to imprecise system and sensor models and external disturbances affecting control and sensor devices have to be reckoned with. Therefore, estimates on the time evolution of the state and the incorporation of sensor observations are subject to different sources of uncertainties. Even the deployment of highly precise sensor technology does not invalidate the necessity to systematically take uncertainties into consideration. In the field of state estimation theory, two directions have primarily been pursued in order to model uncertainties: Stochastic state estimation techniques are most widely used, where uncertain quantities are assumed to follow certain probability distributions. Besides stochastic methods, set-membership state estimation systems are employed that are most appropriate to model unknown but bounded disturbances.

This thesis lays its focus on an intensive and extensive study of state estimation in centralized, distributed, and decentralized systems with uncertainties being modeled and incorporated in a systematic fashion.

Simultaneous stochastic and set-membership state estimation:
As the first principal objective, state estimation techniques are derived that enable us to simultaneously consider stochastic and set-membership uncertainties. The derived concept allows to flexibly model different sources of estimation uncertainty, to profit from the individual advantages of stochastic and set-membership models, and to increase the reliability of the estimation results.

In the first instance, an uncertain quantity that is affected by both a random noise and an unknown but bounded error can be characterized by a set of probability densities, i.e., imprecise probabilities, instead of a single density function. In its simplest case, such a set of densities can be represented by the density of a normal distribution with an imprecise mean that is only defined by its membership to a bounded set. This thesis proposes a generalization of the standard Kalman filtering scheme that simultaneously processes stochastic and set-membership error characteristics by employing sets of Gaussian densities. An alternative version of a combined Kalman filter is attained when the gain is directly derived by minimizing the total mean squared error instead of considering the underlying densities. An additional weighting parameter enables the user to decide whether the stochastic or set-membership uncertainty shall primarily be minimized. The proposed concepts are compared with existing approaches and are applied to linear and nonlinear estimation problems.

Distributed and decentralized Kalman filtering:
Alongside with the aforementioned disturbances that directly affect the system and sensor devices, additional uncertainties may emerge from information processing itself. In particular, implementing estimation algorithms within networked systems entails the difficulty that dependencies among locally computed estimates may arise. As a leading example, a sensor network is considered that consists of a possibly massive amount of sensor nodes, each of which observes the same phenomenon and processes a local estimate of its state. Local estimates are, in general, correlated with each other, because the same state transition model is employed in each sensor node for the prediction of its local estimate. In order to prevent biased estimation results when local estimates are fused, the underlying cross-covariance matrices either have to be kept track of continually or have to be bounded conservatively.

As a key tool in distributed and decentralized estimation problems, the information filter can efficiently process large numbers of sensor measurements, and it can also be generalized to the simultaneous incorporation of stochastic and set-membership uncertainties. The conservative fusion of local estimates, in case of unknown correlations, becomes a simple convex combination in the information form. It is an interesting result that, for the purpose of treating unknown correlations, conservative fusion rules are closely linked with set-membership estimation techniques.

Nonlinear filtering under nonlinear dependencies:
In many situations, state transition and sensor models can only be defined in terms of nonlinear functions. Cross-covariance matrices are then no longer sufficient to characterize dependencies between locally computed estimates in distributed systems. Solutions towards a systematic treatment and parameterization of nonlinear dependencies are scarce. For this reason, this thesis considers state space transformations to higher-dimensional spaces where system and sensor models become linear mappings. Uncertain quantities are represented by pseudo Gaussian densities in the transformed state space, and hence dependencies can again uniquely be paremeterized by means of cross-covariance matrices. Linear fusion methodologies and the notion of covariance consistency can then directly be employed in distributed nonlinear estimation problems. In general, linear concepts like information filtering or federated Kalman filtering, where only the process noise is conservatively bounded, have counterparts in nonlinear estimation theory.

CHAPTER
1

Introduction

Overview of Chapter 1

The problem of determining an estimate of an uncertain quantity is encountered in many research directions including but not limited to the field of engineering and control. Lack of precision is an ever-present danger to countless technical and non-technical fields of application. It is an unfortunate fact that an increasing amount of accessible information does not innately enhance precision and significance—a fact that everyone recognizes in increasingly information-driven daily life. However, interconnected information sources can bring great benefits if exploited properly. Accordingly, the principal challenge addressed by this thesis is to extract meaningful information without being negligent of involved uncertainties.

Imprecise information justifies by no means precise conclusions, which must be borne in mind whenever an estimate of an uncertain quantity is to be computed. In general, an estimate is fed to decision-making or controlling systems, and, of course, an enactment of decisions under severe uncertainties faces considerable risks. Appropriate models of uncertainty can significantly contribute to ensuring reliability of estimation results and to preventing overconfident and unjustified conclusions. The design of a state estimation algorithm generally requires mathematical models of the real system, its internal state, and deployed sensor devices. In conjunction with an uncertainty model, estimation techniques can be derived that are capable of providing informative and uncertainty-aware estimates at once. As the first major focus, this thesis particularly aims at developing an estimation concept that enables us to account for a variety of different sources of uncertainty.

The aforementioned interconnectivity of information sources imposes additional challenges to the development and deployment of state estimation algorithms. For instance, information reported by two different sources may still stem from the same origin, which is possibly unknown to the state estimation system. In general, information can be interdependent for numerous reasons. To be more specific, networks of sensors and actuators have moved to the center of interest in various domains of research. Here, an estimate on an uncertain state is not necessarily computed on a single instance anymore but multiple times on autonomously operating systems. Even so, an independent computation of estimates does by no means imply that they are independent. An estimator that combines the information provided by other estimators must hence be aware of possible underlying dependencies. Essentially, it is the estimation errors that are interdependent: Each estimator may employ an uncertainty model for the same occurring errors, and for fusing their estimation results it must be taken into account that the models do not characterize different independent errors. Hidden dependencies among estimation results can also be viewed as a source of uncertainty, for which again a model can be derived. The treatment of dependent information in networked systems constitutes the second major focus of this thesis.

1.1 Modeling Model and Estimation Uncertainty

Uncertainties do not merely have their source in external disturbances affecting the system's state; they require attention from the very beginning when a model for the system is to be identified. In general, model parameters cannot be specified with precision, and sometimes there are even several models to choose from. These endogenous sources of uncertainty have influence on the entire state estimation process and cannot be left out of consideration. Multiple possibilities can be named to characterize endogenous and exogenous uncertainties, but they differ from each other in their ability to account for different types and sources of estimation errors. The variety ranges from generalizations to alternatives of classical probability theory. However, they generally have in common that uncertainty is interpreted in terms of likeliness and ignorance [107, 162].

In the field of control and engineering, essentially stochastic and set-membership error descriptions are the methods of choice [103, 111, 154, 161]. Stochastic errors are generally characterized by means of probability density functions, as illustrated in Figure 1.1(a). Possibly unbounded errors and especially outliers can easily be captured by stochastic formulations provided that precise probability densities can be stated. The error behavior must be identifiable in terms of probabilities, as otherwise the lack of precise probabilities superimposes a second type of uncertainty on the stochastic uncertainty model. More precisely, imprecise probabilities cannot yield a precise probability density for the state estimate. In contrast, set-membership error models, as indicated in Figure 1.1(b), offer the advantage that no specific error behavior needs to be pinpointed, except for boundedness. In particular, it is irrelevant whether the set represents an unknown systematic error, a bounded random error, or a combination of both. Owing to the requirement of boundedness, set-membership models do not envisage possible outliers.

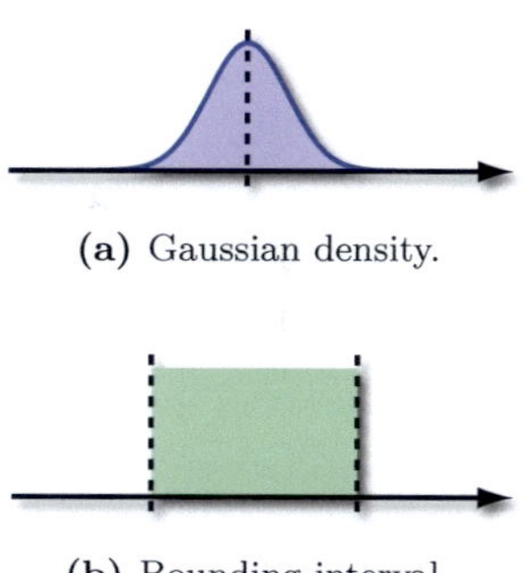

(a) Gaussian density.

(b) Bounding interval.

Figure 1.1: Examples of most-used uncertainty models in state estimation.

Characterizing possible errors by either or both likeliness and boundedness appears to be intuitive. Unfortunately, common estimation techniques compel the user to choose either one or the other uncertainty model. This thesis demonstrates that a simultaneous incorporation of stochastic and set-membership uncertainties is not more challenging than the treatment of purely stochastic or purely set-membership uncertainties. In essence, two directions can be named. The first one is a generalization of classical probability theory that enables us to characterize uncertain quantities by means of imprecise probabilities, i.e., sets of probability densities. Hence, a new, combined error model is defined. The second approach is to simply compute a point estimate by minimizing a measure that is aware of both possibly present stochastic and set-membership errors. The estimate is provided together with stochastic and set-membership error characteristics, and both types of error remain distinguishable throughout the entire state estimation process.

1.2 New Challenges in Sensor Networks

The rapid advances in sensor and communication technology entail an increasing demand for implementing estimation algorithms in distributed networked systems. Networks of sensors and actuators can monitor and control large-scale physical phenomena. Instead of the deployment of highly-accurate, high-resolution, but cost-intensive sensor technology, the advancements in the field of sensor networks reveal a clear tendency towards low-cost, energy-conserving technology that allows the usage of an immense number of sensor nodes. Sensor networks enable a wide range of applications including, inter alia, indoor localization [211], car-2-X communication [56], or forecasting of hazards [104], such as fire detection [124] and monitoring of volcanic eruptions [176]. The most ambitious goal is maybe smart dust [144] with nodes of a few cubic millimeters in size. However, the outstanding spatial resolution of a sensor network is, in general, paid with less accurate, fail-prone sensor data.

State estimation methods provide the means to extract meaningful information from the received sensor data, even in the presence of severe uncertainties. Not only the low accuracy of individual sensors poses a challenge to estimation algorithms, but also constraints on power consumption and communication require attention. Battery-driven nodes compromise their lifetime when continually transmitting sensor data, and autonomous mobile nodes can provide no warranty for reliable communication links. In order to reduce the data volume to be transferred, sensor measurements can be quantized or discretized. In consequence, the state estimator is fed with additional set-membership uncertainties, i.e., quantization and discretization errors. The communication rate can be reduced by only transmitting data when a predefined event occurs, for instance, when the value under observation varies significantly. An event-based state estimator is only aware of a certain interval, which the sensor data belongs to, as long as no new measurement arrives. These examples elucidate that efficient data processing mechanisms in networked systems may be accompanied by additional sources of uncertainties that must be incorporated in the state estimation algorithms.

In many applications, the nodes of a sensor network employ their own state estimation systems in order to be capable of operating independently. A local computation of an estimate often proves to be the best strategy as it also significantly alleviates the need to store and transmit long sequences of sensor data. Managing local estimates in distributed or decentralized systems involves additional challenges, as discussed in the following.

1.3 Distributed and Decentralized Information Processing

Distributing an estimation algorithm over a networked system potentially introduces an additional source of uncertainty. Especially if no central unit is present that can keep track of the information being shared by different nodes, it turns out to be a difficult task to amalgamate information from several nodes and simultaneously avoid biased or overconfident estimates. Without information fusion, a single local estimate is often hardly informative since, for instance, each sensor node can only cover a small area of the monitored phenomenon, as indicated in Figure 1.2. Unfortunately, dependencies that are unknown inflict a systematic error to the fusion problem and require careful consideration. For distributed network architectures, specific fusion methods can be derived that can treat dependencies among adjacent nodes in an optimal manner [71,119]. In the situation of frequently changing network topologies and especially for fully decentralized networks, bookkeeping of dependencies is by far too expensive in terms of required storage and processing power. Conservative strategies [99] then remain the only possibility to prevent fusion results from being overconfident, i.e., from underestimating the underlying uncertainty.

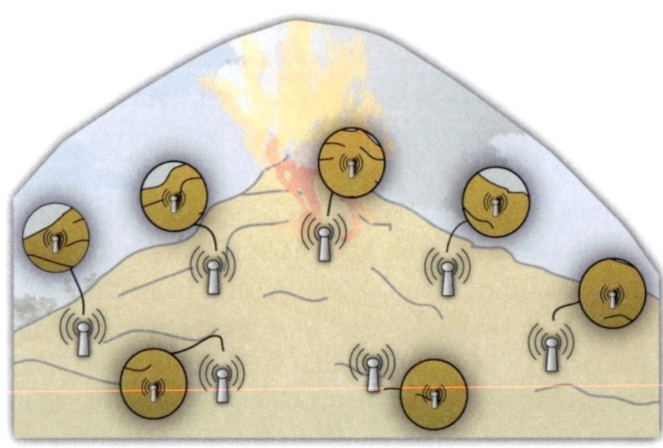

Figure 1.2: Each node observes only a part of the phenomenon, but as a sensor network they achieve a high spatial resolution.

In linear estimation problems, where state transition as well as sensor models are assumed to be linear and disturbances follow Gaussian distributions, a clear and intuitive notion of conservativeness and consistency is attainable. Dependencies do not pose a problem for every type of employed uncertainty model. The treatment of set-membership uncertainties, in particular, remains unaffected by possible dependencies, whereas, for stochastic errors, dependencies must be taken into consideration. However, they can be parameterized in terms of correlations coefficients and can be bounded conservatively. Nonlinear estimation problems, in contrast, neither allow finite-dimensional parameterizations of dependencies nor reveal an intuitive notion of conservativeness.

1.4 Outline and Overview

The aforementioned challenges are addressed by three chapters. The major focus is laid on simultaneous stochastic and set-membership state estimation. With stochastic quantities, knowledge about possible realizations of errors can be incorporated and it can be assessed how likely they are to occur. Set-membership descriptions are most appropriate to take unknown but bounded errors into account. In combination with each other, both directions of describing uncertainties enable a flexible and comprehensive modeling of different sources of estimation uncertainty. A special source of uncertainty is related to possible underlying dependencies between local estimates in distributed or decentralized estimation problems. In this case, the design of estimation algorithms need to be adapted, which is first studied against the background of linear system and sensor models. Subsequently, the insights gained are transferred to nonlinear estimation problem, where a lack of finite parameterizations of dependencies exacerbates the challenges faced by distributed and decentralized state estimation systems.

Chapter 2—Simultaneous Stochastic and Set-membership State Estimation commences with a study of different models for characterizing uncertainties in a systematic fashion. Accordingly, several possibilities are revealed to incorporate these uncertainty models into state estimation systems. Stochastic and set-membership models are most commonly used. In general, they are considered alone but not in combination with each other. Research towards combined stochastic and set-membership estimation has been conducted into different directions, which range from generalizations of classical probability theory that clearly distinguish between stochastic and set-membership uncertainties to concepts that subsume both types of uncertainty under an alternative description, such as constraints. In general, it is clearly preferable that a combined estimation concept reduces to a stochastic estimator in the absence of set-membership errors and vice versa. As already pointed out in Section 1.1, Chapter 2 systematically generalizes the Kalman filtering scheme and ends up with an estimator that reduces the overall MSE in the presence of both types of uncertainties. Furthermore, a weighting parameter allows us to favor either the minimization of the stochastic uncertainty or the minimization of the set-membership uncertainty. It is also discussed how to employ the proposed approach in nonlinear estimation problems and in specific applications.

Chapter 3—Distributed and Decentralized Kalman Filtering: Challenges and Solutions addresses the implementation of estimation algorithms in a distributed or decentralized fashion. Especially, information processing in sensor networks lies in the focus of this chapter. Even in distributed systems, where often a central unit is still available, standard formulations of state estimation algorithms may be not applicable. In the first place, reformulations of the Kalman filter are considered that allow an efficient processing of multisensor data being affecting by stochastic and set-membership errors. In the second place, the computation of local estimates on sensor nodes is examined. Since, in general, it is the same state that is observed, fusion of local estimates can eventually cause double-counting of information and may lead to overconfident results. If no knowledge about the underlying dependencies can be exploited—as it is usually the case in fully decentralized networks—only suboptimal fusion strategies can be pursued. In order to reduce conservatism, several possibilities are discussed to incorporate dependencies that can at least be identified to a certain extent. It is an interesting fact that, for the treatment of unknown dependencies, set-membership techniques are the method of choice.

Chapter 4—Nonlinear State Estimation under Nonlinear Dependencies is dedicated to the fact that distributed and decentralized state estimation requires particular attention when nonlinearities are involved. Nonlinear state estimation is in itself challenging but becomes even more intricate if possible dependencies between local estimates need to be taken into account. Unfortunately, nonlinear dependencies do not allow for simple parameterizations, and hence bookkeeping of dependencies is, in general, not an option. Accordingly, a notion of conservativeness in case of unknown dependencies is difficult to define. In some estimation problems, the state space can be transformed to a higher-dimensional state space, where the system and sensor models can be represented by linear mappings. By means of so-called pseudo Gaussian densities, dependencies then become also linear and can be parameterized in terms of correlation coefficients. Also, the definition of consistency and conservativeness can be borrowed from linear estimation theory. However, state space transformations do not simplify the estimation problem in every case. This chapter therefore discusses a general notion of conservativeness and an according fusion rule is studied that is a direct generalization of a conservative linear fusion rule. It can be shown that this rule counteracts typical sources of dependencies, i.e., common sensor data and common process noise. Fortunately, techniques can be defined to exploit partially known dependencies and hence to reduce conservatism.

CHAPTER
2

Simultaneous Stochastic and Set-membership State Estimation

Overview of Chapter 2

State estimation refers to the task of computing an estimate for the unknown state of a dynamic system. In general, neither a precise description of the underlying system dynamics is attainable nor any sensor observation is accurate. Therefore, the derivation of specific estimates from noisy measurements is of little value if the involved uncertainties are not considered appropriately. That means that imprecise information prevents precise estimation results and conclusions. Thus, a central challenge in state estimation theory is to define suitable models to express the lack of precise information. Employing uncertainty models can significantly contribute to ensuring robustness and reliability in decision and control applications, but for this purpose it is necessary to propagate and update uncertainty characterizations throughout the entire state estimation process. In a dynamic state estimation system, the uncertainty associated with the initial position of an object to be tracked, for example, also affects the position estimate at any later time and therefore needs to be propagated through the motion model.

Devoting attention to uncertainties is necessary from the early beginning when a model for the time evolution of the system's state is to be identified. For the same system, different analytic formulations with different properties may exist, and it can prove difficult to choose one among them. In this work, we investigate state space models whose parameters have generally to be inferred on the basis of observed data from the system. In the prediction phase of an estimation algorithm, a process noise term is commonly utilized to express the uncertainty concerning the identified model as well as to account for any external disturbances affecting the system. Accordingly, in the update phase of an estimator, both uncertain model and sensor properties as well as external influences on the sensor system are characterized in terms of a measurement noise.

In state estimation theory, uncertainty in most instances refers to a probabilistic description like the mean squared error. But it must be recognized that a probabilistic approach requires the selection of precise probability distributions and therefore can considerably restrict the possibilities of representing incomplete knowledge. Especially from the field of economics, it is known that a pure probabilistic description may not suffice for drawing robust conclusions. As an example, the Ellsberg paradox [55] or a similar version [105] are often cited, where balls are drawn from urns and a payout is received based on the color. Only the number of red balls is known, but except for the total number, the composition of blue and green balls remains unknown.

Example 2.1: Ellsberg's three-color urn experiment
An urn contains 90 balls. 30 balls are red and the remaining 60 balls are blue and green in an unspecified proportion. One can choose between the two

gambles A1 and A2 and receives a payout according to the drawn color:

	🔴	🟢	🔵
Gamble A1	£10	£0	£0
Gamble A2	£0	£10	£0

Most persons have shown a clear preference for Gamble A1 since there is no ambiguity involved. After this bet, a second one with different gambles is offered:

	🔴	🟢	🔵
Gamble B1	£10	£0	£10
Gamble B2	£0	£10	£10

In this situation, Gamble B2 is most often preferred, although the preference in the first bet implicitly means that $P(\{\text{green}\}) < P(\{\text{blue}\})$ is assumed. This inequality in turn implies for the second bet that Gamble B1 should be the better choice. Hence, decisions made in this experiment cannot be justified on the basis of purely stochastic assertions.

In the example, only a probability P for drawing a red ball can be stated; the probability for drawing a certain other color lies between 0 and $1 - P$. The latter case, for which no probability can be stated in advance, is often referred to as Knightian uncertainty [107], ambiguity [55], or imprecision [169]. In this and many other situations [15, 23], it can be argued that the principle of indifference, i.e., assigning equal probabilities for drawing a blue or green ball, is not applicable and can even be misleading. In the considered example, the assumption of a balanced ratio of blue and green balls can apparently be very inappropriate to compute the expected payback and to bet for an outcome if the actual urn contains an unfavorable proportion. It is up to the decision maker to assume a certain ratio in order to, for instance, reduce the risk of high loss. A decision maker should therefore not only be provided with purely probabilistic descriptions of uncertainty but also with a model that describes void information. Accordingly, the development of state estimation methods should not become an end in itself but shall rather focus on the task to provide controllers and decision makers with a sound insight into the involved uncertainties.

Errors affecting technical systems are in general categorized as being either random or systematic [70, 164]. Unfortunately, common estimation techniques have difficulty dealing with both types of errors simultaneously. *Systematic errors* are related to flaws in the measurement equipment and, in contrast to *random errors*, cannot simply be identified by repeating measurements since each observation is possibly falsified by the same error, e.g., a bias. Calibration methods are intended to counteract systematic errors, but in general they can only narrow them down to certain intervals. Hence, systematic errors are predestined to be represented by their memberships to bounding sets. The combined stochastic and set-membership error model derived in this chapter therefore appears particularly useful to account for the superposition of random and systematic measurement errors. However, at this point, it is important to emphasize that the set-membership error model is not restricted to unknown systematic errors. A bounded random error can likewise be represented as an unknown but bounded error. It is the advantage of set-membership descriptions that no knowledge about the error behavior within the bounds is required.

This chapter commences in Section 2.1 with the description of stochastic and set-membership estimation principles, which evolved almost independently of each other. As indicated above, Bayesian estimation techniques, which characterize uncertain quantities by means of random variables, are most frequently used. In particular, the Kalman filter is studied as it embodies an optimal solution to the state estimation problem when both system dynamics and measurement models are linear and perturbations are normally distributed. Besides the Bayesian branch, set-membership concepts are investigated that are most appropriate to model an unknown error behavior within certain bounds, i.e., an unknown but bounded error. Thus, these models are useful to mimic the void information we have discussed earlier. In order to profit from the individual advantages of stochastic and set-membership error representations, a combined representation is aspired, which lies in the focus of Section 2.2. In Section 2.3, the generic

Bayesian estimation scheme for sets of probability densities, i.e., imprecise probabilities, is presented that becomes a tractable generalized Kalman filter algorithm (see Section 2.4) in the case of linear system and observation models. The key result is a minimum mean squared error (MMSE) estimator under stochastic and set-membership uncertainties derived in Section 2.5. In Section 2.6 and Section 2.7, an extension to nonlinear estimation problems and a discussion of applications will conclude this chapter.

2.1 Models of Uncertainty and Related Principles of State Estimation

In state estimation theory, two directions have essentially been pursued in order to model disturbances and errors: Either uncertainties are modeled as stochastic quantities or they are characterized by their membership to a set. Both concepts have distinct advantages and disadvantages making each one inherently better suited to model different sources of estimation uncertainty. This section separately explains the principles of stochastic and set-membership estimation while particularly focusing on tractable solutions, namely the Kalman filter for the Bayesian estimation framework and ellipsoidal calculus for the set-membership estimation methodology, respectively.

2.1.1 Stochastic Models, Estimators, and in particular Kalman Filtering

In Bayesian estimation theory, uncertain quantities are regarded as stochastic variables. The unknown state of a system is either considered to be random itself or the incomplete knowledge about the potentially deterministic state is modeled probabilistically. In both cases, the state is represented by a random variable or vector $\underline{x}$ and compiles the smallest set of parameters with the help of which the system's behavior can completely be characterized. The dynamic behavior of a system is a stochastic process that essentially describes a transformation of random variables or vectors [141, 161]. In this work, we confine ourselves to discrete-time continuous-state stochastic models, which specify how the real-valued state vector $\underline{\boldsymbol{x}}_k \in \mathbb{R}^{n_x}$ evolves from a time step $k \in \mathbb{N}$ to the subsequent step $k+1$. This state vector $\underline{\boldsymbol{x}}_k$ is related to $\underline{\boldsymbol{x}}_{k+1}$ according to the state transition model

$$\underline{\boldsymbol{x}}_{k+1} = \underline{a}_k(\underline{\boldsymbol{x}}_k, \hat{\underline{u}}_k, \underline{\boldsymbol{w}}_k) \,, \tag{2.1}$$

where $\underline{a}_k : \mathbb{R}^{n_x} \times \mathbb{R}^{n_u} \times \mathbb{R}^{n_w} \to \mathbb{R}^{n_x}$ is a possibly nonlinear and time-variant function, $\hat{\underline{u}}_k \in \mathbb{R}^{n_u}$ denotes a control-input vector, and $\underline{\boldsymbol{w}}_k$ is an n_w-dimensional process noise.

A second possibly time-variant model $\underline{h}_k : \mathbb{R}^{n_x} \times \mathbb{R}^{n_v} \to \mathbb{R}^{n_z}$ characterizes the relationship of an obtained observation $\hat{\underline{z}}_k \in \mathbb{R}^{n_z}$ to the state $\underline{\boldsymbol{x}}_k$. The measurement $\hat{\underline{z}}_k$ is a realization of

$$\underline{\boldsymbol{z}}_k = \underline{h}_k(\underline{\boldsymbol{x}}_k, \underline{\boldsymbol{v}}_k) \,, \tag{2.2}$$

where $\underline{\boldsymbol{v}}_k$ denotes an n_v-dimensional measurement noise, which embraces any uncertainties associated to the sensor device. In many modern applications, multiple possibly heterogeneous sensor systems are employed that require the derivation of individual measurement models (2.2).

Throughout this work, $\{\underline{\boldsymbol{w}}_k\}_{k \in \mathbb{N}}$ and $\{\underline{\boldsymbol{v}}_k\}_{k \in \mathbb{N}}$ represent white noise terms [141], i.e., $\underline{\boldsymbol{w}}_k$ is independent[1] from $\underline{\boldsymbol{w}}_l$, and $\underline{\boldsymbol{v}}_k$ is independent from $\underline{\boldsymbol{v}}_l$ for every k and $l \neq k$. Furthermore, $\underline{\boldsymbol{w}}_k$ and $\underline{\boldsymbol{v}}_l$ are assumed to be independent from each other for every k and l.

A Bayesian State Estimation

The key challenge to be tackled by a Bayesian estimator is to recursively estimate the probability density of $\underline{\boldsymbol{x}}_k$ conditioned on inputs and received measurements [87]. At a time step k, all input quantities are stacked together into a single vector $\hat{\underline{u}}_k$, but instead of a single measurement vector, different measurements are collected in a set $\mathcal{Z}_k = \{\hat{\underline{z}}_k^1, \ldots, \hat{\underline{z}}_k^N\}$ in order

---

[1]Two random variables $\underline{a}$ and $\underline{b}$ are independent, iff $f(\underline{a}, \underline{b}) = f(\underline{a}) \cdot f(\underline{b})$ holds for the joint density.

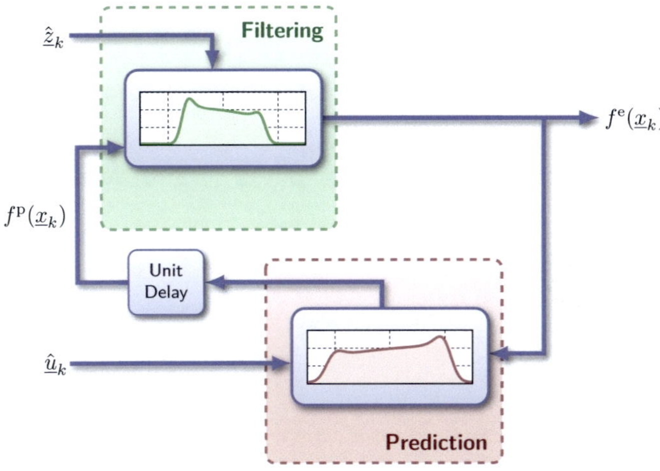

Figure 2.1: Bayesian estimation scheme consisting of prediction and filtering steps.

to emphasize that they may belong to different sensor devices. A current estimate on the state is expressed in terms of the conditional density

$$f^{\mathrm{e}}(\underline{x}_k) := f(\underline{x}_k \,|\, \mathcal{Z}_{0:k}, \hat{\underline{u}}_{0:k-1})$$
$$= f(\underline{x}_k \,|\, \mathcal{Z}_0, \ldots, \mathcal{Z}_k, \hat{\underline{u}}_0, \ldots, \hat{\underline{u}}_{k-1}) \,, \tag{2.3}$$

which can then be predicted and updated with further input and measurement information. Beginning from prior information on the initial state $\underline{x}_0$ represented by a probability density function $f^{\mathrm{p}}(\underline{x}_0) = f(\underline{x}_0)$, the estimator recursively computes the density (2.3), where predicted or prior densities are marked by the superscript p and posterior filtered densities by the superscript e. Figure 2.1 displays an overview of the Bayesian estimation scheme for conditional probability densities.

Prediction For the prediction of the current state estimate, a transition density has to be derived from the state transition model (2.1). For fixed $\underline{x}_k$

and $\underline{w}_k$, the outcome of $\underline{\boldsymbol{x}}_{k+1}$ in (2.1) is deterministic and can be expressed by

$$f(\underline{x}_{k+1} \,|\, \underline{x}_k, \hat{\underline{u}}_k, \underline{w}_k) = \delta\big(\underline{x}_{k+1} - \underline{a}_k(\underline{x}_k, \hat{\underline{u}}_k, \underline{w}_k)\big) \,,$$

where δ denotes the n_x-dimensional Dirac delta function. The transition density is obtained by marginalizing[2] out $\underline{w}_k$, which yields

$$\begin{aligned}
f(\underline{x}_{k+1} \,|\, \underline{x}_k, \hat{\underline{u}}_k) &= \int_{\mathbb{R}^{n_w}} f(\underline{x}_{k+1}, \underline{w}_k \,|\, \underline{x}_k, \hat{\underline{u}}_k) \,\mathrm{d}\underline{w}_k \\
&= \int_{\mathbb{R}^{n_w}} \delta\big(\underline{x}_{k+1} - \underline{a}_k(\underline{x}_k, \hat{\underline{u}}_k, \underline{w}_k)\big) \cdot f^w(\underline{w}_k) \,\mathrm{d}\underline{w}_k \,.
\end{aligned} \tag{2.4}$$

The predicted density f^{P} of the state is finally obtained by the Chapman-Kolmogorov integral

$$\begin{aligned}
f^{\mathrm{P}}(\underline{x}_{k+1}) &:= f(\underline{x}_{k+1} \,|\, \mathcal{Z}_{0:k}, \hat{\underline{u}}_{0:k}) \\
&= \int_{\mathbb{R}^{n_x}} f(\underline{x}_{k+1} \,|\, \underline{x}_k, \hat{\underline{u}}_k) \cdot f^{\mathrm{e}}(\underline{x}_k) \,\mathrm{d}\underline{x}_k \,,
\end{aligned} \tag{2.5}$$

where $\underline{x}_k$ is marginalized out of the joint density $f^{\mathrm{P}}(\underline{x}_{k+1}, \underline{x}_k)$.

Filtering Incoming measurement information is incorporated into the density f^{P} by means of Bayes' theorem for probability densities

$$f^{\mathrm{e}}(\underline{x}_k) = \frac{f(\mathcal{Z}_k \,|\, \underline{x}_k) \cdot f^{\mathrm{P}}(\underline{x}_k)}{\int_{\mathbb{R}^{n_x}} f(\mathcal{Z}_k \,|\, \underline{x}_k) \cdot f^{\mathrm{P}}(\underline{x}_k) \,\mathrm{d}\underline{x}_k} \,, \tag{2.6}$$

where the denominator is a normalization constant. The likelihood $f(\mathcal{Z}_k \,|\, \underline{x}_k)$ is to be derived from the model (2.2) of the sensor system. It is generally assumed that the measurement noise across multiple sensors is independent, and therefore the measurements conditioned on the current state are also independent, i.e., we obtain a product

$$\begin{aligned}
f(\mathcal{Z}_k \,|\, \underline{x}_k) &= f(\hat{\underline{z}}_k^1, \dots, \hat{\underline{z}}_k^N \,|\, \underline{x}_k) \\
&= f(\hat{\underline{z}}_k^1 \,|\, \underline{x}_k) \cdot \,\dots\, \cdot f(\hat{\underline{z}}_k^N \,|\, \underline{x}_k)
\end{aligned} \tag{2.7}$$

---

[2]We assume that all integrals exist, and, where necessary, the delta distribution is applicable.

of likelihoods. Hence, the fusion rule (2.6) can sequentially be applied to each measurement in $\mathcal{Z}_k$ and it suffices to consider the single-sensor case $\mathcal{Z}_k = \{\hat{\underline{z}}_k\}$ in the following. The particular challenges of multisensor estimation problems lie in the focus of chapters 3 and 4. The measurement equation (2.2) can be expressed in terms of the Dirac delta function by

$$f(\hat{\underline{z}}_k \,|\, \underline{x}_k, \underline{v}_k) = \delta\big(\hat{\underline{z}}_k - \underline{h}_k(\underline{x}_k, \underline{v}_k)\big) \ .$$

Analogously to (2.4), the likelihood for $\hat{\underline{z}}_k$ is hence given by

$$
\begin{aligned}
f(\hat{\underline{z}}_k \,|\, \underline{x}_k) &= \int_{\mathbb{R}^{n_v}} f(\hat{\underline{z}}_k \,|\, \underline{x}_k, \underline{v}_k) \cdot f^v(\underline{v}_k) \, \mathrm{d}\underline{v}_k \\
&= \int_{\mathbb{R}^{n_v}} \delta\big(\hat{\underline{z}}_k - \underline{h}_k(\underline{x}_k, \underline{v}_k)\big) \cdot f^v(\underline{v}_k) \, \mathrm{d}\underline{v}_k \ ,
\end{aligned}
\tag{2.8}
$$

which can then be deployed to update the prior density f^{P} by means of (2.6).

Additive Noise Models In many technical applications, the random noise terms in (2.1) and (2.2) represent the sum of many individual and independent disturbances, which externally affect the nonlinearities $\underline{a}_k$ and $\underline{h}_k$. In consequence, the noise terms are, in general, modeled to be, first, additive and, second, normally distributed. As stated in the central limit theorem [161], the latter property owes to the fact that the sum of independent random noise terms tends towards a normal random vector regardless of the probability densities of the individual noise terms. On the basis of the first considerations, the models are simplified to

$$\underline{x}_{k+1} = \underline{a}_k(\underline{x}_k, \hat{\underline{u}}_k) + \underline{w}_k$$

and

$$\underline{z}_k = \underline{h}_k(\underline{x}_k) + \underline{v}_k \ , \tag{2.9}$$

where $\underline{w}_k$ and $\underline{v}_k$ are additive zero-mean white noise terms with probability densities f^w and f^v, respectively. With additive noise, the sifting property of the Dirac delta function can be exploited, i.e., the transition density (2.4) then becomes

$$f(\underline{x}_{k+1} \mid \underline{x}_k, \hat{\underline{u}}_k) = \int_{\mathbb{R}^{n_w}} \delta\big(\underline{x}_{k+1} - \underline{a}_k(\underline{x}_k, \hat{\underline{u}}_k) - \underline{w}_k\big) \cdot f^w(\underline{w}_k) \, \mathrm{d}\underline{w}_k \\ = f^w(\underline{x}_{k+1} - \underline{a}_k(\underline{x}_k, \hat{\underline{u}}_k)) \tag{2.10}$$

and the likelihood (2.8) reads

$$f(\hat{\underline{z}}_k \mid \underline{x}_k) = \int_{\mathbb{R}^{n_v}} \delta\big(\hat{\underline{z}}_k - \underline{h}_k(\underline{x}_k) - \underline{v}_k\big) \cdot f^v(\underline{v}_k) \, \mathrm{d}\underline{v}_k \\ = f^v(\hat{\underline{z}}_k - \underline{h}_k(\underline{x}_k)) \ . \tag{2.11}$$

The second aforementioned assumption that $\underline{w}_k$ and $\underline{v}_k$ are normally distributed further simplifies (2.10) and (2.11), but, despite these simplifications, the generic Bayesian inference scheme, i.e., (2.5) for prediction and (2.6) for filtering, generally suffers from a lack of practical applicability due to the nonlinearities $\underline{a}_k$ and $\underline{h}_k$. For instance, the integral (2.5) has to evaluated at every point $\underline{x}_{k+1}$ and complicated densities need to be approximated and parameterized. This, of course, indicates a strong need for a closed-form solution.

B   The Kalman Filter

The generic Bayesian inference scheme operates on the underlying probability densities and features the ability of providing an optimal solution to the estimation problem. Unfortunately, this scheme is only of conceptual value since, in general, the absence of finite parameterizations prevents efficient and closed-form calculations of the densities. However, the most notable exception is given by the *Kalman filter* formulas [101, 103] in the case of linear system dynamics

$$\underline{x}_{k+1} = \mathbf{A}_k \, \underline{x}_k + \mathbf{B}_k(\hat{\underline{u}}_k + \underline{w}_k) \tag{2.12}$$

and linear sensor models

$$\underline{z}_k = \mathbf{H}_k \, \underline{x}_k + \underline{v}_k \ , \tag{2.13}$$

where $\underline{\boldsymbol{w}}_k \sim \mathcal{N}(\underline{0}, \mathbf{C}_k^u)$ and $\underline{\boldsymbol{v}}_k \sim \mathcal{N}(\underline{0}, \mathbf{C}_k^z)$ are zero-mean white Gaussian perturbations. By $\mathcal{N}(\underline{\hat{x}}, \mathbf{C})$, the normal distribution with mean $\underline{\hat{x}}$ and covariance matrix $\mathbf{C}$ is denoted, and it has the probability density function

$$\mathcal{N}(\underline{x}; \underline{\hat{x}}, \mathbf{C}) := \frac{1}{(2\pi)^{\frac{n_x}{2}} \det(\mathbf{C})^{\frac{1}{2}}} \exp\left(-\frac{1}{2}(\underline{x} - \underline{\hat{x}})^{\mathrm{T}} \mathbf{C}^{-1} (\underline{x} - \underline{\hat{x}})\right). \quad (2.14)$$

If the prior density $f^{\mathrm{p}}(\underline{x}_0) = \mathcal{N}(\underline{x}_0; \underline{\hat{x}}_0^{\mathrm{p}}, \mathbf{C}_0^{\mathrm{p}})$ is Gaussian, the predicted and posterior densities (2.5) and (2.6) remain Gaussian as well. Conditional mean $\underline{\hat{x}}_k^{\mathrm{e}}$ and covariance matrix $\mathbf{C}_k^{\mathrm{e}}$, which uniquely characterize the posterior Gaussian density, can then be calculated in closed form, which yields the Kalman filter formulas.

However, in most derivations of the Kalman filtering scheme, mean $\underline{\hat{x}}_k^{\mathrm{e}}$ and covariance matrix $\mathbf{C}_k^{\mathrm{e}}$ are less regarded as the parameters of a Gaussian density but rather characterize a point estimate and its corresponding *mean squared error* (MSE) matrix [116]. Given the observations $\underline{\hat{z}}_{0:k}$, the objective in this case is to compute an estimate $\underline{\hat{x}}_k^{\mathrm{e}} = \underline{\hat{x}}(\underline{\hat{z}}_{0:k})$ that minimizes the MSE

$$\begin{aligned}
\mathrm{E}[(\underline{\hat{x}}_k^{\mathrm{e}} - \underline{\boldsymbol{x}}_k)^{\mathrm{T}} &(\underline{\hat{x}}_k^{\mathrm{e}} - \underline{\boldsymbol{x}}_k) \,|\, \underline{\hat{z}}_{0:k}] \\
&= \int_{\mathbb{R}^{n_x}} (\underline{\hat{x}}_k^{\mathrm{e}} - \underline{x}_k)^{\mathrm{T}} (\underline{\hat{x}}_k^{\mathrm{e}} - \underline{x}_k) \cdot f(\underline{x} \,|\, \underline{\hat{z}}_{0:k}) \; \mathrm{d}\underline{x}_k \,,
\end{aligned} \quad (2.15)$$

which is more precisely the a posteriori MSE. By setting $\underline{\hat{x}}_k^{\mathrm{e}} - \underline{\boldsymbol{x}}_k = \underline{\hat{x}}_k^{\mathrm{e}} - \underline{c} + \underline{c} - \underline{\boldsymbol{x}}_k$, this conditional MSE becomes

$$\begin{aligned}
\mathrm{E}[(\underline{\hat{x}}_k^{\mathrm{e}} &- \underline{\boldsymbol{x}}_k)^{\mathrm{T}} (\underline{\hat{x}}_k^{\mathrm{e}} - \underline{\boldsymbol{x}}_k) \,|\, \underline{\hat{z}}_{0:k}] \\
&= (\underline{\hat{x}}_k^{\mathrm{e}} - \underline{c})^{\mathrm{T}} (\underline{\hat{x}}_k^{\mathrm{e}} - \underline{c}) + (\underline{\hat{x}}_k^{\mathrm{e}} - \underline{c})^{\mathrm{T}} (\underline{c} - \mathrm{E}[\underline{\boldsymbol{x}}_k \,|\, \underline{\hat{z}}_{0:k}]) \\
&\quad + (\underline{c} - \mathrm{E}[\underline{\boldsymbol{x}}_k \,|\, \underline{\hat{z}}_{0:k}])^{\mathrm{T}} (\underline{\hat{x}}_k^{\mathrm{e}} - \underline{c}) + \mathrm{E}[(\underline{c} - \underline{\boldsymbol{x}}_k)^{\mathrm{T}} (\underline{c} - \underline{\boldsymbol{x}}_k) \,|\, \underline{\hat{z}}_{0:k}]
\end{aligned}$$

which reduces to

$$\begin{aligned}
\mathrm{E}[(\underline{\hat{x}}_k^{\mathrm{e}} - \underline{\boldsymbol{x}}_k)^{\mathrm{T}} (\underline{\hat{x}}_k^{\mathrm{e}} - \underline{\boldsymbol{x}}_k) \,|\, \underline{\hat{z}}_{0:k}] &= (\underline{\hat{x}}_k^{\mathrm{e}} - \underline{c})^{\mathrm{T}} (\underline{\hat{x}}_k^{\mathrm{e}} - \underline{c}) \\
&\quad + \mathrm{E}[(\underline{c} - \underline{\boldsymbol{x}}_k)^{\mathrm{T}} (\underline{c} - \underline{\boldsymbol{x}}_k) \,|\, \underline{\hat{z}}_{0:k}]
\end{aligned}$$

for $\underline{c} := \mathrm{E}[\underline{\boldsymbol{x}}_k \,|\, \underline{\hat{z}}_{0:k}]$. We can now easily accept that (2.15) becomes minimal for the conditional mean

$$\underline{\hat{x}}_k^{\mathrm{e}} := \underline{c} = \mathrm{E}[\underline{\boldsymbol{x}}_k \,|\, \underline{\hat{z}}_{0:k}] \,.$$

Because of

$$\mathrm{E}[\hat{\underline{x}}_k^{\mathrm{e}}] = \mathrm{E}\left[\mathrm{E}[\underline{x}_k \mid \underline{z}_{0:k}]\right] = \mathrm{E}[\underline{x}_k] , \tag{2.16}$$

the estimator $\hat{\underline{x}}_k^{\mathrm{e}} = \mathrm{E}[\underline{x}_k \mid \cdot\,]$ is unbiased and minimizes (2.15) for each particular value $\underline{z}_{0:k}$. This means it minimizes the unconditional MSE

$$\mathrm{MSE}(\hat{\underline{x}}_k^{\mathrm{e}}) = \mathrm{E}[(\hat{\underline{x}}_k^{\mathrm{e}} - \underline{x}_k)^{\mathrm{T}}(\hat{\underline{x}}_k^{\mathrm{e}} - \underline{x}_k)] \tag{2.17}$$

due to equation

$$\mathrm{E}[(\hat{\underline{x}}_k^{\mathrm{e}} - \underline{x}_k)^{\mathrm{T}}(\hat{\underline{x}}_k^{\mathrm{e}} - \underline{x}_k)] = \mathrm{E}\left[\mathrm{E}[(\hat{\underline{x}}_k^{\mathrm{e}} - \underline{x}_k)^{\mathrm{T}}(\hat{\underline{x}}_k^{\mathrm{e}} - \underline{x}_k) \mid \underline{z}_{0:k}]\right]$$

$$= \int_{\mathbb{R}^{n_z}} \cdots \int_{\mathbb{R}^{n_z}} \mathrm{E}[(\hat{\underline{x}}_k^{\mathrm{e}} - \underline{x}_k)^{\mathrm{T}}(\hat{\underline{x}}_k^{\mathrm{e}} - \underline{x}_k) \mid \underline{z}_{0:k}]\, f(\underline{z}_{0:k})\ \mathrm{d}\underline{z}_{0:k}$$

and nonnegativity of $f(\underline{z}_{0:k})$. It is important to notice that, in the considered linear and Gaussian case, the conditional MSE (2.15) is the same for all possible realizations $\hat{\underline{z}}_0, \ldots, \hat{\underline{z}}_k$. As a significant result, the conditional MSE (2.15) is then equal to the unconditional MSE (2.17).

Alongside the estimate $\hat{\underline{x}}_k^{\mathrm{e}}$, the Kalman filter algorithm recursively computes the MSE matrix[3]

$$\mathbf{C}_k^{\mathrm{e}} := \mathrm{E}[(\hat{\underline{x}}_k^{\mathrm{e}} - \underline{x}_k)(\hat{\underline{x}}_k^{\mathrm{e}} - \underline{x}_k)^{\mathrm{T}}] \in \mathbb{R}^{n_x \times n_x} , \tag{2.18}$$

whose trace exactly yields the MSE (2.17). Because of the unbiasedness (2.16), i.e., $\mathrm{E}[\hat{\underline{x}}_k^{\mathrm{e}} - \underline{x}] = 0$, the matrix (2.18) also represents the covariance matrix $\mathrm{Cov}(\hat{\underline{x}}_k^{\mathrm{e}} - \underline{x}_k)$ for the error $(\hat{\underline{x}}_k^{\mathrm{e}} - \underline{x}_k)$ and has therefore been denoted by $\mathbf{C}_k^{\mathrm{e}}$. Although it can also be expressed in terms of a single equation, the Kalman filter algorithm is composed of a time update (*prediction*) and a measurement update (*filtering*) phase, as shown in the schematic overview provided by Figure 2.2.

Prediction   The prediction step employs the system model (2.12) in order to compute the predicted state estimate

$$\hat{\underline{x}}_{k+1}^{\mathrm{p}} := \mathrm{E}[\underline{x}_{k+1} \mid \hat{\underline{z}}_{0:k}]$$

$$= \mathrm{E}[\mathbf{A}_k\,\underline{x}_k + \mathbf{B}_k(\hat{\underline{u}}_k + \underline{w}_k) \mid \hat{\underline{z}}_{0:k}] \tag{2.19}$$

$$= \mathbf{A}_k\,\hat{\underline{x}}_k^{\mathrm{e}} + \mathbf{B}_k\,\hat{\underline{u}}_k$$

---

[3]This simple definition can only be used for the linear and Gaussian case. For the general (e.g., non-Gaussian) case, it has to be $\mathbf{C}_k^{\mathrm{e}} := \mathrm{E}[(\hat{\underline{x}}_k^{\mathrm{e}} - \underline{x}_k)(\hat{\underline{x}}_k^{\mathrm{e}} - \underline{x}_k)^{\mathrm{T}} \mid \underline{z}_{0:k}]$.

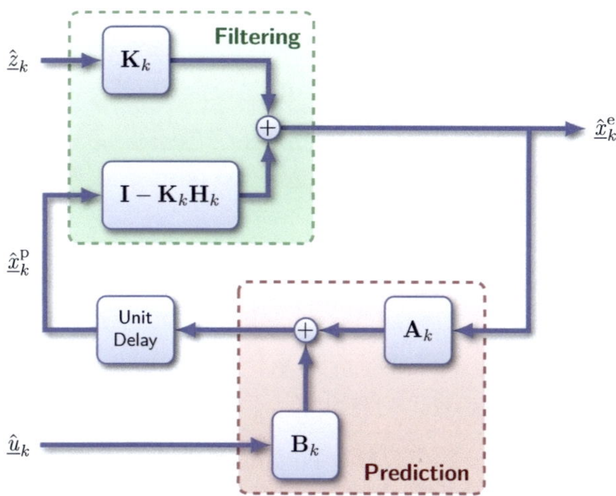

Figure 2.2: The Kalman filter is a recursive estimation scheme consisting of prediction and filtering steps.

and its corresponding MSE matrix

$$
\begin{aligned}
\mathbf{C}_{k+1}^{\mathrm{P}} &= \mathrm{E}[(\hat{\underline{x}}_{k+1}^{\mathrm{P}} - \underline{x}_{k+1})(\hat{\underline{x}}_{k+1}^{\mathrm{P}} - \underline{x}_{k+1})^{\mathrm{T}}] \\
&= \mathrm{E}[(\mathbf{A}_k(\hat{\underline{x}}_k^{\mathrm{e}} - \underline{x}_k) + \mathbf{B}_k\underline{w}_k)(\mathbf{A}_k(\hat{\underline{x}}_k^{\mathrm{e}} - \underline{x}_k) + \mathbf{B}_k\underline{w}_k)^{\mathrm{T}}] \qquad (2.20) \\
&= \mathbf{A}_k\mathbf{C}_k^{\mathrm{e}}\mathbf{A}_k^{\mathrm{T}} + \mathbf{B}_k\mathbf{C}_k^{u}\mathbf{B}_k^{\mathrm{T}}\ ,
\end{aligned}
$$

where the independence of $\underline{w}_k$ from the estimation error $(\hat{\underline{x}}_k^{\mathrm{e}} - \underline{x}_k)$ at time step k has been exploited.

Filtering In the filtering step, the prior or predicted estimate $\hat{\underline{x}}_k^{\mathrm{p}}$ and a received measurement $\hat{\underline{z}}_k$ are linearly combined in order to obtain an updated estimate

$$
\hat{\underline{x}}_k^{\mathrm{e}} = \mathbf{K}_1\,\hat{\underline{x}}_k^{\mathrm{p}} + \mathbf{K}_2\,\hat{\underline{z}}_k\ .
$$

In the first place, we figure out that we, of course, have to require the linear estimator to be unbiased: As for the derivation of (2.16), a filtering result that introduces a bias $\Delta\underline{x}_k^{\mathrm{e}}$, i.e.,

$$
\bar{\underline{x}}_k^{\mathrm{e}} = \hat{\underline{x}}_k^{\mathrm{e}} + \Delta\underline{x}_k^{\mathrm{e}}\ ,
$$

leads to the MSE

$$\mathrm{E}[(\bar{\hat{x}}_k^{\mathrm{e}} - \underline{x}_k)^{\mathrm{T}}(\bar{\hat{x}}_k^{\mathrm{e}} - \underline{x}_k)] = \mathrm{E}[(\hat{\underline{x}}_k^{\mathrm{e}} - \underline{x}_k)^{\mathrm{T}}(\hat{\underline{x}}_k^{\mathrm{e}} - \underline{x}_k)] + (\Delta\underline{x}_k^{\mathrm{e}})^{\mathrm{T}}(\Delta\underline{x}_k^{\mathrm{e}})$$

where $(\Delta\underline{x}_k^{\mathrm{e}})^{\mathrm{T}}(\Delta\underline{x}_k^{\mathrm{e}})$ is a positive constant. Hence, a biased estimator cannot provide a lower MSE than an unbiased estimator. The unbiasedness

$$\begin{aligned}
0 &= \mathrm{E}\left[\hat{\underline{x}}_k^{\mathrm{e}} - \mathrm{E}[\underline{x}_k \,|\, \hat{\underline{z}}_{0:k}]\right] \\
&= \mathrm{E}[\mathbf{K}_1\,\hat{\underline{x}}_k^{\mathrm{p}} + \mathbf{K}_2\,\hat{\underline{z}}_k - \underline{x}_k] \\
&= \mathrm{E}[\mathbf{K}_1(\hat{\underline{x}}_k^{\mathrm{p}} - \underline{x}_k) + \mathbf{K}_1\,\underline{x}_k + \mathbf{K}_2(\mathbf{H}_k\,\underline{x}_k + \underline{v}_k) - \underline{x}_k] \\
&= \mathrm{E}[\mathbf{K}_1\,\underline{x}_k + \mathbf{K}_2\,\mathbf{H}_k\,\underline{x}_k - \underline{x}_k] \\
&= (\mathbf{K}_1 + \mathbf{K}_2\mathbf{H}_k - \mathbf{I})\,\mathrm{E}[\underline{x}_k]\;,
\end{aligned}$$

then implies that $\mathbf{K}_1 = \mathbf{I} - \mathbf{K}_2\mathbf{H}_k$. Let $\mathbf{K}_k := \mathbf{K}_2$ in the following. In the second place, the trace of the MSE matrix

$$\begin{aligned}
\mathbf{C}_k^{\mathrm{e}} &= \mathrm{E}[(\hat{\underline{x}}_k^{\mathrm{e}} - \underline{x}_k)(\hat{\underline{x}}_k^{\mathrm{e}} - \underline{x}_k)^{\mathrm{T}}] \\
&= \mathrm{E}\left[\left((\mathbf{I} - \mathbf{K}_k\mathbf{H}_k)\hat{\underline{x}}_k^{\mathrm{p}} + \mathbf{K}_k\hat{\underline{z}}_k - \underline{x}_k\right)(\ldots)^{\mathrm{T}}\right] \\
&= \mathrm{E}\left[\left((\mathbf{I} - \mathbf{K}_k\mathbf{H}_k)\hat{\underline{x}}_k^{\mathrm{p}} + \mathbf{K}_k(\mathbf{H}_k\,\underline{x}_k + \underline{v}_k) - \underline{x}_k\right)(\ldots)^{\mathrm{T}}\right] \quad (2.21) \\
&= \mathrm{E}\left[\left((\mathbf{I} - \mathbf{K}_k\mathbf{H}_k)(\hat{\underline{x}}_k^{\mathrm{p}} - \underline{x}_k) + \mathbf{K}_k\,\underline{v}_k)\right)(\ldots)^{\mathrm{T}}\right] \\
&= (\mathbf{I} - \mathbf{K}_k\mathbf{H}_k)\mathbf{C}_k^{\mathrm{p}}(\mathbf{I} - \mathbf{K}_k\mathbf{H}_k)^{\mathrm{T}} + \mathbf{K}_k\mathbf{C}_k^{z}\mathbf{K}_k^{\mathrm{T}}
\end{aligned}$$

is to be minimized, where the measurement noise $\underline{v}_k$ is independent of the prediction error $(\hat{\underline{x}}_k^{\mathrm{p}} - \underline{x}_k)$. In order to compute the optimal gain $\mathbf{K}_k$, the derivative of the trace of $\mathbf{C}_k^{\mathrm{e}}$ is considered. More precisely, by applying the derivative rules [143]

$$\frac{\partial}{\partial\mathbf{K}_k}\,\mathrm{trace}\left(\mathbf{K}_k\mathbf{A}\right) = \mathbf{A}^{\mathrm{T}}\;, \quad \frac{\partial}{\partial\mathbf{K}_k}\,\mathrm{trace}\left(\mathbf{A}\mathbf{K}_k^{\mathrm{T}}\right) = \mathbf{A} \quad (2.22)$$

and

$$\frac{\partial}{\partial\mathbf{K}_k}\,\mathrm{trace}\left(\mathbf{K}_k\mathbf{A}\mathbf{K}_k^{\mathrm{T}}\right) = \mathbf{K}_k(\mathbf{A}^{\mathrm{T}} + \mathbf{A}) \quad (2.23)$$

for the trace to (2.21) and setting the derivative to zero, the optimal Kalman gain can finally be derived as

$$\mathbf{K}_k = \mathbf{C}_k^{\mathrm{p}}\mathbf{H}_k^{\mathrm{T}}(\mathbf{C}_k^{z} + \mathbf{H}_k\mathbf{C}_k^{\mathrm{p}}\mathbf{H}_k^{\mathrm{T}})^{-1}\;. \quad (2.24)$$

The second derivative is a positive definite Hessian matrix, and hence the MSE attains a local minimum. The final filtering result consists of the estimate

$$\hat{\underline{x}}_k^e = \hat{\underline{x}}_k^p + \mathbf{K}_k(\hat{\underline{z}}_k - \mathbf{H}_k\hat{\underline{x}}_k^p)$$
$$= (\mathbf{I} - \mathbf{K}_k\mathbf{H}_k)\hat{\underline{x}}_k^p + \mathbf{K}_k\,\hat{\underline{z}}_k \tag{2.25}$$

and error covariance matrix

$$\mathbf{C}_k^e = (\mathbf{I} - \mathbf{K}_k\mathbf{H}_k)\mathbf{C}_k^p(\mathbf{I} - \mathbf{K}_k\mathbf{H}_k)^{\mathrm{T}} + \mathbf{K}_k\mathbf{C}_k^z\mathbf{K}_k^{\mathrm{T}}$$
$$= \mathbf{C}_k^p - \mathbf{K}_k\mathbf{H}_k\mathbf{C}_k^p\;. \tag{2.26}$$

The first part of the latter formula is often referred to as *Joseph form* of the updated covariance matrix, and the second part is a simplification, which only holds for the particular Kalman gain (2.24) and shows that the covariance matrix decreases in the sense that $\mathbf{C}_k^p - \mathbf{C}_k^e$ is positive semidefinite.

In the considered problem setup, the Kalman filter represents a *minimum mean squared error* (MMSE) estimator, but also, in case of arbitrary noise densities, the Kalman filter can still be applied, if mean and covariance matrix of the noise are known, and encompasses the *best linear unbiased estimator* (BLUE), also referred to as *linear minimum mean squared error* (LMMSE) estimator. The scheme of the Kalman filter summarized in Figure 2.2 constitutes the basis for the considerations in the following sections.

2.1.2 Set-membership Models, Estimators, and in particular Ellipsoidal Calculus

Besides stochastic approaches, set-membership uncertainty models have been intensely investigated, which are most appropriate to cope with *unknown but bounded* perturbations [44, 125]. This involves the major advantage that no knowledge about the error behavior within the bounds is required. A set-membership error characterization can be utilized for uncertainties that are difficult to identify in terms of a probability distribution or that even do not reveal any probabilistic nature[4]. In particular, unknown systematic errors are well-suited to be represented through their membership to sets.

Compared to a stochastic system modeling (2.1), set-membership systems are similar in structure. The quantities in the state transition model

$$\underline{x}_{k+1} = \underline{a}_k(\underline{x}_k, \hat{\underline{u}}_k, \underline{d}_k) \tag{2.27}$$

are specified by their membership to sets $\underline{x}_k \in \mathcal{X}_k^{\mathrm{e}}$ and $\underline{d}_k \in \mathcal{X}_k^u$ instead of probability densities. The superscript u in $\mathcal{X}_k^u$ indicates that $\underline{d}_k$ can be considered as a set of additional unknown inputs. Applying the mapping $\underline{a}_k$ to these sets then defines the predicted set $\mathcal{X}_{k+1}^{\mathrm{P}}$ that encloses the true but unknown state $\underline{x}_{k+1}$. Beginning from an initial set $\underline{x}_0 \in \mathcal{X}_0^{\mathrm{P}}$, such set-valued models can be deployed to explore which values can be attained by the state in future. These so-called *reachable sets* are, for instance, a useful tool in many control applications [4, 39, 111].

Sensor devices are commonly affected by several set-membership uncertainties, which include, inter alia, calibration errors, biases, defects, and symptoms of fatigue. A measurement is thus perturbed by a set-membership quantity $\underline{e}_k \in \mathcal{X}_k^z$ according to

$$\underline{z}_k = \underline{h}_k(\underline{x}_k, \underline{e}_k) \; . \tag{2.28}$$

A set-membership state estimator provides the means to draw conclusions from a set-membership observation about the underlying state. The filtering step of an estimator involves the computation of the set of all possible values of the system state that are consistent with the prior set and the sensor data, i.e., the computation of intersections. The derived estimation results constitute sets to which the actual system state is considered to certainly belong.

---

[4]It is a controversial discussion whether every kind of error follows a probability distribution or can at least be modeled as such. However, the true distribution may not be identifiable and therefore may remain unknown forever.

A Set-membership State Estimation

Except for a bounding set, a set-membership estimator does not rely on the definition of certain error statistics or certain probability distributions. Essentially, set-membership estimators differ in the way the bounding sets are represented. The most well-known implementations of set-membership estimation algorithms employ interval-based and polytopic bounds [65, 85] or ellipsoidal sets [24, 40, 111, 154]. Both representations enable efficient computations of, for instance, linear transformations, Minkowski sums, and intersections. For the selected representation of set-membership uncertainty, the challenge of predicting information resides in finding a tight outer approximation

$$
\begin{aligned}
\mathcal{X}_{k+1}^{\mathrm{p}} &\supseteq \underline{a}_k(\mathcal{X}_k^{\mathrm{e}}, \hat{\underline{u}}_k, \mathcal{X}_k^u) \\
&= \left\{ \underline{x}_{k+1} = \underline{a}_k(\underline{x}_k, \hat{\underline{u}}_k, \underline{d}_k) \,|\, \underline{x}_k \in \mathcal{X}_k^{\mathrm{e}}, \underline{d}_k \in \mathcal{X}_k^u \right\}
\end{aligned}
\tag{2.29}
$$

that has the same representation as $\mathcal{X}_k^{\mathrm{e}}$ and $\mathcal{X}_k^u$. In order to update a set with a received measurement, the intersection of the prior set and the set of all values that are compatible to the measurement has to be computed. More precisely, a tight outer approximation

$$
\mathcal{X}_k^{\mathrm{e}} \supseteq \mathcal{X}_k^{\mathrm{p}} \cap \left\{ \underline{x}_k \,|\, \exists \underline{e}_k \in \mathcal{X}_k^z : \hat{\underline{z}}_k = \underline{h}_k(\underline{x}_k, \underline{e}_k) \right\}
\tag{2.30}
$$

is to be determined. Often, also inner approximations are computed in order to keep track of approximation errors.

Set-membership estimation can also prove to be very challenging. For example, the use of rigid bounds prevents a simple treatment of outliers and, in contrast to the Kalman filter, a high number of measurements does not necessarily imply better filtering results. Analogously to stochastic estimation problems, closed-form solutions are generally not available in case of nonlinear system and sensor models, but again, for linear models, efficient estimation algorithms exist.

B Ellipsoidal Estimation

It is not without reason that *ellipsoidal representations* of set-membership uncertainties are widely used [40, 111, 151, 154]—for they pose several advantages such as simple parameterizations and good algebraic properties. In this work, an ellipsoid

$$\mathcal{E}(\hat{\underline{x}}, \mathbf{X}) = \left\{ \underline{x} \,\middle|\, (\hat{\underline{x}} - \underline{x})^{\mathrm{T}} \mathbf{X}^{-1} (\hat{\underline{x}} - \underline{x}) \leq 1 \right\} \tag{2.31}$$

is defined by a midpoint $\hat{\underline{x}} \in \mathbb{R}^{n_x}$ and a nonnegative definite shape matrix $\mathbf{X} \in \mathbb{R}^{n_x \times n_x}$. Apparently, these parameters allow for an interpretation similar to mean and covariance matrix of a Gaussian density (2.14), where n-sigma bounds are often graphically illustrated by covariance ellipsoids.

For linear system models

$$\underline{x}_{k+1} = \mathbf{A}_k \, \underline{x}_k + \mathbf{B}_k \left(\hat{\underline{u}}_k + \underline{d}_k \right) \tag{2.32}$$

and linear sensor models

$$\hat{\underline{z}} = \mathbf{H}_k \, \underline{x}_k + \underline{e}_k \ , \tag{2.33}$$

where the perturbations $\underline{d}_k$ and $\underline{e}_k$ are characterized by their membership to $\mathcal{E}(\underline{0}, \mathbf{X}_k^u)$ and $\mathcal{E}(\underline{0}, \mathbf{X}_k^z)$, an efficient estimation algorithm can be stated. As explained in the following, it consists of a prediction and a filtering phase.

Prediction For the time update of a current state estimate given by an ellipsoid $\mathcal{X}_k^{\mathrm{e}} = \mathcal{E}(\hat{\underline{x}}_k^{\mathrm{e}}, \mathbf{X}_k^{\mathrm{e}})$, the system model (2.32) has to be applied to each element of the sets, which can then be expressed in terms of the Minkowski sum

$$\mathcal{X}_{k+1}^{\mathrm{p}} = \mathbf{A}_k \, \mathcal{E}(\hat{\underline{x}}_k^{\mathrm{e}}, \mathbf{X}_k^{\mathrm{e}}) \oplus \mathbf{B}_k \left(\hat{\underline{u}}_k + \mathcal{E}(\underline{0}, \mathbf{X}_k^u) \right) , \tag{2.34}$$

i.e., the elementwise sum of the sets, which is denoted by $\oplus$. Affine transformations $\underline{x} \mapsto \mathbf{A}\, \underline{x} + \underline{b}$ can easily be computed by the corresponding transformations of the parameters, i.e,.

$$\mathbf{A}\mathcal{E}(\hat{\underline{x}}, \mathbf{X}) + \underline{b} = \mathcal{E}(\mathbf{A}\, \hat{\underline{x}} + \underline{b}, \mathbf{A}\mathbf{X}\mathbf{A}^{\mathrm{T}}) \ . \tag{2.35}$$

The elementwise sum of two ellipsoids does in general not yield an ellipsoid anymore, but it can be enclosed by a bounding ellipsoid

$$\mathcal{E}(\hat{\underline{x}}_1, \mathbf{X}_1) \oplus \mathcal{E}(\hat{\underline{x}}_2, \mathbf{X}_2) \subseteq \mathcal{E}(\hat{\underline{x}}_1 + \hat{\underline{x}}_2, \mathbf{X}(p)) \tag{2.36}$$

with shape matrix

$$\mathbf{X}(p) = (1 + p^{-1})\mathbf{X}_1 + (1 + p)\mathbf{X}_2 \ . \tag{2.37}$$

The inclusion (2.36) holds for every $p > 0$ [111]. In order to select an ellipsoid out of the parametric family (2.36), the parameter p is often determined to minimize the determinant of (2.37), which yields the enclosing ellipsoid with minimum volume. In this work, we choose p to minimize the trace of (2.37), which corresponds to the sum of the squared lengths of semiaxes. The inequality of arithmetic and geometric means can be applied in order to obtain

$$\begin{aligned}
\mathrm{trace}\left(\mathbf{X}(p)\right) &= \mathrm{trace}\left(\mathbf{X}_1\right) + \mathrm{trace}\left(\mathbf{X}_2\right) + p^{-1}\,\mathrm{trace}\left(\mathbf{X}_1\right) + p\,\mathrm{trace}\left(\mathbf{X}_1\right) \\
&\geq \mathrm{trace}\left(\mathbf{X}_1\right) + \mathrm{trace}\left(\mathbf{X}_1\right) + 2\sqrt{\mathrm{trace}\left(\mathbf{X}_1\right)\mathrm{trace}\left(\mathbf{X}_2\right)} \\
&= \left(\sqrt{\mathrm{trace}\left(\mathbf{X}_1\right)} + \sqrt{\mathrm{trace}\left(\mathbf{X}_2\right)}\right)^2 \ .
\end{aligned} \tag{2.38}$$

Hence, the trace is minimal when equality holds, i.e.,

$$p^{\mathrm{opt}} = \mathrm{trace}(\mathbf{X}_1)^{\frac{1}{2}} \cdot \mathrm{trace}(\mathbf{X}_2)^{-\frac{1}{2}} \ . \tag{2.39}$$

Choosing the trace as optimality criterion is reasonable since selfsame criterion is used to determine the Kalman gain in (2.21). With (2.35) and (2.36), the prediction step (2.34) can now be computed.

Filtering For updating the currently estimated ellipsoid, the intersection of a prior ellipsoid and a measured ellipsoid has to be computed. The measurement $\hat{\underline{z}}_k$ and the associated set-membership uncertainty $\underline{e}_k \in \mathcal{E}(\underline{0}, \mathbf{X}_k^z)$ in (2.33) allow for a combined interpretation as an ellipsoid $\mathcal{E}(\hat{\underline{z}}, \mathbf{X}_k^z) = \hat{\underline{z}} - \mathcal{E}(\underline{0}, \mathbf{X}_k^z)$ of possible measurements, where the symmetry of ellipsoids has been exploited. For the intersection, an outer ellipsoidal approximation $\mathcal{E}(\hat{\underline{x}}_k^e, \mathbf{X}_k^e)$ has to be computed, which is again related to a parameterized family of outer approximations

$$\mathcal{E}(\hat{\underline{x}}_k^e(w), \mathbf{X}_k^e(w)) \supseteq \mathcal{E}(\hat{\underline{x}}_k^{\mathrm{p}}, \mathbf{X}_k^{\mathrm{p}}) \cap \{\underline{x} \,|\, \mathbf{H}_k\,\underline{x} \in \mathcal{E}(\hat{\underline{z}}, \mathbf{X}_k^z)\}$$

with midpoint

$$\hat{\underline{x}}_k^{\mathrm{e}} = \tilde{\mathbf{X}}_k^{\mathrm{e}} \left((1-\omega)(\mathbf{X}_k^{\mathrm{p}})^{-1}\hat{\underline{x}}_k^{\mathrm{e}} + \omega \mathbf{H}_k^{\mathrm{T}}(\mathbf{X}_k^z)^{-1}\hat{\underline{z}}_k \right)$$

and shape matrix

$$\mathbf{X}_k^{\mathrm{e}} = \lambda \cdot \tilde{\mathbf{X}}_k^{\mathrm{e}} \, ,$$

where $\tilde{\mathbf{X}}_k^{\mathrm{e}}$ and λ are given by

$$\tilde{\mathbf{X}}_k^{\mathrm{e}} = \left((1-\omega)(\mathbf{X}_k^{\mathrm{p}})^{-1} + \omega \mathbf{H}_k^{\mathrm{T}}(\mathbf{X}_k^z)^{-1}\mathbf{H}_k \right)^{-1} \tag{2.40}$$

and

$$\lambda = 1 - \left((1-\omega)(\hat{\underline{x}}_k^{\mathrm{p}})^{\mathrm{T}}(\mathbf{X}_k^{\mathrm{p}})^{-1}\hat{\underline{x}}_k^{\mathrm{p}} + \omega\hat{\underline{z}}_k^{\mathrm{T}}(\mathbf{X}_k^z)^{-1}\hat{\underline{z}}_k - (\hat{\underline{x}}_k^{\mathrm{e}})^{\mathrm{T}}(\tilde{\mathbf{X}}_k^{\mathrm{e}})^{-1}\hat{\underline{x}}_k^{\mathrm{e}} \right) \, ,$$

$$\tag{2.41}$$

respectively. The parameter $\omega \in [0,1]$ can, for instance, be determined to minimize the determinant or trace of $\mathbf{X}_k^{\mathrm{e}}$. In both case, this is only possible numerically. The ellipsoid $\mathcal{E}(\hat{\underline{x}}_k^{\mathrm{e}}, \mathbf{X}_k^{\mathrm{e}})$ is an outer conservative approximation of the intersection but never contains any additional point not included either in the prior set or in the measured set [151].

Example 2.2: Set-membership filtering
The results of two different filtering steps are depicted in Figure 2.3. In both cases, the prior ellipsoid is drawn in green and the measurement is associated to the red ellipsoid. In Figure 2.3(a), a two-dimensional measurement is received and the intersection is enclosed by the blue ellipsoid, which represents the outer approximation with minimum determinant. In Figure 2.3(b), the interval $[-1,1]$ for component x_1 is measured, which corresponds to $\hat{z} = 0$, $X^z = 1$, and $H = [1,0]$. The corresponding ellipsoid in the considered two-dimensional state space is degenerate.

Set-membership estimators often encounter difficulties in treating outliers that may even lead to non-intersecting sets. Thus, the set-membership uncertainty description should be large enough in order to account for outliers, but also small enough such that still new insights are gained.

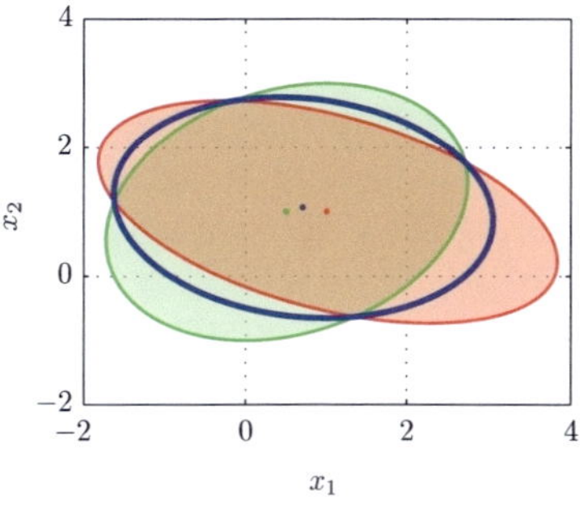

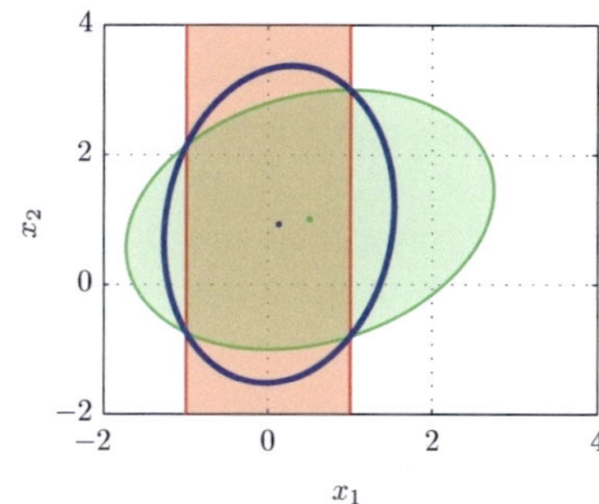

(a) Intersection of two nondegenerate ellipsoids.

(b) Intersection with a degenerate ellipsoid.

Figure 2.3: The blue ellipsoids are outer approximations of the intersections of the green and red ellipsoids. In Fig. 2.3(a), the intersection of two nondegenerate ellipsoids is approximated. In Fig. 2.3(b), the red degenerate ellipsoid corresponds to a one-dimensional measurement of component $\underline{x}_1$ and intersects with the green ellipsoid.

2.2 Combined Models of Stochastic and Set-membership Uncertainties

Imperfect information can take various forms and has to be taken into account properly by a state estimation system in order to provide a complete and correct picture of the imperfect knowledge about the real world. In literature, imperfect information is often subdivided into the terms uncertainty and imprecision [162]. This definition relates uncertainty to the incapability of deciding how certain a statement is and essentially corresponds to stochastic uncertainty. Imprecision refers to the lack of knowledge to decide between two or more items that are compatible with the available information and can therefore be regarded as set-membership uncertainty. In general, a survey of literature does not reveal a unified notion of different types of uncertainty.

The previous section has illustrated that stochastic and set-membership uncertainty models have distinct advantages and disadvantages. A simultaneous consideration of stochastic and set-membership uncertainties therefore appears promising in order to allow to flexibly model different sources of estimation uncertainty, to profit from the individual advantages, and to increase the reliability of the estimation results. Research towards combined stochastic and set-membership estimation has been conducted into different directions, which range from generalizations of classical probability theory that clearly distinguish between stochastic and set-membership uncertainties to concepts that subsume both types of uncertainty under an alternative notion of uncertainty.

Especially, *random sets* [66, 121, 135] have received significant attention in multitarget tracking estimation problems, where the number of targets encompasses the unknown component. By means of random sets, additional set-membership uncertainties can be incorporated into a *generalized likelihood function* [122, 149]. The idea behind general Bayes filtering is similar and rests upon the use of quantization cells, which the measurements belong to [52, 122]. Another approach models the combination of stochastic and set-membership uncertainties by means of sets of probability density functions, so-called *imprecise probabilities* [19, 126, 169]. The random set and the imprecise probability approach encompass the most widely used concepts for state estimation under stochastic and set-membership uncertainties in continuous state spaces and lie in the focus of this section. The examined approaches are discussed based on the example of computing a likelihood for an erroneous measurement affected by both uncertainties.

2.2.1   Random Set Perspective

As the name already suggests, random sets, i.e., set-valued random variables, embody a combination of stochastic and set-membership uncertainties [121, 135]. Analogously to (2.9), we consider a measurement model

$$\underline{z}_k = \underline{h}_k(\underline{x}_k) + \underline{v}_k \tag{2.42}$$

with additive stochastic noise $\underline{v}_k$, but in contrast to (2.9) an observation $\hat{\underline{z}}_k$ is not known with complete precision but by its membership to a closed subset $\mathcal{E}_{\hat{\underline{z}}_k}$, so, instead of $\underline{h}_k(\underline{\boldsymbol{x}}_k) + \underline{\boldsymbol{v}}_k$, a closed set $\boldsymbol{\mathcal{E}}_{\underline{h}_k(\underline{\boldsymbol{x}}_k)+\underline{\boldsymbol{v}}_k}$ is randomly observed [122, 149]. More specifically, the identity $\underline{h}_k(\underline{\boldsymbol{x}}_k) = \hat{\underline{z}}_k - \underline{\boldsymbol{v}}_k$ does not hold anymore, but the relation

$$\underline{h}_k(\underline{\boldsymbol{x}}_k) \in \boldsymbol{\mathcal{E}}_{\hat{\underline{z}}_k-\underline{\boldsymbol{v}}_k} = \left\{ \underline{z}_k - \underline{\boldsymbol{v}}_k \mid \underline{z}_k \in \mathcal{E}_{\hat{\underline{z}}_k} \right\} \tag{2.43}$$

has to be considered. The *generalized likelihood* is defined by

$$
\begin{aligned}
f(\mathcal{E}_{\hat{\underline{z}}_k} \mid \underline{x}_k) &= P(\hat{\underline{z}}_k \in \boldsymbol{\mathcal{E}}_{\underline{z}_k} \mid \underline{x}_k) = P(\underline{h}_k(\underline{\boldsymbol{x}}_k) \in \boldsymbol{\mathcal{E}}_{\hat{\underline{z}}_k-\underline{\boldsymbol{v}}_k} \mid \underline{x}_k) \\
&= P(\underline{\boldsymbol{v}}_k \in \boldsymbol{\mathcal{E}}_{\hat{\underline{z}}_k-\underline{h}_k(\underline{\boldsymbol{x}}_k)} \mid \underline{x}_k) = P(\underline{\boldsymbol{v}}_k \in \mathcal{E}_{\hat{\underline{z}}_k-\underline{h}_k(\underline{\boldsymbol{x}}_k)}) \\
&= \int_{\mathcal{E}_{\hat{\underline{z}}_k-\underline{h}_k(\underline{x}_k)}} f^v(\underline{v}_k)\,\mathrm{d}\underline{v}_k \ .
\end{aligned}
\tag{2.44}
$$

Apparently, the considered measurement model is a stochastic model that incorporates set-membership uncertainties in terms of imprecise observations, i.e., $\hat{\underline{z}}_k$ cannot be associated to a precise value. The generalized likelihood (2.44) can also be viewed as a fuzzy membership function for the membership of $\underline{h}_k(\underline{x})$ to the observed set $\mathcal{E}_{\hat{\underline{z}}_k}$ [66, 149]. It is important to emphasize that (2.44) generally does not allow for a simple parameterization, even if (2.42) is linear and $\underline{\boldsymbol{v}}_k$ is normally distributed. Due to the integral (2.44), a linear relationship anyway turns out to be a nonlinear estimation problem, as further discussed in Subsection 2.2.2.

Fusing Random Set Estimates

An interesting special case arises when two or more random sets are measured. In the simplest case, two estimates

$$\hat{\underline{x}}_1 = \underline{\boldsymbol{x}} + \tilde{\underline{\boldsymbol{x}}}_1 \quad \text{and} \quad \hat{\underline{x}}_2 = \underline{\boldsymbol{x}} + \tilde{\underline{\boldsymbol{x}}}_2$$

are given with probabilistic errors $\tilde{\underline{\boldsymbol{x}}}_1$ and $\tilde{\underline{\boldsymbol{x}}}_2$. Both estimates $\hat{\underline{x}}_1$ and $\hat{\underline{x}}_1$ are only imprecisely characterized by their memberships to $\mathcal{E}_{\hat{\underline{x}}_1}$ and $\mathcal{E}_{\hat{\underline{x}}_2}$, respectively. The joint likelihood is then obtained by

$$
\begin{aligned}
f(\mathcal{E}_{\hat{\underline{x}}_1}, \mathcal{E}_{\hat{\underline{x}}_2} \mid \underline{x}) &= P(\hat{\underline{x}}_1 \in \boldsymbol{\mathcal{E}}_{\underline{x}+\tilde{\underline{x}}_1}, \hat{\underline{x}}_2 \in \boldsymbol{\mathcal{E}}_{\underline{x}+\tilde{\underline{x}}_2} \mid \underline{x}) \\
&= P(\underline{x} \in \boldsymbol{\mathcal{E}}_{\hat{\underline{x}}_1-\tilde{\underline{x}}_1}, \underline{x} \in \boldsymbol{\mathcal{E}}_{\hat{\underline{x}}_2-\tilde{\underline{x}}_2} \mid \underline{x}) \\
&= P(\underline{x} \in \boldsymbol{\mathcal{E}}_{\hat{\underline{x}}_1-\tilde{\underline{x}}_1} \cap \boldsymbol{\mathcal{E}}_{\hat{\underline{x}}_2-\tilde{\underline{x}}_2} \mid \underline{x}) \\
&= P(\underline{x} \in \boldsymbol{\mathcal{E}}_{\hat{\underline{x}}_1-\tilde{\underline{x}}_1} \cap \boldsymbol{\mathcal{E}}_{\hat{\underline{x}}_2-\tilde{\underline{x}}_2}) \; .
\end{aligned}
$$

Evidently, if the state $\underline{x}$ is associated to two or more random set estimates, an intersection of these random sets has to be computed. This fusion methodology is widely utilized to establish estimators for mixed stochastic and set-membership uncertainties, for instance, by the *statistical and set-theoretic information* (SSI) filter [76–78]. Again, these methods encounter the practical difficulty that even, in the case of linear models and Gaussian noise, complicated densities have to be dealt with and a nonlinear estimation problem turns up that requires approximate solutions.

2.2.2 General Bayes Filtering (of Quantized Measurements)

An imprecise observation originating from (2.42) can also be viewed as a quantized measurement of the state [46, 51, 52, 122]. The idea relies on the measure-theoretic identity

$$
\mathrm{E}\left[g(\underline{\boldsymbol{x}}_k) \mid \underline{z} \in \mathcal{Q}_{\hat{\underline{z}}_k}\right] = \mathrm{E}\left[\mathrm{E}[g(\underline{\boldsymbol{x}}_k) \mid \underline{z}] \mid \underline{z} \in \mathcal{Q}_{\hat{\underline{z}}_k}\right] \tag{2.45}
$$

for any quantization cell $\mathcal{Q}_{\hat{\underline{z}}_k}$ of the measurement space and test function g [46, 48]. In [122], it has been shown that the expectation (2.45) yields

$$
\begin{aligned}
\mathrm{E}\left[\mathrm{E}[g(\underline{\boldsymbol{x}}_k) \mid \underline{z}] \mid \underline{z} \in \mathcal{Q}_{\hat{\underline{z}}_k}\right] &= \int_{\mathbb{R}^{n_x}} \mathrm{E}[g(\underline{\boldsymbol{x}}_k) \mid \underline{z}] \cdot f(\underline{z} \mid \underline{z} \in \mathcal{Q}_{\hat{\underline{z}}_k}) \, \mathrm{d}\underline{z} \\
&= \int_{\mathbb{R}^{n_x}} g(\underline{x}_k) \cdot \frac{f(\mathcal{Q}_{\hat{\underline{z}}_k} \mid \underline{x}_k) f^{\mathrm{P}}(\underline{x}_k)}{f(\mathcal{Q}_{\hat{\underline{z}}_k})} \, \mathrm{d}\underline{x}_k
\end{aligned}
\tag{2.46}
$$

with

$$
f(\mathcal{Q}_{\hat{\underline{z}}_k} \mid \underline{x}_k) = \int_{\mathcal{Q}_{\hat{\underline{z}}_k}} f(\underline{z}_k \mid \underline{x}_k) \, \mathrm{d}\underline{z}_k \tag{2.47}
$$

and

$$f(\mathcal{Q}_{\hat{\underline{z}}_k}) = \int_{\mathbb{R}^{n_x}} f(\mathcal{Q}_{\hat{\underline{z}}_k} \,|\, \underline{x}_k) \cdot f^{\mathrm{P}}(\underline{x}_k) \, \mathrm{d}\underline{x}_k \ . \tag{2.48}$$

For linear system and sensor models, the Kalman filter equations can be generalized to quantized measurements [46, 52]. The filtering (2.25) step then alters to the computation of the conditional mean[5]

$$\hat{\underline{x}}_k^{\mathrm{quant}} = \mathrm{E}\left[\underline{x}_k \,|\, \underline{z} \in \mathcal{Q}_{\hat{\underline{z}}_k}\right] = \hat{\underline{x}}_k^{\mathrm{P}} + \mathbf{K}_k \left(\mathrm{E}\left[\underline{z} \,|\, \underline{z} \in \mathcal{Q}_{\hat{\underline{z}}_k}\right] - \mathbf{H}_k \,\hat{\underline{x}}_k^{\mathrm{P}}\right)$$

and MSE matrix

$$\begin{aligned}
\mathbf{C}_k^{\mathrm{quant}} &= \mathrm{E}\left[(\hat{\underline{x}}_k^{\mathrm{e}} - \underline{x}_k)(\hat{\underline{x}}_k^{\mathrm{e}} - \underline{x}_k)^{\mathrm{T}} \,|\, \underline{z} \in \mathcal{Q}_{\hat{\underline{z}}_k}\right] \\
&= \mathbf{C}_k^{\mathrm{e}} + \mathbf{K}_k \, \mathrm{E}\left[\underline{z}\,\underline{z}^{\mathrm{T}} \,|\, \underline{z} \in \mathcal{Q}_{\hat{\underline{z}}_k}\right]\mathbf{K}_k^{\mathrm{T}} \\
&= \mathbf{C}_k^{\mathrm{e}} + \mathbf{K}_k \, \mathrm{Cov}\left[\underline{z} \,|\, \underline{z} \in \mathcal{Q}_{\hat{\underline{z}}_k}\right]\mathbf{K}_k^{\mathrm{T}} \ ,
\end{aligned}$$

where the matrix $\mathbf{K}_k$ is the standard Kalman gain (2.24) and $\mathbf{C}_k^{\mathrm{e}}$ is the standard Kalman MSE matrix (2.21). The mean $\mathrm{E}\left[\underline{z} \,|\, \underline{z} \in \mathcal{Q}_{\hat{\underline{z}}_k}\right]$ and covariance matrix $\mathrm{Cov}\left[\underline{z} \,|\, \underline{z} \in \mathcal{Q}_{\hat{\underline{z}}_k}\right]$ can be computed by means of (2.46). At this point, the algorithm emerges as a nonlinear estimation problem since the densities (2.47) and (2.48) are, in general, non-Gaussian and their computation is a difficult issue in the multi-dimensional case.

The quantization of measurement information can also be viewed as an additional nonlinear many-to-few mapping $\underline{z}_k^* = Q(\underline{z}), \underline{z} \in \mathcal{Q}_{\hat{\underline{z}}_k}$ that is applied after the measurement is generated and that yields the quantization cell or an identifier $\underline{z}_k^*$ of it, for example, the center of the cell. This also states the reason why the concept of quantized measurements completely resides in the Bayesian estimation framework, and the fusion result is given by a single probability density. Although the explained idea has also been generalized to arbitrary set-membership measurements [53], it relies on the definition of a certain quantization mapping. Irrespective of the actual unknown but bounded error, the corrupted measurement is reduced to the quantization cell $\underline{z}_k^*$. The following Subsection 2.2.3 will accord more attention to this issue.

---

[5]For a shorter notation, we write $\mathrm{E}\left[\underline{x}_k \,|\, \underline{z} \in \mathcal{Q}_{\hat{\underline{z}}_k}\right]$ conditioned only on the current observation.

Relation to Random Sets

The filtering of quantized measurements directly corresponds to the generalized likelihood for random sets in Subsection 2.2.1. Let the randomly observed set $\mathcal{E}_{\hat{\underline{z}}_k}$ represent the quantization cell $\mathcal{Q}_{\hat{\underline{z}}_k}$. Then, the generalized likelihood

$$f(\mathcal{E}_{\hat{\underline{z}}_k} \mid \underline{x}_k) \overset{(2.44)}{=} P(\underline{v}_k \in \mathcal{E}_{\hat{\underline{z}}_k - \underline{h}_k(\underline{x}_k)})$$

$$= \int_{\mathcal{E}_{\hat{\underline{z}}_k - \underline{h}_k(\underline{x}_k)}} f^v(\underline{v}_k)\, \mathrm{d}\underline{v}_k = \int_{\mathcal{E}_{\hat{\underline{z}}_k}} f^v(\tilde{\underline{z}}_k - \underline{h}_k(\underline{x}_k))\, \mathrm{d}\tilde{\underline{z}}_k$$

$$\overset{(2.11)}{=} \int_{\mathcal{E}_{\hat{\underline{z}}_k}} f(\underline{z}_k \mid \underline{x})\, \mathrm{d}\underline{z}_k \ ,$$

is identical to (2.47) and the posterior density yields

$$f(\underline{x} \mid \mathcal{E}_{\hat{\underline{z}}_k}) = \frac{f(\mathcal{Q}_{\hat{\underline{z}}_k} \mid \underline{x}) \cdot f^{\mathrm{P}}(\underline{x}_k)}{f(\mathcal{Q}_{\hat{\underline{z}}_k})} \ .$$

Inserting this density in (2.46) directly leads to the equality

$$\mathrm{E}\left[h_k(\underline{x}_k) \mid \underline{z} \in \mathcal{Q}_{\hat{\underline{z}}_k}\right] = \mathrm{E}\left[h_k(\underline{x}_k) \mid \mathcal{E}_{\hat{\underline{z}}_k}\right] \ . \tag{2.49}$$

In conclusion, the filtering of quantized measurement can be regarded as an instance of the generalized likelihood approach for random sets.

2.2.3 Sets as Additional Unknown But Bounded Error Terms and Imprecise Probabilities

In the previous subsections, the set-membership uncertainty refers to the incapability to associate an observation with a certain value. An alternative way is to consider a set-membership uncertainty as an additional error [198–201] that affects the observed quantity $\hat{\underline{z}}_k$ in the same fashion as the stochastic noise, i.e., $\hat{\underline{z}}_k$ is related to the state via

$$\underline{z}_k = \underline{h}_k(\underline{x}_k) + \underline{v}_k + \underline{e}_k \ , \tag{2.50}$$

where $\underline{e}_k \in \mathcal{X}_k^z$ denotes the unknown but bounded error. As mentioned at the beginning of this chapter, we do not make any assumption about the error behavior within the bounds. However, the error might also be random with unknown probability density f^e and the total measurement error is then given by

$$\underline{\tilde{z}}_k = \underline{z}_k - h_k(\underline{x}_k) = \underline{v}_k + \underline{e}_k \ .$$

The density of the measurement noise $\underline{\tilde{z}}_k$ is obtained by the convolution

$$f^{\tilde{z}}(\underline{\tilde{z}}_k) = f^{v+e}(\underline{\tilde{z}}_k) = \int_{\mathbb{R}^{n_v}} f^v(\underline{\tilde{z}}_k - \underline{e}_k) \cdot f^e(\underline{e}_k) \, \mathrm{d}\underline{e}_k \ .$$

Note that f^e has compact support $\mathrm{supp}(f^e) \subseteq \mathcal{X}_k^z$. Analogously to (2.11), the corresponding likelihood yields

$$\begin{aligned}
f(\hat{\underline{z}}_k \,|\, \underline{x}_k) &= f^{\tilde{z}}(\hat{\underline{z}}_k - \underline{h}_k(\underline{x}_k)) \\
&= \int_{\mathbb{R}^{n_v}} f^v(\hat{\underline{z}}_k - \underline{h}_k(\underline{x}_k) - \underline{e}_k) \cdot f^e(\underline{e}_k) \, \mathrm{d}\underline{e}_k \ .
\end{aligned} \tag{2.51}$$

Since f^e is unknown except for its support, a set of possible densities has to be taken into consideration for $\underline{e}_k$. The likelihood (2.51) can also be regarded as a compound or mixture density with unknown weights $\omega(\underline{i}) = f^e(\underline{i})$ and an infinite number of components $f_{\underline{i}}^v(\hat{\underline{z}}_k - \underline{h}_k(\underline{x}_k) - \underline{i})$. Therefore, the true but unknown likelihood lies in the convex closure

$$f(\,\cdot\,|\,\underline{x}_k) \in \overline{\mathrm{conv}\big\{f^v(\,\cdot\, - \underline{h}_k(\underline{x}_k) - \underline{\nu})\,\big|\,\underline{\nu} \in \mathcal{X}_k^z\big\}} =: \boldsymbol{\mathcal{F}}^{\mathrm{L}} \tag{2.52}$$

of $\{f^v(\,\cdot\, - \underline{h}_k(\underline{x}_k) - \underline{\nu})\,|\,\underline{\nu} \in \mathcal{X}_k^z\}$, which apparently is a set of translated versions of f^v due to the additive model (2.50). The entire set $\boldsymbol{\mathcal{F}}^{\mathrm{L}}$ must be taken into account by an estimator in order to provide reliable results, as elucidated in detail in Section 2.3.

A   Comparison to Random Sets

Interestingly, by setting $f^e := 1_{\mathcal{X}_k^z}$, the imprecise likelihood (2.51) becomes precisely the generalized likelihood (2.44), i.e.,

$$
\begin{aligned}
f(\hat{\underline{z}}_k \,|\, \underline{x}_k) &= \int_{\mathbb{R}^{n_v}} f^v(\hat{\underline{z}}_k - \underline{h}_k(\underline{x}_k) - \underline{e}_k) \cdot 1_{\mathcal{X}_k^z}(\underline{e}_k) \, \mathrm{d}\underline{e}_k \\
&= \int_{\mathcal{X}_k^z} f^v(\hat{\underline{z}}_k - \underline{h}_k(\underline{x}_k) - \underline{e}_k) \, \mathrm{d}\underline{e}_k \\
&= \int_{\mathcal{E}_{\hat{\underline{z}}_k}} f^v(\underline{z}_k - \underline{h}_k(\underline{x}_k)) \, \mathrm{d}\underline{z}_k
\end{aligned}
\tag{2.53}
$$

with $\mathcal{E}_{\hat{\underline{z}}_k} = \{\hat{\underline{z}}_k - \underline{e}_k \,|\, \underline{e}_k \in \mathcal{X}_k^z\}$. Contrary to the deliberations in subsections 2.2.1 and 2.2.2, the focus is now laid on an explicit source $\underline{e}_k$ of set-membership uncertainty. This case can also be expressed in terms of a random set by means of

$$
\underline{h}_k(\underline{x}_k) + \underline{e}_k \in \mathcal{E}_{\hat{\underline{z}}_k - \underline{v}_k} = \{\underline{z}_k - \underline{v}_k \,|\, \underline{z}_k \in \mathcal{E}_{\hat{\underline{z}}_k}\} \, ,
$$

which replaces relation (2.43) in Subsection 2.2.1. For a moment, we suppose that $\underline{e}_k$ is random with probability density f^e. The generalized likelihood then results in

$$
\begin{aligned}
f(\mathcal{E}_{\hat{\underline{z}}_k} \,|\, \underline{x}_k) &= P(\underline{z} \in \mathcal{E}_{\hat{\underline{z}}_k} \,|\, \underline{x}_k) \\
&= P(\underline{h}_k(\underline{x}_k) + \underline{e}_k \in \mathcal{E}_{\hat{\underline{z}}_k - \underline{v}_k} \,|\, \underline{x}_k) \\
&= P(\underline{v}_k + \underline{e}_k \in \mathcal{E}_{\hat{\underline{z}}_k - \underline{h}_k(\underline{x}_k)}) \\
&= \int_{\mathcal{E}_{\hat{\underline{z}}_k - \underline{h}_k(\underline{x}_k)}} f^{v+e}(\underline{v}_k) \, \mathrm{d}\underline{v}_k \\
&= \int_{\mathcal{E}_{\hat{\underline{z}}_k - \underline{h}_k(\underline{x}_k)}} \int_{\mathcal{X}_k^z} f^v(\underline{v}_k - \underline{e}_k) f^e(\underline{e}_k) \, \mathrm{d}\underline{e}_k \, \mathrm{d}\underline{v}_k \, .
\end{aligned}
$$

However, since the true f^e is unknown, it is not a precise generalized likelihood that is an eligible probabilistic representation of the observation model (2.50). In terms of fuzzy sets (see Subsection 2.2.1), this imprecise generalized likelihood can be seen as a vague fuzzy membership function and only an interval for the membership to $\mathcal{E}_{\hat{\underline{z}}_k}$ can be stated, and similarly the expectation (2.49) is now imprecise lying between a lower and upper expectation. The reason for the imprecise likelihood is that, instead of an uncertain association of the measurement to a precise value, the set-membership uncertainty is here considered as an additional error term that may induce some unknown additional error behavior to the estimation problem.

In [23], two different interpretations of the set-membership uncertainty associated to a random set are discussed. More specifically, the authors describe two ways to derive probability distributions compatible with a random set: On the one hand, the principle of indifference can be applied and a uniform distribution is associated with the set. This then results into the precise generalized likelihood (2.53). On the other hand, the entire set of compatible distributions, called *measurable selection* or *selectors*, can be considered, which then yields the set (2.52) of likelihoods. In this work, we strongly prefer the second viewpoint since ignorance expressed by a set implies that every outcome is possible and might even contradict the current observations. The random set perspective from Subsection 2.2.1 indicates to trust or distrust incoming observations proportionally to the extend of the set-membership uncertainty, but excludes the case that unknown information may strongly contradict currently available information. So, the approach to use sets of likelihoods proves to be more conservative and cautious and resides in the theory of imprecise probabilities.

B Imprecise Probabilities as a Comprehensive Concept

Although Bayesian state estimation techniques have evolved into a key tool in many research directions, numerous situations and applications point out that estimation results are questionable if precise probability distributions for the involved uncertainties cannot unambiguously be identified. In particular, the choice of appropriate prior probabilities is a major challenge and is often difficult to justify.

Early investigations have started to scrutinize the impact of varying input distributions on the answer of a Bayesian estimator, which is a central topic of *robust Bayesian analysis* [20, 21, 83]. For the purpose of testing the sensitivity of an estimator, several possibilities have been presented. By the ϵ-contamination model, the family of all convex combinations of a known nominal distribution with any arbitrary distribution is denoted and the resulting set of posterior distributions is analyzed [22]. In [10, 11], distribution bands are employed as lower and upper bounds for the set of reasonable distributions and, in [57, 58], p-boxes are constructed that enclose the cumulative distribution function. In [112], density-bounded neighborhoods, i.e., sets of densities bounded by lower and upper functions, are considered. Similarly, intervals of measures are used in [47], which are investigated further in [45], where imprecise prior probability measures as well as imprecise likelihoods are modeled by convex sets of probability measures. A result of these studies surely is that an estimator often cannot provide valid and reliable conclusions if only one probability density is chosen among many equally reasonable densities. These examinations evidently support the view to apply an estimator to all possible probability densities.

All the mentioned considerations are strongly related to the concept of *imprecise probabilities* [169] since, in each case, uncertain quantities are not modeled by means of precise probability densities but by means of sets of probability densities. In general, these sets are continuous and connected. The probability of a certain outcome A thus yields an *interval probability* $[\underline{P}(A), \overline{P}(A)] \subseteq [0, 1]$ [173–175], i.e., lies between the *lower and upper probabilities* $\underline{P}(A)$ and $\overline{P}(A)$ [19, 126, 169]. More generally, imprecise probabilities can be defined by means of *coherent lower and upper previsions* [18, 19, 126, 169], which can be regarded as lower and upper expectation functionals

$$\underline{\mathrm{E}}[h_k] = \inf_{f \in \mathcal{F}} \int h_k(\underline{x}_k) f(\underline{x}_k) \, \mathrm{d}\underline{x}_k$$

and

$$\overline{\mathrm{E}}[h_k] = \sup_{f \in \mathcal{F}} \int h_k(\underline{x}_k) f(\underline{x}_k) \, \mathrm{d}\underline{x}_k \, ,$$

where $\mathcal{F}$ denotes the set of possible candidates for the unknown density. With the relationship $\underline{\mathrm{E}}[h_k] = -\overline{\mathrm{E}}[-h_k]$, it suffices to consider the lower prevision, i.e., the lower envelope of all expectations. In general, as in (2.52), $\mathcal{F}$ is assumed to be a closed convex set since any arbitrary set and its convex closure yields the same lower prevision. A closed convex set of probability densities is often referred to as *credal set* [117, 130, 131].

As mentioned earlier, also random set theory is linked to imprecise probabilities through measurable selections, which induce the set of probability distributions that are compatible to the random set. This relation is intensely studied in [127, 128]. In conclusion, the concept of imprecise probabilities renders a comprehensive theory for incorporating set-membership uncertainty into probabilistic uncertainty descriptions.

2.2.4 Is That All?

Apparently, the previous considerations are not an all-encompassing survey of combined stochastic and set-membership uncertainty representations. Many approaches exist that focus on discrete domains and propose alternative representations of uncertainty. Well-known is the Dempster-Shafer theory [155] that models the degree of belief for a particular proposition instead of using a probability distribution. The degree of belief is represented by an interval. More precisely, belief denotes the lower bound that directly supports a proposition. Plausibility is the upper bound on the possibility the proposition can be true—up to that amount of evidence that does not contradict the proposition. Dempster's rule of combining evidences is often criticized to contradict the results of a Bayesian fusion [17, 179] if a probabilistic interpretation is imposed on the degree of belief.

Uncertain and incomplete knowledge in logical reasoning can be covered by means subjective logic [88, 89]. The opinion about the truth of a proposition is here defined by belief, disbelief, uncertainty, and a base rate, where the latter two parameters represent ignorance and prior belief, respectively. Subjective logic is a powerful tool to derive a logical formulation from verbal descriptions that are in general imprecise.

An already mentioned concept is fuzzy set theory [178], where the membership of an element to a set is modeled to be potentially imprecise. Although the generalized likelihood in Subsection 2.2.1 can be interpreted as a fuzzy membership function, fuzzy set theory is rather suitable to represent vagueness than uncertainty and ignorance.

The approaches in info-gap decision theory [14, 15] seek the same objective as the considered concepts, but focus on decision making under severe uncertainty and are detached from probabilistic or probability-like uncertainty descriptions. In this theory, a family of nested subsets around point estimates serves as an uncertainty model and a parameter, the horizon of uncertainty, measures the error between the estimate and the universe of all possible true values.

Although providing convincing results, alternatives to particularly probabilistic methods are often criticized by showing that they contradict probability theory or reside in it. The objective pursued by this thesis therefore is to consequently and consistently develop a combined estimation method that does not contradict the classical stochastic and set-membership methodologies, but that includes them as special cases.

2.3   Bayesian State Estimation with Sets of Probability Densities

As figured out in Section 2.2.3, the purpose of a cautious and conservative incorporation of stochastic and set-membership uncertainty supports the idea of imprecise probabilities and points to an elementwise processing of the underlying densities within the Bayesian estimation framework, as outlined in [198, 200, 201]. This and the following sections pursue the objective to provide a unified framework of stochastic and set-membership state estimation, each of which has been discussed in sections 2.1.1 and 2.1.2, respectively.

Instead of focusing on additive perturbation terms as in (2.50), we start our studies with the more general discrete-time nonlinear dynamic state transition model

$$\underline{x}_{k+1} = \underline{a}_k(\underline{x}_k, \hat{\underline{u}}_k, \underline{w}_k, \underline{d}_k) \,, \tag{2.54}$$

which can be recognized as a combination of the stochastic model (2.1) and the set-valued model (2.27). The transition from $\underline{x}_k$ to $\underline{x}_{k+1}$ is corrupted by a probabilistic process noise $\underline{w}_k$ with probability density f^w and an unknown but bounded error $\underline{d}_k \in \mathcal{X}_k^u \subset \mathbb{R}^{n_d}$. The outcome of a measurement $\hat{\underline{z}}_k$ is related to a possibly nonlinear sensor model

$$\underline{z}_k = \underline{h}_k(\underline{x}_k, \underline{v}_k, \underline{e}_k) \tag{2.55}$$

with respect to the state $\underline{x}_k$, a stochastic measurement noise $\underline{v}_k$ with density f^v, and an unknown but bounded error $\underline{e}_k \in \mathcal{X}_k^z \subset \mathbb{R}^{n_e}$. Again, the model (2.55) unites the stochastic observation model (2.2) and the set-membership model (2.28). In the following, we assume that $\underline{x}_0, \underline{w}_1, \ldots, \underline{w}_0,$ $\underline{v}_0, \ldots, \underline{v}_k$ are mutually independent and that their outcomes are also independent from the unknown but bounded errors $\underline{d}_0, \ldots, \underline{d}_k, \underline{e}_0, \ldots, \underline{e}_k.$

2.3.1 Generic Prediction and Filtering of Sets of Probability Densities

In order to derive an estimation algorithm, we follow the line of thought discussed in Section 2.2.3. In the presence of both stochastic and set-membership uncertainties, only imprecise probability densities can be stated. The task is thus to recursively compute a set of estimated conditional densities for the state, i.e., a credal set. The following considerations are based on [198].

Prediction If we first assume $\underline{d}_k$ to be known, the time update or prediction step can be expressed, analogously to (2.4) and (2.5), as the Chapman-Kolmogorov integral

$$f^{\mathrm{P}}(\underline{x}_{k+1}) = \int_{\mathbb{R}^{n_x}} f(\underline{x}_{k+1} \,|\, \underline{x}_k, \hat{\underline{u}}_k, \underline{d}_k) \cdot f^{\mathrm{e}}(\underline{x}_k) \, \mathrm{d}\underline{x}_k \,, \tag{2.56}$$

where

$$f(\underline{x}_{k+1} \,|\, \underline{x}_k, \hat{\underline{u}}_k, \underline{d}_k) = \int_{\mathbb{R}^{n_w}} \delta\big(\underline{x}_{k+1} - \underline{a}_k(\underline{x}_k, \hat{\underline{u}}_k, \underline{w}_k, \underline{d}_k)\big) \cdot f^w(\underline{w}_k) \, \mathrm{d}\underline{w}_k \tag{2.57}$$

is the transition density deduced from the system model (2.54). However, since $\underline{d}_k$ is unknown, we have to consider the entire set of eligible transition densities

$$\boldsymbol{\mathcal{F}}_k^{\mathrm{T}} = \left\{ f^{\mathrm{T}}(\underline{x}_{k+1} \,|\, \underline{x}_k) = f(\underline{x}_{k+1} \,|\, \underline{x}_k, \hat{\underline{u}}_k, \underline{d}_k) \,\Big|\, \underline{d}_k \in \mathcal{X}_k^u \right\} ,$$

each of which is a seriously potential candidate for the true state transition density. Consequently, the integral (2.56) is applied elementwise to the entire set in order to obtain the set

$$\boldsymbol{\mathcal{F}}_{k+1}^{\mathrm{P}} = \left\{ f^{\mathrm{P}}(\underline{x}_{k+1}) = \int_{\mathbb{R}^{n_x}} f^{\mathrm{T}}(\underline{x}_{k+1} \,|\, \underline{x}_k) \cdot f^{\mathrm{e}}(\underline{x}_k) \,\mathrm{d}\underline{x}_k \,\bigg|\, f^{\mathrm{e}} \in \boldsymbol{\mathcal{F}}_k^{\mathrm{e}}, \ f^{\mathrm{T}} \in \boldsymbol{\mathcal{F}}_k^{\mathrm{T}} \right\}$$

$$(2.58)$$

of predicted densities, where also the current estimated density f^{e} might only be defined imprecisely through its membership to $\boldsymbol{\mathcal{F}}_k^{\mathrm{e}}$.

Filtering For fixed $\underline{e}_k$, the prior density f^{P} can be updated in the filtering step according to Bayes' rule (2.6), i.e.,

$$f^{\mathrm{e}}(\underline{x}_k) = \frac{f(\hat{\underline{z}}_k \,|\, \underline{x}_k, \underline{e}_k) \cdot f^{\mathrm{P}}(\underline{x}_k)}{\int_{\mathbb{R}^{n_x}} f(\hat{\underline{z}}_k \,|\, \underline{x}_k, \underline{e}_k) \cdot f^{\mathrm{P}}(\underline{x}_k) \,\mathrm{d}\underline{x}_k} ,$$

where $f(\hat{\underline{z}}_k \,|\, \underline{x}_k, \underline{e}_k)$ denotes the likelihood

$$f(\hat{\underline{z}}_k \,|\, \underline{x}_k, \underline{e}_k) = \int_{\mathbb{R}^n} \delta\big(\hat{\underline{z}}_k - \underline{h}_k(\underline{x}_k, \underline{v}_k, \underline{e}_k)\big) \cdot f^v(\underline{v}_k) \,\mathrm{d}\underline{v}_k \qquad (2.59)$$

being derived from model (2.55) similarly to the computation of (2.8). Again, this function cannot be identified uniquely since we only know $\underline{e}_k \in \mathcal{X}_k^z$, and hence the filtering step is carried out elementwise for every likelihood in

$$\boldsymbol{\mathcal{F}}_k^{\mathrm{L}} = \left\{ f^{\mathrm{L}}(\hat{\underline{z}}_k \,|\, \underline{x}_k) = f(\hat{\underline{z}}_k \,|\, \underline{x}_k, \underline{e}_k) \,\Big|\, \underline{e}_k \in \mathcal{X}_k^z \right\}$$

and, if f^{P} is not precise either, also for every $f^{\mathrm{P}} \in \boldsymbol{\mathcal{F}}_k^{\mathrm{P}}$. The filtering step yields the set

$$\boldsymbol{\mathcal{F}}_k^{\mathrm{e}} = \left\{ f^{\mathrm{e}}(\underline{x}_k) = \frac{f^{\mathrm{L}}(\hat{\underline{z}}_k \,|\, \underline{x}_k) \cdot f^{\mathrm{P}}(\underline{x}_k)}{\int_{\mathbb{R}^{n_x}} f^{\mathrm{L}}(\hat{\underline{z}}_k \,|\, \underline{x}_k) \cdot f^{\mathrm{P}}(\underline{x}_k) \,\mathrm{d}\underline{x}_k} \,\bigg|\, f^{\mathrm{P}} \in \boldsymbol{\mathcal{F}}_k^{\mathrm{P}}, \ f^{\mathrm{L}} \in \boldsymbol{\mathcal{F}}_k^{\mathrm{L}} \right\}$$

$$(2.60)$$

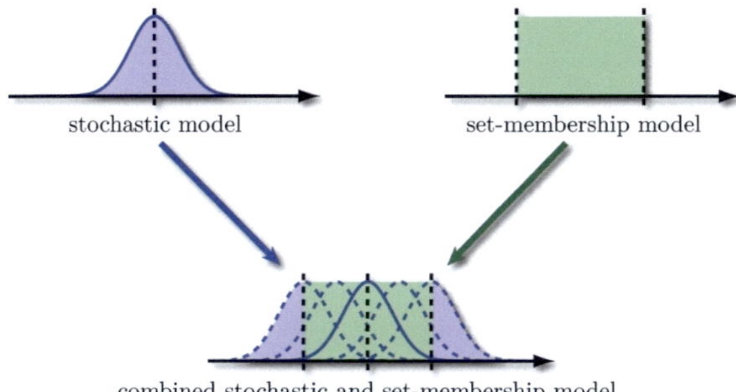

Figure 2.4: Stochastic and set-membership uncertainties can be considered simultaneously by means of sets of probability densities. For instance, an additive combination of a normally distributed and an interval-based error characterization becomes a set of translated Gaussian densities.

of estimated densities, which can then be further processed elementwise with the set of transition densities or likelihoods in subsequent prediction or filtering steps, respectively.

Additive Noise Models   In the majority of estimation problems, the noise is assumed to additively affect the system and measurement model so that (2.54) and (2.55) are changed to

$$\underline{x}_{k+1} = \underline{a}_k(\underline{x}_k, \hat{\underline{u}}_k) + \underline{w}_k + \underline{d}_k \tag{2.61}$$

and

$$\underline{z}_k = \underline{h}_k(\underline{x}_k) + \underline{v}_k + \underline{e}_k \ , \tag{2.62}$$

respectively. The sifting property of the delta distribution can now be employed to simplify the imprecise transition density (2.57) and likelihood (2.59) to

$$f^{\mathrm{T}}(\underline{x}_{k+1} \,|\, \underline{x}_k, \hat{\underline{u}}_k, \underline{d}_k) = f^w(\underline{x}_{k+1} - \underline{a}_k(\underline{x}_k, \hat{\underline{u}}_k) - \underline{d}_k)$$

and

$$f^{\mathrm{L}}(\underline{x}_k \,|\, \hat{\underline{z}}_k, \underline{e}_k) = f^v(\hat{\underline{z}}_k - \underline{h}_k(\underline{x}_k) - \underline{e}_k) \ ,$$

respectively. Accordingly, we obtain the set

$$\boldsymbol{\mathcal{F}}_k^{\mathrm{T}} = \left\{ f^w(\underline{x}_{k+1} - \underline{a}_k(\underline{x}_k, \hat{\underline{u}}_k) - \underline{d}_k) \,\middle|\, \underline{d}_k \in \mathcal{X}_k^u \right\} \tag{2.63}$$

of transition densities and the set

$$\boldsymbol{\mathcal{F}}_k^{\mathrm{L}} = \left\{ f^v(\hat{\underline{z}}_k - \underline{h}_k(\underline{x}_k) - \underline{e}_k) \,\middle|\, \underline{e}_k \in \mathcal{X}_k^z \right\} \tag{2.64}$$

of likelihoods. Each set contains translated versions of a density, where the set-membership vectors $\underline{d}_k$ and $\underline{e}_k$ represent the corresponding shift parameters. This way of combining probabilistic and set-membership uncertainty is depicted in Figure 2.4. A set like $\boldsymbol{\mathcal{F}}_k^{\mathrm{T}}$ or $\boldsymbol{\mathcal{F}}_k^{\mathrm{L}}$ can be interpreted as a density (or likelihood) with imprecisely known mean and can therefore be parameterized by the according set $\mathcal{X}_k^u$ or $\mathcal{X}_k^z$ of possible means.

2.3.2 Special Cases, Consistency, and Convexity

Up to now, we have achieved that classical probabilistic methods are a special case of the proposed estimation concept, which occurs when each set collapses to a single density. The opposite direction is more intricate and unintuitive as the examinations in [193] have shown. The special case of vanishing stochastic uncertainty implies that all involved densities reduce to Dirac delta functions. Since the product of delta functions is not defined, the filtered set (2.60) cannot be computed. From a simplistic view point, which is far from mathematically sound, we may constitute that the multiplication of delta distributions at the same positions is defined to yield again the same delta function. As a result, the set (2.60) represents an intersection of sets of delta functions, which corresponds to the set-membership filtering step (2.30). The predicted set (2.58) may be computed if the sifting property is generalized to the product of delta distributions and is then comparable to the set-membership prediction step (2.29). However, the linear estimation algorithm derived later in Section 2.5 will effectively include set-membership estimation as a special case, while we can achieve this special case only artificially for the presented generic estimation framework.

In general, Bayes' rule can only be applied if the involved densities have intersecting supports, i.e., prior and measurement information are compatible. Of course, it is in any case reasonable to assume that sensor observations are supported by prior information, which means that the product of the prior density and the likelihood is a nonzero function. For imprecise probabilities, the application of Bayes' theorem and the definition—or generalization—of independence is more sophisticated as stated in [19, 126, 129, 169] or [190], and further prerequisites must be met in order to circumvent inconsistencies. Thus, we require that each density in (2.58) and (2.60) can be computed, which is definitely fulfilled for additive models (2.61) and (2.62) with Gaussian noise. In the following sections, where generalizations of the Kalman filtering scheme are considered, we therefore do not need to make any explicit assumption.

In Section 2.2.3, it has been stated that the set of densities, representing an imprecise random variable, is commonly convex, i.e., a credal set. As investigated in [198], the resulting sets (2.58) and (2.60) are, in general, not convex even if the inputs are convex sets. However, each set processed in prediction and filtering steps can be considered as a *generator set*: For a certain credal set, every set whose convex hull yields this credal set is called a generator set. As depicted schematically in Figure 2.5, it suffices to process the generator sets, and the resulting set is still a generator set of the actual estimation result, i.e., the convex closure of the resulting set encloses the actual estimation result [198]. Consequently, we can confine ourselves to the processing of the set of extremal densities or any other generator set, which generally allow for simpler representations, and we are still able to obtain the estimated credal set.

2.4   Linear State Estimation under Stochastic and Set-membership Uncertainties

Linear estimation problems have been studied most intensively, and the Kalman filtering scheme can be viewed as the most well-established estimation algorithm. Although being optimal for linear systems and Gaussian

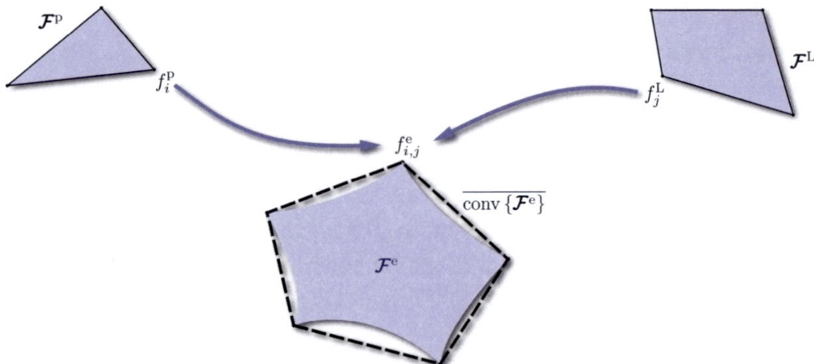

Figure 2.5: The set $\mathcal{F}^{\mathrm{P}}$ of prior densities and the set $\mathcal{F}^{\mathrm{L}}$ of likelihoods are each the convex hull of a finite number of densities, i.e., a convex polytope. It suffices to process the extremal densities f_i^{P} and f_j^{L} elementwise [198]. The set $\mathcal{F}^{\mathrm{e}}$ of filtered densities (2.60) is not convex anymore but lies in the convex hull of the extremal densities $f_{i,j}^{\mathrm{e}}$.

noise characteristics, the Kalman filter may severely diverge under influences of nonlinearities, non-Gaussian perturbations, and systematic errors. In Section 2.2, we have discussed extensions to the standard models in order to be capable of encompassing these additional influences. Of course, many approaches can be named that aspire a robust Kalman filter design. For example, uncertainties can be described by means of sum quadratic [102] or relative entropy constraints [142], irrespective of their actual error characteristics. A robust filter is here obtained by solving a nonlinear optimal control problem or finding approximate solutions to it. These constraints can generally be regarded to be defined on a distance measure between a nominal distribution and an unknown contaminating distribution, and these approaches are thereby also related to imprecise probabilities. In [61] and [156], the system matrix is perturbed by an unknown but norm-bounded block, and linear matrix inequalities [28] are employed to derive a robust filter. The method in [80] explicitly distinguishes between stochastic and set-membership uncertainties, and the gain is computed by minimizing a cost function that depends on both types of errors. The minimization is also done under linear matrix inequality conditions.

Our considerations in the preceding section provide the spadework for a straightforward generalization of the Kalman filtering scheme to a conservative and reliable linear estimator. This generalization is inferred from the elementwise processing of sets of Gaussian densities in case of linear models. This idea has been published in [199], and this section investigates also a second derivation of a Kalman filter algorithm for the simultaneous treatment of stochastic and set-membership uncertainties. It essentially represents an MMSE estimator in the presence of additional set-membership deviations [205]. Both approaches can be regarded as combinations of the standard Kalman filter from Section 2.1.1 and ellipsoidal calculus from Section 2.1.2.

The linear stochastic and set-membership models (2.12) and (2.32) of Section 2.2 are merged into the linear system dynamics

$$\underline{x}_{k+1} = \mathbf{A}_k \, \underline{x}_k + \mathbf{B}_k \left(\underline{\hat{u}} + \underline{w}_k + \underline{d}_k \right) \tag{2.65}$$

of the n_x-dimensional state vector $\underline{x}_k$ with system matrix $\mathbf{A}_k \in \mathbb{R}^{n_x \times n_x}$ and control-input matrix $\mathbf{B}_k \in \mathbb{R}^{n_x \times n_u}$. The input vector $\underline{\hat{u}}_k \in \mathbb{R}^{n_u}$ is affected by the white zero-mean noise $\underline{w}_k \sim \mathcal{N}(\underline{0}, \mathbf{C}_k^u)$ with covariance matrix $\mathbf{C}_k^u \in \mathbb{R}^{n_u \times n_u}$ and the unknown but bounded perturbation $\underline{d}_k \in \mathcal{E}(\underline{0}, \mathbf{X}_k^u)$ with shape matrix $\mathbf{X}_k^u \in \mathbb{R}^{n_u \times n_u}$. W.l.o.g., the set-membership errors are assumed to be centered at $\underline{0}$. Otherwise, we consider a shifted version $\underline{\hat{u}}_k' = \underline{\hat{u}}_k + \underline{\hat{c}}_k$ of the input quantity when $\underline{d}_k \in \mathcal{E}(\underline{\hat{c}}_k, \mathbf{X}_k^u)$.

A measurement $\underline{\hat{z}}_k$ is related to the state $\underline{x}_k$ according to a linear sensor model

$$\underline{z}_k = \mathbf{H}_k \, \underline{x}_k + \underline{v}_k + \underline{e}_k \; , \tag{2.66}$$

with the measurement matrix $\mathbf{H}_k \in \mathbb{R}^{n_z \times n_x}$. It combines the stochastic linear model (2.13) and the set-membership model (2.33) and is hence affected by both a white zero-mean noise $\underline{v}_k \sim \mathcal{N}(\underline{0}, \mathbf{C}_k^z)$ with covariance matrix $\mathbf{C}_k^z \in \mathbb{R}^{n_x \times n_x}$ and an unknown error $\underline{e}_k$ bounded by the ellipsoid $\mathcal{E}(\underline{0}, \mathbf{X}_k^z)$ with shape matrix $\mathbf{X}_k^z \in \mathbb{R}^{n_x \times n_x}$. In the following,

$\underline{x}_0, \underline{w}_0, \ldots, \underline{w}_k, \underline{v}_0, \ldots, \underline{v}_k$ are mutually independent and, furthermore, do not depend on the values of $\underline{d}_0, \ldots \underline{d}_k, \underline{e}_0, \ldots, \underline{e}_k$.

2.4.1 Kalman Filtering with Sets of Gaussian Densities

The generic Bayesian estimation framework for sets of probability densities proposed in Section 2.3 becomes a tractable estimation algorithm in case of linear models and Gaussian noise characteristics. In earlier studies [131,132], the first assumption that has been relaxed is that a precise prior estimate must exist for the Kalman filter. The prior has been characterized by an ellipsoidal set of means and a single covariance matrix, i.e., by a set of translated versions a Gaussian density. The further developments in [199–201] allow to also model transition densities and likelihoods to be imprecise, which is necessary to incorporate the set-membership uncertainties in (2.65) and (2.66). The objective is then to recursively predict and update the set

$$\mathcal{F}_k^{\mathrm{e}} = \left\{ f^{\mathrm{e}}(\underline{x}_k) = \mathcal{N}(\underline{x}_k; \underline{c}, \mathbf{C}_k^{\mathrm{e}}) \,\middle|\, \underline{c} \in \mathcal{E}(\hat{\underline{x}}_k^{\mathrm{e}}, \mathbf{X}_k^{\mathrm{e}}) \right\} \tag{2.67}$$

of conditional estimated densities, which can uniquely be parameterized by the mean or midpoint $\hat{\underline{x}}_k^{\mathrm{e}}$, the shape matrix $\mathbf{X}_k^{\mathrm{e}}$, and the covariance matrix $\mathbf{C}_k^{\mathrm{e}}$.

Prediction The input vector $\hat{\underline{u}}_k$ being additively affected by the unknown but bounded error $\underline{d}_k \in \mathcal{E}(\underline{0}, \mathbf{X}_k^u)$ can be regarded as a set

$$\mathcal{U}_k := \left\{ \hat{\underline{u}}_k + \underline{d}_k \,\middle|\, \underline{d}_k \in \mathcal{E}(\underline{0}, \mathbf{X}_k^u) \right\} = \mathcal{E}(\hat{\underline{u}}_k, \mathbf{X}_k^u)$$

of imprecise input quantities[6]. With the Gaussian process noise density $f^w = \mathcal{N}(\,\cdot\,; \underline{0}, \mathbf{C}_k^u)$, the set (2.63) of transition densities is simplified to

$$\mathcal{F}_k^{\mathrm{T}} = \Big\{ f(\underline{x}_{k+1} | \underline{x}_k)$$

$$= \mathcal{N}\Big(\underline{x}_{k+1} - \big(\mathbf{A}_k\,\underline{x}_k + \mathbf{B}_k\,\underline{u}_k\big); \underline{0}, \mathbf{C}_k^u\Big) \,\Big|\, \underline{u}_k \in \mathcal{E}(\hat{\underline{u}}_k, \mathbf{X}_k^u) \Big\},$$

---

[6]This interpretation states the reason why we denote set-membership uncertainties affecting the prediction step with the superscript u from the early beginning of this chapter. Analogously, the superscript z is used for measurement uncertainties.

which has to be processed elementwise with the set (2.67) of estimated densities according to the prediction step (2.58). Obviously, the prediction result $\mathcal{F}^{\mathrm{p}}_{k+1}$ then solely contains Gaussian densities. From the predicted covariance matrix (2.20) of the standard Kalman filter (see Section 2.1.1), we recognize that its computation is independent of the mean. This implies that every Gaussian density in the predicted set $\mathcal{F}^{\mathrm{p}}_{k+1}$ has the same covariance matrix

$$\mathbf{C}^{\mathrm{p}}_{k+1} \stackrel{(2.20)}{=} \mathbf{A}_k \mathbf{C}^{\mathrm{e}}_k \mathbf{A}^{\mathrm{T}}_k + \mathbf{B}_k \mathbf{C}^{u}_k \mathbf{B}^{\mathrm{T}}_k \ . \tag{2.68}$$

The mean of each density in $\mathcal{F}^{\mathrm{p}}_{k+1}$ can be calculated by means of the standard Kalman filter formula (2.19), i.e., (2.19) has to be applied to every possible combination of an estimate $\underline{x}^{\mathrm{e}}_k \in \mathcal{E}(\hat{\underline{x}}^{\mathrm{e}}_k, \mathbf{X}^{\mathrm{e}}_k)$ and imprecise input $\underline{u}_k \in \mathcal{E}(\hat{\underline{u}}_k, \mathbf{X}^{u}_k)$. Thus, the set $\mathcal{X}^{\mathrm{p}}_{k+1}$ of possible predicted means is obtained by

$$\begin{aligned} \mathcal{X}^{\mathrm{p}}_{k+1} &\stackrel{(2.19)}{=} \mathbf{A}_k\, \mathcal{E}(\hat{\underline{x}}^{\mathrm{e}}_k, \mathbf{X}^{\mathrm{e}}_k) \oplus \mathbf{B}_k \mathcal{E}(\hat{\underline{u}}_k, \mathbf{X}^{u}_k) \\ &\stackrel{(2.35)}{=} \mathcal{E}(\mathbf{A}_k\, \hat{\underline{x}}^{\mathrm{e}}_k, \mathbf{A}_k \mathbf{X}^{\mathrm{e}}_k \mathbf{A}^{\mathrm{T}}_k) \oplus \mathcal{E}(\mathbf{B}_k\, \hat{\underline{u}}_k, \mathbf{B}_k \mathbf{X}^{u}_k \mathbf{B}^{\mathrm{T}}_k) \\ &\stackrel{(2.36)}{\subseteq} \mathcal{E}(\hat{\underline{x}}^{\mathrm{p}}_{k+1}, \mathbf{X}^{\mathrm{p}}_{k+1}) \ , \end{aligned} \tag{2.69}$$

where the latter ellipsoid is an outer approximation of the Minkowski sum. Evidently, the computation of the set of means corresponds exactly to the ellipsoidal precition step (2.34) in Section 2.1.2, where the midpoint is given by

$$\hat{\underline{x}}^{\mathrm{p}}_{k+1} \stackrel{(2.36)}{=} \mathbf{A}_k\, \hat{\underline{x}}^{\mathrm{p}}_{k+1} + \mathbf{B}_k\, \hat{\underline{u}}_k \ ,$$

and $\mathbf{X}^{\mathrm{p}}_{k+1}$ is chosen out of the family of possible shape matrices

$$\mathbf{X}^{\mathrm{p}}_{k+1}(p) \stackrel{(2.37)}{=} (1 + p^{-1})\, \mathbf{A}_k \mathbf{X}^{\mathrm{e}}_k \mathbf{A}^{\mathrm{T}}_k + (1 + p)\, \mathbf{B}_k \mathbf{X}^{u}_k \mathbf{B}^{\mathrm{T}}_k$$

by minimizing the trace with

$$p^{\mathrm{opt}} \stackrel{(2.39)}{=} \mathrm{trace}(\mathbf{A}_k \mathbf{X}^{\mathrm{e}}_k \mathbf{A}^{\mathrm{T}}_k)^{\frac{1}{2}} \cdot \mathrm{trace}(\mathbf{B}_k \mathbf{X}^{u}_k \mathbf{B}^{\mathrm{T}}_k)^{-\frac{1}{2}} \ . \tag{2.70}$$

The result of the prediction step is finally parameterized by the midpoint $\hat{\underline{x}}^{\mathrm{P}}_{k+1}$, the shape matrix $\mathbf{X}^{\mathrm{P}}_{k+1}$, and the covariance matrix $\mathbf{C}^{\mathrm{P}}_{k+1}$, which define the set

$$\mathcal{F}^{\mathrm{P}}_{k+1} = \left\{ f^{\mathrm{P}}(\underline{x}_{k+1}) = \mathcal{N}(\underline{x}_{k+1}; \underline{c}, \mathbf{C}^{\mathrm{P}}_{k+1}) \,\middle|\, \underline{c} \in \mathcal{E}(\hat{\underline{x}}^{\mathrm{P}}_{k+1}, \mathbf{X}^{\mathrm{P}}_{k+1}) \right\} \qquad (2.71)$$

of predicted Gaussian densities.

Filtering In order to follow the same line of reasoning as in the prediction step, the measurement equation (2.66) is rewritten to

$$\underline{z}_k - \underline{e}_k = \mathbf{H}_k \, \underline{x}_k + \underline{v}_k$$

so that a concrete measurement $\hat{\underline{z}}_k$ together with the unknown but bounded error $\underline{e}_k \in \mathcal{E}(\underline{0}, \mathbf{X}^z_k)$ can be considered as a set

$$\mathcal{Z}_k := \left\{ \hat{\underline{z}}_k - \underline{e}_k \,\middle|\, \underline{e}_k \in \mathcal{E}(\underline{0}, \mathbf{X}^z_k) \right\} = \mathcal{E}(\hat{\underline{z}}_k, \mathbf{X}^z_k)$$

of possible measurements. The last part of the equation is due to the symmetry of the ellipsoid, so it is irrelevant whether $\underline{e}_k$ is added or subtracted. The set (2.64) of translated likelihoods then becomes

$$\mathcal{F}^{\mathrm{L}}_k = \left\{ f(\underline{z}_k \,|\, \underline{x}_k) = \mathcal{N}\left(\underline{z}_k - \mathbf{H}_k \, \underline{x}_k; \underline{0}, \mathbf{C}^z_k\right) \,\middle|\, \underline{z}_k \in \mathcal{E}(\hat{\underline{z}}_k, \mathbf{X}^z_k) \right\}$$

with $f^v = \mathcal{N}(\,\cdot\,; \underline{0}, \mathbf{C}^z_k)$ being the density of the measurement noise $\underline{v}_k$. In order to compute the set (2.60) of filtered densities, the set (2.71) of prior or predicted densities has to be updated elementwise with $\mathcal{F}^{\mathrm{L}}_k$. Again, the computation of the filtered covariance matrix is independent from the computation of the mean, as can be seen from (2.26). Since every density in $\mathcal{F}^{\mathrm{P}}_k$ and every density in $\mathcal{F}^{\mathrm{L}}_k$ has the same covariance matrix, every filtered density in the set $\mathcal{F}^{\mathrm{e}}_k$ therefore has also the same covariance matrix

$$\mathbf{C}^{\mathrm{e}}_k \overset{(2.26)}{=} \mathbf{C}^{\mathrm{P}}_k - \mathbf{K}_k \mathbf{H}_k \mathbf{C}^{\mathrm{P}}_k \,, \qquad (2.72)$$

which also implies that in each case the same Kalman gain (2.24) has been used. Consequently, the same Kalman gain is also employed to combine every prior mean $\underline{x}_k^{\mathrm{P}} \in \mathcal{E}(\hat{\underline{x}}_k^{\mathrm{P}}, \mathbf{X}_k^{\mathrm{P}})$ with every possible measurement $\underline{z}_k \in \mathcal{E}(\hat{\underline{z}}_k, \mathbf{X}_k^{z})$. This combination can again be written as a Minkowski sum

$$
\begin{aligned}
\mathcal{X}_k^{\mathrm{e}} &\overset{(2.25)}{=} (\mathbf{I} - \mathbf{K}_k\mathbf{H}_k)\,\mathcal{E}(\hat{\underline{x}}_k^{\mathrm{P}}, \mathbf{X}_k^{\mathrm{P}}) \oplus \mathbf{K}_k\,\mathcal{E}(\hat{\underline{z}}_k, \mathbf{X}_k^{z}) \\
&\overset{(2.35)}{=} \mathcal{E}\big((\mathbf{I} - \mathbf{K}_k\mathbf{H}_k)\,\hat{\underline{x}}_k^{\mathrm{P}}, (\mathbf{I} - \mathbf{K}_k\mathbf{H}_k)\mathbf{X}_k^{\mathrm{P}}(\mathbf{I} - \mathbf{K}_k\mathbf{H}_k)^{\mathrm{T}}\big) \\
&\qquad \oplus \mathcal{E}(\hat{\underline{z}}_k, \mathbf{K}_k\mathbf{X}_k^{z}\mathbf{K}_k^{\mathrm{T}}) \\
&\overset{(2.36)}{\subseteq} \mathcal{E}(\hat{\underline{x}}_k^{\mathrm{e}}, \mathbf{X}_k^{\mathrm{e}}) \,,
\end{aligned}
\tag{2.73}
$$

where the latter ellipsoid denotes an outer approximation with the midpoint

$$
\hat{\underline{x}}_k^{\mathrm{e}} \overset{(2.36)}{=} (\mathbf{I} - \mathbf{K}_k\mathbf{H}_k)\hat{\underline{x}}_k^{\mathrm{P}} + \mathbf{K}_k\,\hat{\underline{z}}_k
$$

and the shape matrix

$$
\mathbf{X}_k^{\mathrm{e}} \overset{(2.37)}{=} (1+p^{-1})\,(\mathbf{I} - \mathbf{K}_k\mathbf{H}_k)\mathbf{X}_k^{\mathrm{P}}(\mathbf{I} - \mathbf{K}_k\mathbf{H}_k)^{\mathrm{T}} + (1+p)\,\mathbf{K}_k\mathbf{X}_k^{z}\mathbf{K}_k^{\mathrm{T}} \,.
\tag{2.74}
$$

An enclosing ellipsoid with minimal trace of $\mathbf{X}_k^{\mathrm{e}}$ can be calculated by employing

$$
p^{\mathrm{opt}} \overset{(2.39)}{=} \operatorname{trace}((\mathbf{I} - \mathbf{K}_k\mathbf{H}_k)\mathbf{X}_k^{\mathrm{P}}(\mathbf{I} - \mathbf{K}_k\mathbf{H}_k)^{\mathrm{T}})^{\frac{1}{2}} \cdot \operatorname{trace}(\mathbf{K}_k\mathbf{X}_k^{z}\mathbf{K}_k^{\mathrm{T}})^{-\frac{1}{2}} \,.
\tag{2.75}
$$

It is particularly important to note that the estimated set of means is not, as one might expect, an intersection of ellipsoids but again a weighted Minkowski sum. An intersection can only be attained by adapting the gain $\mathbf{K}_k$, which is done in the following Subsection 2.4.2.

The generic Bayesian estimation framework for sets of densities discussed in Section 2.3 generally does not allow to explicitly distinguish between stochastic and set-membership uncertainties as it can be impossible to discern how the set-membership error description is incorporated in an arbitrary set of densities, especially when no parameterization is known. Contrary to nonlinear models and arbitrary densities, stochastic and set-membership error characteristics remain distinguishable from one another in the presented generalization of the Kalman filter: The former is

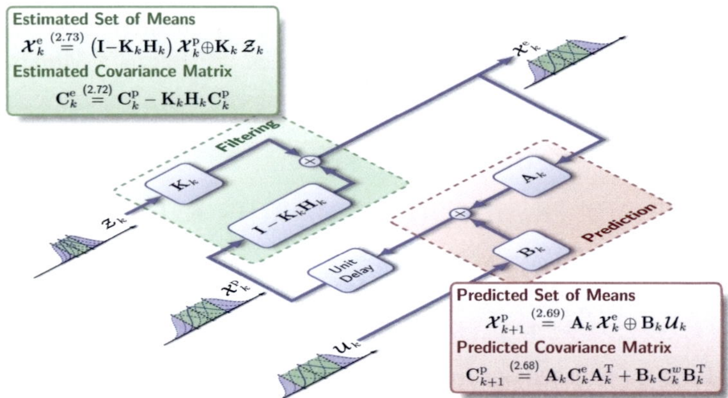

Figure 2.6: Uncertain quantities are characterized by sets of translated Gaussian densities. Each set can be parameterized by a set of means and a single covariance matrix. In comparison to the standard Kalman filter, only the computation of the means needs to be generalized to the Minkowski sums (2.69) and (2.73).

represented by the covariance matrix $\mathbf{C}_k^{\mathrm{e}}$ and the latter is characterized by the shape matrix $\mathbf{X}_k^{\mathrm{e}}$ of the ellipsoidal bound. Figure 2.6 summarizes the generalized Kalman filtering scheme and, compared to the purely stochastic scheme in Section 2.1.1, only the calculations of the means are replaced by Minkowski sums.

2.4.2 Linear Minimum Mean Squared Error Estimator

In the previous subsection, the Kalman filter has been generalized to set-membership uncertainties by considering the underlying sets of Gaussian probability densities. As stated earlier in Section 2.1.1, it is a misconception to assume that the Kalman filter strictly relies on Bayes' rule applied to underlying Gaussian densities. The derivation of the purely stochastic Kalman filter does not require the exploitation of specific probabilistic information other than mean and covariance matrix. Analogously, we can also derive a linear minimum mean squared error (LMMSE) estimator in the simultaneous presence of set-membership uncertainties. This concept is based on the results published in [205] and is studied more closely in the following. A different derivation of the optimal gain is provided, and the discussion of special and limiting cases is expanded.

According to (2.16) and (2.17), the optimal estimate $\hat{\underline{x}}_k^{\mathrm{e}}$ has the property that the random deviation

$$\tilde{\underline{x}}_k^{\mathrm{stoc}} = \hat{\underline{x}}_k^{\mathrm{e}} - \underline{x}_k$$

from the uncertain state $\underline{x}_k$ has zero mean and the corresponding MSE is given by $\mathrm{E}[(\tilde{\underline{x}}_k^{\mathrm{stoc}})^{\mathrm{T}}(\tilde{\underline{x}}_k^{\mathrm{stoc}})]$. By following the considerations in [184] and [205], we aspire to derive a Kalman filtering scheme for estimation problems where also an unknown but bounded deviation may affect the state estimate and the total deviation yields

$$\tilde{\underline{x}}_k^{\mathrm{stoc}} + \tilde{\underline{x}}_k^{\mathrm{set}} = \hat{\underline{x}}_k^{\mathrm{e}} - \underline{x}_k \ . \tag{2.76}$$

The stochastic part $\tilde{\underline{x}}_k^{\mathrm{stoc}}$ has zero mean and the error covariance matrix $\mathbf{C}_k^{\mathrm{e}} = \mathrm{Cov}(\tilde{\underline{x}}_k^{\mathrm{stoc}})$, and the set-membership $\tilde{\underline{x}}_k^{\mathrm{set}}$ deviation is bounded by the ellipsoid $\mathcal{E}(\underline{0}, \mathbf{X}_k^{\mathrm{e}})$. As a result, the unbiasedness (2.16) does not hold anymore due to

$$\mathrm{E}[\hat{\underline{x}}_k^{\mathrm{e}}] - \mathrm{E}[\underline{x}_k] = \mathrm{E}[\hat{\underline{x}}_k^{\mathrm{e}} - \underline{x}_k] = \mathrm{E}[\tilde{\underline{x}}^{\mathrm{set}}] \in \mathcal{E}(\underline{0}, \mathbf{X}_k^{\mathrm{e}}) \ . \tag{2.77}$$

The MSE matrix in this case becomes

$$\begin{aligned}
\mathrm{E}\left[(\hat{\underline{x}}_k^{\mathrm{e}} - \underline{x}_k)(\hat{\underline{x}}_k^{\mathrm{e}} - \underline{x}_k)^{\mathrm{T}}\right] &= \mathrm{E}\left[(\tilde{\underline{x}}_k^{\mathrm{stoc}} + \tilde{\underline{x}}_k^{\mathrm{set}})(\tilde{\underline{x}}_k^{\mathrm{stoc}} + \tilde{\underline{x}}_k^{\mathrm{set}})^{\mathrm{T}}\right] \\
&= \mathrm{E}\left[(\tilde{\underline{x}}_k^{\mathrm{stoc}})(\tilde{\underline{x}}_k^{\mathrm{stoc}})^{\mathrm{T}}\right] + \mathrm{E}\left[(\tilde{\underline{x}}_k^{\mathrm{set}})(\tilde{\underline{x}}_k^{\mathrm{set}})^{\mathrm{T}}\right] \\
&= \mathbf{C}_k^{\mathrm{e}} + \mathrm{E}\left[(\tilde{\underline{x}}_k^{\mathrm{set}})(\tilde{\underline{x}}_k^{\mathrm{set}})^{\mathrm{T}}\right] \\
&= \mathbf{C}_k^{\mathrm{e}} + (\tilde{\underline{x}}_k^{\mathrm{set}})(\tilde{\underline{x}}_k^{\mathrm{set}})^{\mathrm{T}} \ ,
\end{aligned} \tag{2.78}$$

where the latter equality holds when $\tilde{\underline{x}}_k^{\mathrm{set}}$ is non-stochastic[7], e.g., a systematic error. The trace of this matrix then yields the MSE

$$\mathrm{E}\left[(\hat{\underline{x}}_k^{\mathrm{e}} - \underline{x}_k)^{\mathrm{T}}(\hat{\underline{x}}_k^{\mathrm{e}} - \underline{x}_k)\right] = \underbrace{\mathrm{E}\left[(\tilde{\underline{x}}_k^{\mathrm{stoc}})^{\mathrm{T}}(\tilde{\underline{x}}_k^{\mathrm{stoc}})\right]}_{= \mathrm{trace}(\mathbf{C}^{\mathrm{e}})} + \underbrace{(\tilde{\underline{x}}_k^{\mathrm{set}})^{\mathrm{T}}(\tilde{\underline{x}}_k^{\mathrm{set}})}_{\leq \mathrm{trace}(\mathbf{X}^{\mathrm{e}})} \tag{2.79}$$

$$\leq \mathrm{trace}(\mathbf{C}_k^{\mathrm{e}} + \mathbf{X}_k^{\mathrm{e}}) \ .$$

---

[7]In the following, we assume $\tilde{\underline{x}}_k^{\mathrm{set}}$ to be non-stochastic in order to simplify matters. Otherwise, we define $\tilde{\underline{x}}_k^{\mathrm{set}} := \mathrm{E}[\tilde{\underline{x}}_k^{\mathrm{set}}] \in \mathcal{E}(\underline{0}, \mathbf{X}_k^{\mathrm{e}})$, and the conclusions remain the same.

The latter inequality holds because the Euclidean distance of any element $\underline{e}_k$ in an ellipsoid $\mathcal{E}(\underline{\hat{c}}, \mathbf{X})$ to the center $\underline{\hat{c}}$ is related to the trace of $\mathbf{X}$ via the inequality

$$
\begin{aligned}
\|\underline{\hat{c}} - \underline{e}\|^2 &= \operatorname{trace}\left((\underline{\hat{c}} - \underline{e})(\underline{\hat{c}} - \underline{e})^{\mathrm{T}}\right) \\
&= (\underline{\hat{c}} - \underline{e})^{\mathrm{T}}(\underline{\hat{c}} - \underline{e}) \\
&\leq \operatorname{trace}\left(\mathbf{X}\right) ,
\end{aligned}
\tag{2.80}
$$

as stated in [111]. Thus, the relation $(\underline{\tilde{x}}_k^{\mathrm{set}})^{\mathrm{T}}(\underline{\tilde{x}}_k^{\mathrm{set}}) = \|\underline{\tilde{x}}_k^{\mathrm{set}}\|^2 \leq \operatorname{trace}(\mathbf{X}_k^{\mathrm{e}})$ immediately justifies the use of ellipsoidal error bounds since the Euclidean length of the error is directly bounded by the sum of the squared lengths of the semiaxes, i.e., the trace of $\mathbf{X}_k^{\mathrm{e}}$. Consequently, the trace of the covariance matrix characterizes the mean squared error of the stochastic term and the trace of the shape matrix bounds the maximum squared error of the set-membership term. Before we derive the Kalman filter algorithm for the considered situation, we extend the notion of unbiasedness to set-membership uncertainties.

A   On the Unbiased Condition

As explained in [205], set-membership perturbations may directly affect the mean of a linear estimator, according to

$$
\begin{aligned}
\mathrm{E}\left[\mathbf{K}'\underline{\hat{x}}_1^{\mathrm{e}} + \mathbf{K}\underline{\hat{x}}_2^{\mathrm{e}}\right] &= \mathbf{K}'(\mathrm{E}[\underline{x}] + \underline{\tilde{x}}_1^{\mathrm{set}}) + \mathbf{K}(\mathrm{E}[\underline{x}] + \underline{\tilde{x}}_2^{\mathrm{set}}) \\
&= (\mathbf{K}' + \mathbf{K})\,\mathrm{E}[\underline{x}] + \mathbf{K}'\underline{\tilde{x}}_1^{\mathrm{set}} + \mathbf{K}\underline{\tilde{x}}_2^{\mathrm{set}} ,
\end{aligned}
$$

where $\mathbf{K}'$ and $\mathbf{K}$ are the gains that are used to fuse the two estimates $\underline{\hat{x}}_1^{\mathrm{e}}$ and $\underline{\hat{x}}_2^{\mathrm{e}}$. For known deviations $\underline{\tilde{x}}_1^{\mathrm{set}}$ and $\underline{\tilde{x}}_2^{\mathrm{set}}$, the gains can be determined to eliminate the bias and to minimize (2.79), i.e.,

$$
(\mathbf{K}' + \mathbf{K})\,\mathrm{E}[\underline{x}] = \mathrm{E}[\underline{x}] - \mathbf{K}'\underline{\tilde{x}}_1^{\mathrm{set}} - \mathbf{K}\underline{\tilde{x}}_2^{\mathrm{set}} .
$$

Since the deviations are unknown and are bounded by ellipsoids centered at the origin, even $-\underline{\tilde{x}}_1^{\mathrm{set}}$ and $-\underline{\tilde{x}}_2^{\mathrm{set}}$ are possible. The mean would then yield

$$
\mathrm{E}\left[\mathbf{K}'\underline{\hat{x}}_1^{\mathrm{e}} + \mathbf{K}\underline{\hat{x}}_2^{\mathrm{e}}\right] = (\mathbf{K}' + \mathbf{K})\,\mathrm{E}[\underline{x}] - 2\mathbf{K}'\underline{\tilde{x}}_1^{\mathrm{set}} - 2\mathbf{K}\underline{\tilde{x}}_2^{\mathrm{set}} ,
$$

which could significantly increase the actual MSE. Therefore, $\mathbf{K}' = \mathbf{I} - \mathbf{K}$ minimizes the risk of a high error, which otherwise must be bounded and results into a larger non-optimal bound for the total MSE. A second argument is that the set-membership error bound also includes zero-mean stochastic perturbations, for which we expect an unbiased estimation result.

B   Recursive LMMSE Estimation

In order to compute an estimate that minimizes the MSE (2.79), the formulas for the prediction and filtering step need to be adapted appropriately. While it turns out that the prediction steps remains the same as in Section 2.4.1, the main challenge is to derive the Kalman gain.

Prediction   In order to compute a predicted estimate for the evolved state $\underline{x}_{k+1}$, we consider the conditional mean

$$
\begin{aligned}
\hat{\underline{x}}_{k+1}^{\mathrm{p}} &\overset{(2.65)}{=} \mathrm{E}[\mathbf{A}_k\,\underline{x}_k + \mathbf{B}_k(\hat{\underline{u}}_k + \underline{w}_k + \underline{d}_k)\,|\,\hat{\underline{z}}_{0:k}] \\
&= \mathbf{A}_k(\hat{\underline{x}}_k^{\mathrm{e}} + \tilde{\underline{x}}_k^{\mathrm{set}}) + \mathbf{B}_k(\hat{\underline{u}}_k + \underline{d}_k) \\
&= \mathbf{A}_k\,\hat{\underline{x}}_k^{\mathrm{e}} + \mathbf{B}_k\,\hat{\underline{u}}_k + \mathbf{A}_k\,\tilde{\underline{x}}_k^{\mathrm{set}} + \mathbf{B}_k\,\underline{d}_k\;,
\end{aligned}
$$

which is in compliance with the standard Kalman prediction step (2.19) in Section 2.1.1-B. It is affected by the set-membership error[8] $\tilde{\underline{x}}_k^{\mathrm{set}}$ of the previous estimate $\hat{\underline{x}}_k^{\mathrm{e}}$ and the unknown input error $\underline{d}_k$. At the moment, we therefore cannot state a specific value for $\hat{\underline{x}}_{k+1}^{\mathrm{p}}$. For the following considerations, we choose the midpoint

$$
\hat{\underline{x}}_{k+1}^{\mathrm{p}} = \mathbf{A}_k\,\hat{\underline{x}}_k^{\mathrm{e}} + \mathbf{B}_k\,\hat{\underline{u}}_k \tag{2.81}
$$

of this set of possible estimates and compute the MSE matrix of the error

$$
\begin{aligned}
\hat{\underline{x}}_{k+1}^{\mathrm{p}} - \underline{x}_{k+1} &= \mathbf{A}_k(\hat{\underline{x}}_k^{\mathrm{e}} - \underline{x}_k^{\mathrm{e}}) + \mathbf{B}_k(\underline{w}_k + \underline{d}_k) \\
&\overset{(2.76)}{=} \mathbf{A}_k(\tilde{\underline{x}}_k^{\mathrm{stoc}} + \tilde{\underline{x}}_k^{\mathrm{set}}) + \mathbf{B}_k(\underline{w}_k + \underline{d}_k)\;,
\end{aligned}
$$

---

[8]Note that, according to (2.77), it should say $-\tilde{\underline{x}}_k^{\mathrm{set}}$, but the set of possible errors is modeled to be symmetric around origin.

i.e.,

$$
\begin{aligned}
\mathrm{E}[(\hat{\underline{x}}^{\mathrm{p}}_{k+1} &- \underline{x}_{k+1})(\hat{\underline{x}}^{\mathrm{p}}_{k+1} - \underline{x}_{k+1})^{\mathrm{T}}] \\
&= \mathrm{E}\left[\left(\mathbf{A}_k(\hat{\underline{x}}^{\mathrm{e}}_k - \underline{x}_k) + \mathbf{B}_k(\underline{w}_k + \underline{d}_k)\right)\left(\mathbf{A}_k(\hat{\underline{x}}^{\mathrm{e}}_k - \underline{x}_k) + \mathbf{B}_k(\underline{w}_k + \underline{d}_k)\right)^{\mathrm{T}}\right] \\
&= \mathbf{A}_k\mathbf{C}^{\mathrm{e}}_k\mathbf{A}^{\mathrm{T}}_k + \mathbf{B}_k\mathbf{C}^{u}_k\mathbf{B}^{\mathrm{T}}_k + (\mathbf{A}_k\,\tilde{\underline{x}}^{\mathrm{set}}_k + \mathbf{B}_k\,\underline{d}_k)(\mathbf{A}_k\,\tilde{\underline{x}}^{\mathrm{set}}_k + \mathbf{B}_k\,\underline{d}_k)^{\mathrm{T}} \ .
\end{aligned}
\tag{2.82}
$$

By comparing the result with (2.78), we realize that the predicted stochastic uncertainty is, as expected, represented by

$$
\mathbf{C}^{\mathrm{p}}_{k+1} = \mathbf{A}_k\mathbf{C}^{\mathrm{e}}_k\mathbf{A}^{\mathrm{T}}_k + \mathbf{B}_k\mathbf{C}^{u}_k\mathbf{B}^{\mathrm{T}}_k \ ,
\tag{2.83}
$$

and the set-membership deviations are characterized by a Minkowski sum according to

$$
\mathbf{A}_k\,\tilde{\underline{x}}^{\mathrm{set}}_k + \mathbf{B}_k\,\underline{d}_k \in \mathbf{A}_k\,\mathcal{E}(\underline{0}, \mathbf{X}^{\mathrm{e}}_k) \oplus \mathbf{B}_k\,\mathcal{E}(\underline{0}, \mathbf{X}^{z}_k) \subseteq \mathcal{E}(\underline{0}, \mathbf{X}^{\mathrm{p}}_{k+1}) \ ,
$$

for which the latter ellipsoid is an external approximation with the shape matrix

$$
\mathbf{X}^{\mathrm{p}}_{k+1}(p) \stackrel{(2.37)}{=} (1 + p^{-1})\,\mathbf{A}_k\mathbf{X}^{\mathrm{e}}_k\mathbf{A}^{\mathrm{T}}_k + (1 + p)\,\mathbf{B}_k\mathbf{X}^{u}_k\mathbf{B}^{\mathrm{T}}_k \ .
\tag{2.84}
$$

In compliance with (2.79), the MSE is related to the inequality

$$
\begin{aligned}
\mathrm{E}\left[(\hat{\underline{x}}^{\mathrm{e}}_k - \underline{x}_k)^{\mathrm{T}}(\hat{\underline{x}}^{\mathrm{e}}_k - \underline{x}_k)\right] &\leq \mathrm{trace}(\mathbf{C}^{\mathrm{p}}_{k+1} + \mathbf{X}^{\mathrm{p}}_{k+1}(p)) \\
&= \mathrm{trace}(\mathbf{C}^{\mathrm{p}}_{k+1}) + \mathrm{trace}(\mathbf{X}^{\mathrm{p}}_{k+1}(p)) \ .
\end{aligned}
\tag{2.85}
$$

The matrix $\mathbf{X}^{\mathrm{p}}_{k+1} = \mathbf{X}^{\mathrm{p}}_{k+1}(p^{\mathrm{opt}})$ must have minimal trace by setting p^{opt} to (2.70) so as to minimize the bound (2.85).

The error matrices $\mathbf{C}^{\mathrm{p}}_{k+1}$ and $\mathbf{X}^{\mathrm{p}}_{k+1}$ have been derived for the chosen predicted estimate (2.81). The reason for this choice is the same as in Paragraph 2.4.2-A. For any other point estimate $\hat{\underline{x}}^{\mathrm{p}}_{k+1} = \mathbf{A}_k\,\hat{\underline{x}}^{\mathrm{e}}_k + \mathbf{B}_k\,\hat{\underline{u}}_k + \underline{b}_k$, the actual MSE matrix (2.82) might become $\mathbf{C}^{\mathrm{p}}_{k+1} + 2 \cdot \underline{b}_k\underline{b}^{\mathrm{T}}_k$. Hence, in order to avoid the risk of a high MSE, we have to choose (2.81). Apparently, we have derived the same predicted parameters $\hat{\underline{x}}^{\mathrm{p}}_{k+1}$, $\mathbf{C}^{\mathrm{p}}_{k+1}$, and $\mathbf{X}^{\mathrm{p}}_{k+1}$ as for the prediction step in the previous Subsection 2.4.1. Now, the shape matrix $\mathbf{X}^{\mathrm{p}}_{k+1}$ does not characterize a set of possible means but rather the set-membership uncertainty associated to the point estimate $\hat{\underline{x}}^{\mathrm{p}}_{k+1}$.

Filtering Based upon the measurement data and system dynamics up to a time instance k, we are looking for the Kalman gain $\mathbf{K}_k$ that combines the prior state estimate $\hat{\underline{x}}_k^{\mathrm{p}}$ with the measurement information $\hat{\underline{z}}_k$ according to

$$\hat{\underline{x}}_k^{\mathrm{e}} = (\mathbf{I} - \mathbf{K}_k \mathbf{H}_k)\hat{\underline{x}}_k^{\mathrm{p}} + \mathbf{K}_k\hat{\underline{z}}_k = \hat{\underline{x}}_k^{\mathrm{p}} + \mathbf{K}_k\left(\hat{\underline{z}}_k - \mathbf{H}_k\,\hat{\underline{x}}_k^{\mathrm{p}}\right) , \tag{2.86}$$

and concurrently minimizes the trace of (2.78), i.e., the MSE (2.79). As discussed in Paragraph 2.4.2-A, we require the estimator to be unbiased. With $\tilde{\underline{x}}_k^{\mathrm{stoc}} + \tilde{\underline{x}}_k^{\mathrm{set}}$ denoting the errors perturbing the prior estimate $\hat{\underline{x}}_k^{\mathrm{p}}$, the MSE matrix yields

$$\begin{aligned}
\mathrm{E}&\left[(\hat{\underline{x}}_k^{\mathrm{e}} - \underline{x}_k)(\hat{\underline{x}}_k^{\mathrm{e}} - \underline{x}_k)^{\mathrm{T}}\right]\\[2pt]
&= \mathrm{E}\left[\left(\hat{\underline{x}}_k^{\mathrm{p}} + \mathbf{K}_k(\hat{\underline{z}}_k - \mathbf{H}_k\,\hat{\underline{x}}_k^{\mathrm{p}}) - \underline{x}_k\right)\left(\hat{\underline{x}}_k^{\mathrm{p}} + \mathbf{K}_k(\hat{\underline{z}}_k - \mathbf{H}_k\,\hat{\underline{x}}_k^{\mathrm{p}}) - \underline{x}_k\right)^{\mathrm{T}}\right]\\[2pt]
&= \mathrm{E}\left[\left(\tilde{\underline{x}}_k^{\mathrm{stoc}} + \tilde{\underline{x}}_k^{\mathrm{set}} + \mathbf{K}_k(\mathbf{H}_k\,\underline{x}_k + \underline{v}_k + \underline{e}_k - \mathbf{H}_k\,\hat{\underline{x}}_k^{\mathrm{p}})\right)\left(\ldots\right)^{\mathrm{T}}\right]\\[2pt]
&= \mathrm{E}\left[\left((\mathbf{I} - \mathbf{K}_k\mathbf{H}_k)(\tilde{\underline{x}}_k^{\mathrm{stoc}} + \tilde{\underline{x}}_k^{\mathrm{set}}) + \mathbf{K}_k(\underline{v}_k + \underline{e}_k)\right)\left(\ldots\right)^{\mathrm{T}}\right]\\[2pt]
&= (\mathbf{I} - \mathbf{K}_k\mathbf{H}_k)\,\mathrm{E}[(\tilde{\underline{x}}_k^{\mathrm{stoc}})(\tilde{\underline{x}}_k^{\mathrm{stoc}})^{\mathrm{T}}](\mathbf{I} - \mathbf{K}_k\mathbf{H}_k)^{\mathrm{T}} + \mathbf{K}_k\,\mathrm{E}[(\underline{v}_k)(\underline{v}_k)^{\mathrm{T}}]\mathbf{K}_k^{\mathrm{T}}\\[2pt]
&\quad + \mathrm{E}\left[\left((\mathbf{I} - \mathbf{K}_k\mathbf{H}_k)\tilde{\underline{x}}_k^{\mathrm{set}} + \mathbf{K}_k\underline{e}_k\right)(\ldots)^{\mathrm{T}}\right]\\[2pt]
&= (\mathbf{I} - \mathbf{K}_k\mathbf{H}_k)\mathbf{C}_k^{\mathrm{p}}(\mathbf{I} - \mathbf{K}_k\mathbf{H}_k)^{\mathrm{T}} + \mathbf{K}_k\mathbf{C}_k^{z}\mathbf{K}_k^{\mathrm{T}}\\[2pt]
&\quad + \left((\mathbf{I} - \mathbf{K}_k\mathbf{H}_k)\tilde{\underline{x}}_k^{\mathrm{set}} + \mathbf{K}_k\underline{e}_k\right)\left((\mathbf{I} - \mathbf{K}_k\mathbf{H}_k)\tilde{\underline{x}}_k^{\mathrm{set}} + \mathbf{K}_k\underline{e}_k\right)^{\mathrm{T}} .
\end{aligned}$$
$$\tag{2.87}$$

Compared to (2.78), this sum is again a combination of the estimated error covariance matrix

$$\mathbf{C}_k^{\mathrm{e}} = (\mathbf{I} - \mathbf{K}_k\mathbf{H}_k)\mathbf{C}_k^{\mathrm{p}}(\mathbf{I} - \mathbf{K}_k\mathbf{H}_k)^{\mathrm{T}} + \mathbf{K}_k\mathbf{C}_k^{z}\mathbf{K}_k^{\mathrm{T}} \tag{2.88}$$

and the estimated set-membership error matrix, which is computed with the help of a bounding ellipsoid. Due to the set-membership of $\tilde{\underline{x}}_k^{\mathrm{set}}$ and $\underline{e}_k$, the latter sum is related to a Minkowki sum according to

$$(\mathbf{I} - \mathbf{K}_k\mathbf{H}_k)\,\underline{\tilde{x}}_k^{\mathrm{set}} + \mathbf{K}_k\,\underline{e}_k \in (\mathbf{I} - \mathbf{K}_k\mathbf{H}_k)\mathcal{E}(\underline{0}, \mathbf{X}_k^{\mathrm{p}}) \oplus \mathbf{K}_k\mathcal{E}(\underline{0}, \mathbf{X}_k^{z})$$

$$\stackrel{(2.35)}{=} \mathcal{E}(\underline{0}, (\mathbf{I} - \mathbf{K}_k\mathbf{H}_k)\mathbf{X}_k^{\mathrm{p}}(\mathbf{I} - \mathbf{K}_k\mathbf{H}_k)^{\mathrm{T}})$$

$$\oplus \mathcal{E}(\underline{0}, \mathbf{K}_k\mathbf{X}_k^{z}\mathbf{K}_k^{\mathrm{T}})$$

$$\stackrel{(2.36)}{\subseteq} \mathcal{E}(\underline{0}, \mathbf{X}_k^{\mathrm{e}}(p)) \,,$$

where

$$\mathbf{X}_k^{\mathrm{e}}(p) \stackrel{(2.37)}{=} (1 + p^{-1})\,(\mathbf{I} - \mathbf{K}_k\mathbf{H}_k)\mathbf{X}_k^{\mathrm{p}}(\mathbf{I} - \mathbf{K}_k\mathbf{H}_k)^{\mathrm{T}}$$
$$+ (1 + p)\,\mathbf{K}_k\mathbf{X}_k^{z}\mathbf{K}_k^{\mathrm{T}} \tag{2.89}$$

denotes the shape matrix of the bounding ellipsoid. For this matrix, we have

$$\mathrm{trace}\left(\left((\mathbf{I} - \mathbf{K}_k\mathbf{H}_k)\underline{\tilde{x}}_k^{\mathrm{set}} + \mathbf{K}_k\,\underline{e}_k\right)\left((\mathbf{I} - \mathbf{K}_k\mathbf{H}_k)\underline{\tilde{x}}_k^{\mathrm{set}} + \mathbf{K}_k\,\underline{e}_k\right)^{\mathrm{T}}\right)$$
$$\leq\ \mathrm{trace}\left(\mathbf{X}_k^{\mathrm{e}}(p)\right)$$

for all $p > 0$, due to inequality (2.80). The actual MSE can now be bounded from above by

$$\mathrm{E}\left[(\hat{\underline{x}}_k^{\mathrm{e}} - \underline{x}_k)^{\mathrm{T}}(\hat{\underline{x}}_k^{\mathrm{e}} - \underline{x}_k)\right] = \mathrm{trace}\left(\mathrm{E}\left[(\hat{\underline{x}}_k^{\mathrm{e}} - \underline{x}_k)(\hat{\underline{x}}_k^{\mathrm{e}} - \underline{x}_k)^{\mathrm{T}}\right]\right)$$

$$\stackrel{(2.87)}{=} \mathrm{trace}\left(\mathbf{C}_k^{\mathrm{e}}\right) + \mathrm{trace}\left(\left((\mathbf{I} - \mathbf{K}_k\mathbf{H}_k)\underline{\tilde{x}}_k^{\mathrm{set}} + \mathbf{K}_k\underline{e}_k\right)\left(\ldots\right)^{\mathrm{T}}\right)$$

$$\stackrel{(2.80)}{\leq} \mathrm{trace}\left(\mathbf{C}_k^{\mathrm{e}}\right) + \mathrm{trace}\left(\mathbf{X}_k^{\mathrm{e}}(p)\right)$$

$$= \mathrm{trace}\left((\mathbf{I} - \mathbf{K}_k\mathbf{H}_k)\mathbf{C}_k^{\mathrm{p}}(\mathbf{I} - \mathbf{K}_k\mathbf{H}_k)^{\mathrm{T}}\right) + \mathrm{trace}\left(\mathbf{K}_k\mathbf{C}_k^{z}\mathbf{K}_k^{\mathrm{T}}\right)$$
$$+ (1 + p^{-1})\,\mathrm{trace}\left((\mathbf{I} - \mathbf{K}_k\mathbf{H}_k)\mathbf{X}_k^{\mathrm{p}}(\mathbf{I} - \mathbf{K}_k\mathbf{H}_k)^{\mathrm{T}}\right)$$
$$+ (1 + p)\,\mathrm{trace}\left(\mathbf{K}_k\mathbf{X}_k^{z}\mathbf{K}_k^{\mathrm{T}}\right)$$
$$=: \overline{\mathrm{E}}\left[(\hat{\underline{x}}_k^{\mathrm{e}} - \underline{x}_k)^{\mathrm{T}}(\hat{\underline{x}}_k^{\mathrm{e}} - \underline{x}_k)\right] \,. \tag{2.90}$$

The optimal parameter p that minimizes the upper bound $\overline{\mathrm{E}}\left[(\hat{\underline{x}}_k^{\mathrm{e}} - \underline{x}_k)^{\mathrm{T}}(\hat{\underline{x}}_k^{\mathrm{e}} - \underline{x}_k)\right]$ is given by (2.39), i.e.,

$$p = \frac{\sqrt{\mathrm{trace}\left((\mathbf{I} - \mathbf{K}_k\mathbf{H}_k)\mathbf{X}_k^{\mathrm{p}}(\mathbf{I} - \mathbf{K}_k\mathbf{H}_k)^{\mathrm{T}}\right)}}{\sqrt{\mathrm{trace}\left(\mathbf{K}_k\mathbf{X}_k^{z}\mathbf{K}_k^{\mathrm{T}}\right)}} \,,$$

and, analogously to (2.38), the right-hand side of (2.90) then reads

$$
\begin{aligned}
\overline{\mathrm{E}}&\big[(\hat{\underline{x}}_k^{\mathrm{e}} - \underline{x}_k)^{\mathrm{T}}(\hat{\underline{x}}_k^{\mathrm{e}} - \underline{x}_k)\big] \\
&= \mathrm{trace}\left((\mathbf{I} - \mathbf{K}_k\mathbf{H}_k)\mathbf{C}_k^{\mathrm{p}}(\mathbf{I} - \mathbf{K}_k\mathbf{H}_k)^{\mathrm{T}}\right) + \mathrm{trace}\left(\mathbf{K}_k\mathbf{C}_k^{z}\mathbf{K}_k^{\mathrm{T}}\right) \\
&\quad + \Big(\underbrace{\sqrt{\mathrm{trace}\left((\mathbf{I} - \mathbf{K}_k\mathbf{H}_k)\mathbf{X}_k^{\mathrm{p}}(\mathbf{I} - \mathbf{K}_k\mathbf{H}_k)^{\mathrm{T}}\right)}}_{=:A} + \underbrace{\sqrt{\mathrm{trace}\left(\mathbf{K}_k\mathbf{X}_k^{z}\mathbf{K}_k^{\mathrm{T}}\right)}}_{=:B}\Big)^2 .
\end{aligned}
$$

The derivative rules (2.22) and (2.23) for the trace can be utilized to compute

$$
\begin{aligned}
\frac{\partial}{\partial \mathbf{K}_k}&\overline{\mathrm{E}}\big[(\hat{\underline{x}}_k^{\mathrm{e}} - \underline{x}_k)^{\mathrm{T}}(\hat{\underline{x}}_k^{\mathrm{e}} - \underline{x}_k)\big] \\
&= -2\mathbf{C}_k^{\mathrm{p}}\mathbf{H}_k^{\mathrm{T}} + 2\mathbf{K}_k\mathbf{H}_k\mathbf{C}_k^{\mathrm{p}}\mathbf{H}_k^{\mathrm{T}} + 2\mathbf{K}_k\mathbf{C}_k^{z} \\
&\quad + \left(-\mathbf{X}_k^{\mathrm{p}}\mathbf{H}_k^{\mathrm{T}} + \mathbf{K}_k\mathbf{H}_k\mathbf{X}_k^{\mathrm{p}}\mathbf{H}_k^{\mathrm{T}}\right) \cdot \frac{1}{A} \cdot 2(A + B) \\
&\quad + 2\mathbf{K}_k\mathbf{X}_k^{z} \cdot \frac{1}{B} \cdot (A + B) \\
&= -2\mathbf{C}_k^{\mathrm{p}}\mathbf{H}_k^{\mathrm{T}} + 2\mathbf{K}_k\mathbf{H}_k\mathbf{C}_k^{\mathrm{p}}\mathbf{H}_k^{\mathrm{T}} + 2\mathbf{K}_k\mathbf{C}_k^{z} \\
&\quad + 2\left(1 + \frac{B}{A}\right)\left(-\mathbf{X}_k^{\mathrm{p}}\mathbf{H}_k^{\mathrm{T}} + \mathbf{K}_k\mathbf{H}_k\mathbf{X}_k^{\mathrm{p}}\mathbf{H}_k^{\mathrm{T}}\right) \\
&\quad + 2\left(1 + \frac{A}{B}\right)\mathbf{K}_k\mathbf{X}_k^{z}
\end{aligned}
$$

Substituting $p := \frac{A}{B}$, setting the above derivative to zero, and rearranging the equation yield the gain

$$
\begin{aligned}
\mathbf{K}_k(p) = &\left((1 + p^{-1})\mathbf{X}_k^{\mathrm{p}}\mathbf{H}_k^{\mathrm{T}} + \mathbf{C}_k^{\mathrm{p}}\mathbf{H}_k^{\mathrm{T}}\right) \cdot \\
&\left((1 + p^{-1})\mathbf{H}_k\mathbf{X}_k^{\mathrm{p}}\mathbf{H}_k^{\mathrm{T}} + (1 + p)\mathbf{X}_k^{z} + \mathbf{H}_k\mathbf{C}_k^{\mathrm{p}}\mathbf{H}^{\mathrm{T}} + \mathbf{C}_k^{z}\right)^{-1}
\end{aligned}
\tag{2.91}
$$

According to this result, the minimization of (2.90) only depends on the scalar parameter $p > 0$. Hence, the n_x^2-dimensional optimization problem of finding the optimal gain $\mathbf{K}_k$ has been reduced to a one-dimensional one. Unfortunately, a convex optimization is needed to find that value p^{opt} which minimizes the MSE (2.90). However, the need for such an optimization is a usual issue of ellipsoidal approximations. With $\mathbf{K}_k = \mathbf{K}_k(p^{\mathrm{opt}})$, the point estimate $\hat{\underline{x}}_k^{\mathrm{e}}$ is obtained by (2.86). The updated covariance matrix for the stochastic estimation error and the updated shape matrix for the unknown but bounded error can be computed by (2.88) and (2.89), respectively. These parameters, $\mathbf{C}_k^{\mathrm{e}}$ and $\mathbf{X}_k^{\mathrm{e}}$, then characterize the uncertainty associated to the point estimate $\hat{\underline{x}}_k^{\mathrm{e}}$.

Remark 1 For the derived gain (2.91), the MSE matrix that corresponds to the bound (2.90) can be rewritten to

$$\overline{\mathrm{E}}\big[(\hat{\underline{x}}_k^{\mathrm{e}} - \underline{x}_k)(\hat{\underline{x}}_k^{\mathrm{e}} - \underline{x}_k)^{\mathrm{T}}\big]$$
$$= \Big((\mathbf{C}_k^{\mathrm{p}} + (1 + p^{-1})\mathbf{X}_k^{\mathrm{p}})^{-1} + \mathbf{H}^{\mathrm{T}}(\mathbf{C}_k^z + (1 + p)\mathbf{X}_k^z)^{-1}\mathbf{H}\Big)^{-1}.$$

Remark 2 The scalar factors $(1 + p^{-1})$ and $(1 + p)$ can be replaced according to

$$(1 + p) = \frac{1}{\omega} \quad \text{and} \quad (1 + p^{-1}) = \frac{1}{1 - \omega}$$

with $\omega \in (0, 1)$. In doing so, a bisection method can be employed to find the optimal parameter in the interval $(0, 1)$.

C Special Cases

In Figure 2.7, different estimates to be fused are depicted, i.e., $\mathbf{H} = \mathbf{I}$ and the measurement can itself be seen as a state estimate (see Chapter 3). In special cases [205], the proposed combined estimator reduces to the purely stochastic and purely set-membership estimation principles from Section 2.2, namely, in the situation of vanishing set-membership or vanishing stochastic uncertainty, respectively.

Standard Kalman Filter The Kalman filter appears in its standard formulation of Section 2.1.1-B, if $\mathbf{X}_k^{\mathrm{p}} = \mathbf{X}_k^z = \mathbf{0}$. The gain (2.91) simply becomes

$$\mathbf{K}_k = \mathbf{C}_k^{\mathrm{p}}\mathbf{H}_k^{\mathrm{T}}(\mathbf{H}_k\mathbf{C}_k^{\mathrm{p}}\mathbf{H}^{\mathrm{T}} + \mathbf{C}_k^z)^{-1},$$

which is identical to (2.24). This is an expected result since we have strictly followed and generalized the derivation of the standard Kalman gain. The fusion result is depicted in Figure 2.7(b) by the dashed blue ellipsoid.

Centered Intersection   More surprising is the special case of a centered intersection of ellipsoids, if $\mathbf{C}_k^{\mathrm{p}} = \mathbf{C}_k^z = \mathbf{0}$. The gain (2.91) then reduces to

$$\mathbf{K}_k(p) = (1 + p^{-1})\mathbf{X}_k^{\mathrm{P}}\mathbf{H}_k^{\mathrm{T}}\left((1 + p^{-1})\mathbf{H}_k\mathbf{X}_k^{\mathrm{P}}\mathbf{H}_k^{\mathrm{T}} + (1 + p)\mathbf{X}_k^z\right)^{-1}.$$

With this gain, formula (2.89) for the shape matrix of the bounding ellipsoid can be simplified[9] to

$$\begin{aligned}
\mathbf{X}_k^{\mathrm{e}}(p) &= (1 + p^{-1})\mathbf{X}_k^{\mathrm{P}} - \mathbf{K}_k(p)\mathbf{H}_k\left((1 + p^{-1})\mathbf{X}_k^{\mathrm{P}}\right) \\
&= (1 + p^{-1})\mathbf{X}_k^{\mathrm{P}} - \left((1 + p^{-1})\mathbf{X}_k^{\mathrm{P}}\right)\mathbf{H}_k^{\mathrm{T}} \cdot \\
&\quad \left((1 + p^{-1})\mathbf{H}_k\mathbf{X}_k^{\mathrm{P}}\mathbf{H}_k^{\mathrm{T}} + (1 + p)\mathbf{X}_k^z\right)^{-1}\mathbf{H}_k\left((1 + p^{-1})\mathbf{X}_k^{\mathrm{P}}\right) \\
&= \left(\frac{1}{1 + p^{-1}}(\mathbf{X}_k^{\mathrm{P}})^{-1} + \frac{1}{1 + p}\mathbf{H}_k^{\mathrm{T}}(\mathbf{X}_k^z)^{-1}\mathbf{H}_k\right)^{-1},
\end{aligned}$$

where, for the last equation, the Woodbury matrix identity [177] has been applied. As in Remark 2, the formula can be rewritten to

$$\mathbf{X}_k^{\mathrm{e}}(\omega) = \left((1 - \omega)(\mathbf{X}_k^{\mathrm{P}})^{-1} + \omega\mathbf{H}_k^{\mathrm{T}}(\mathbf{X}_k^z)^{-1}\mathbf{H}_k\right)^{-1} \qquad (2.92)$$

with $\omega = \frac{1}{1+p} \in [0, 1]$ and is identical to the shape matrix (2.40) for the ellipsoidal bound of an intersection. By letting $\mathbf{H}_k = \mathbf{I}$ in order to simplify matters, this matrix characterizes the centered intersection

$$\mathcal{E}(\underline{0}, \mathbf{X}_k^{\mathrm{P}}) \cap \mathcal{E}(\underline{0}, \mathbf{X}_k^z) \subset \mathcal{E}(\underline{0}, \mathbf{X}_k^{\mathrm{e}}(\omega)),$$

which is analogous to the results of covariance intersection [99] (see Section 3.5.3). The obtained bounding ellipsoid is shown in Figure 2.7(c) and is compared to a tight bound for the intersection. The difference to Section 2.1.2-B is that the parameter (2.41) which scales the shape matrix

---

[9]In the same way, the Joseph form (2.26) is simplified.

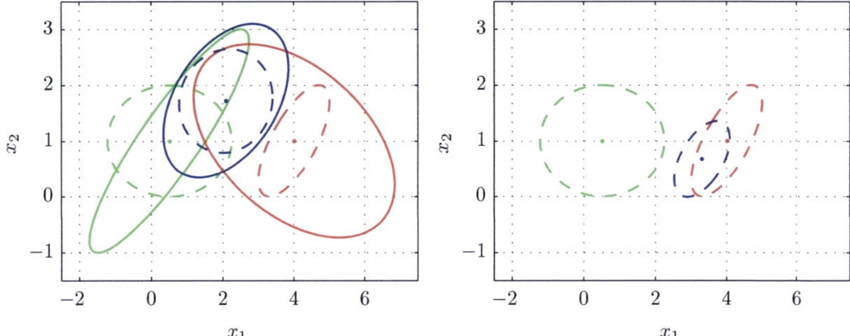

(a) Fusion of two estimates with both stochastic and set-membership error characteristics.

(b) Fusion of the same estimates with zero set-membership uncertainty. Result is the same result as for the standard Kalman filter.

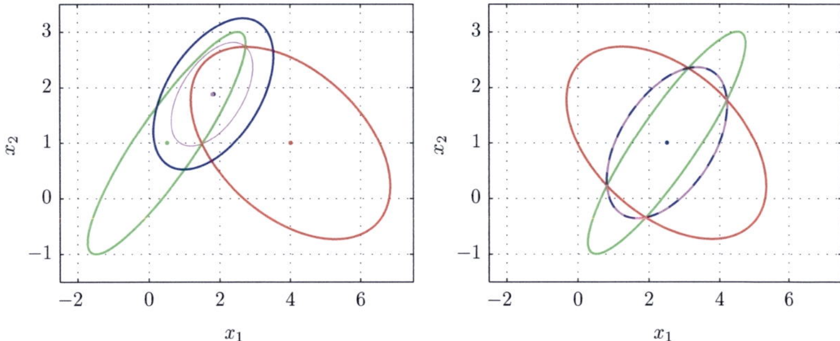

(c) Fusion of the same estimates with zero stochastic uncertainty. Result is a bound for the centered intersection. The magenta ellipsoid is a tight trace-optimal bound for the intersection.

(d) Centralized Intersection: If the estimates are identical, then the blue error bound in Fig. (c) is a tight approximation of the intersection and coincides with the magenta bound.

Figure 2.7: The blue ellipsoids are the uncertainty characteristics for the fused red and green estimates. The set-membership bounds are bold and the covariance ellipsoids are drawn in dashed lines.

down is omitted. Thus, the maximum possible intersection is bounded, which occurs when the estimates to be fused are identical, as illustrated in Figure 2.7(d).

Finally, a generalization of the well-known Kalman filtering scheme is attained that reduces to the standard Kalman filter in the absence of set-membership uncertainty and that otherwise becomes the intersection of sets in case of vanishing stochastic uncertainty.

2.5 Advanced Simultaneous Stochastic and Set-membership Estimation

This section constitutes the main result of this chapter, although not much work is left to be done with keeping in mind the previous results. We provide a generalization of the estimation concept proposed in Section 2.4.2 in order to enable the user to decide whether the stochastic or set-membership uncertainty shall primarily be minimized. The idea behind it rests upon the fact that for the standard Kalman gain the same result is attained when instead of $\mathrm{trace}(\mathbf{C}_k^{\mathrm{e}})$ a scaled version

$$
\begin{aligned}
\mathrm{trace}\left(S \cdot \mathbf{C}_k^{\mathrm{e}}\right) = \mathrm{trace}\Big(&(\mathbf{I} - \mathbf{K}_k \mathbf{H}_k)(S \cdot \mathbf{C}_k^{\mathrm{p}})(\mathbf{I} - \mathbf{K}_k \mathbf{H}_k)^{\mathrm{T}} \\
&+ \mathbf{K}_k (S \cdot \mathbf{C}_k^z) \mathbf{K}_k^{\mathrm{T}} \Big)
\end{aligned}
\tag{2.93}
$$

with $S > 0$ is considered, which corresponds to a scaled covariance ellipsoid around the current state estimate $\hat{\underline{x}}_k^{\mathrm{e}}$.

The basic idea behind the scaling parameter is that, in contrast to enclosing ellipsoids for set-membership errors, covariance ellipsoids do not represent rigid bounds. In a one-dimensional state space, a 95.4% confidence level around a point estimate, for instance, corresponds to the 2-sigma ellipsoid (i.e., interval), and setting $S = 3$ even yields a 99.7% confidence level. With increasing scaling parameter, the bounds can be considered to be more rigid and more similar to a set-membership error description. To express it in a simple way, both types of errors become more comparable with larger values of S. If for ensuing decision making or control tasks a certain confidence level is known in advance, it is particularly desirable to choose a Kalman gain that ensures a minimum confidence set. The proposed concept, which is based on [205], therefore allows to flexibly balance between the minimization of the stochastic uncertainty and the minimization of the set-membership uncertainty.

As in the previous considerations, we consider the combined linear models (2.65) and (2.66). It is easy to recognize from the predicted MSE matrix (2.82) in Section 2.4.2-B that the estimate $\hat{x}^{\mathrm{p}}_{k+1}$, the predicted covariance matrix $\mathbf{C}^{\mathrm{p}}_{k+1}$, and shape matrix $\mathbf{X}^{\mathrm{p}}_{k+1}$, i.e., equations (2.81), (2.83), and (2.84), are not altered by the scaling parameter S. In the following, we examine its effect to the filtering step.

2.5.1 Adjustable Gains

Instead of minimizing the MSE (2.90) in Section 2.4.2, we determine the gain to be optimal for the sum of the modified trace (2.93) and the trace of the ellipsoidal shape matrix , i.e.,

$$\mathbf{K}_k(p) = \arg\min \left\{ \operatorname{trace}\left(S \cdot \mathbf{C}^{\mathrm{e}}_k\right) + \operatorname{trace}\left(\mathbf{X}^{\mathrm{e}}_k(p)\right) \right\} .$$

for $p > 0$. By following the derivation of the filtering step in Section 2.4.2-B, we obtain the Kalman gain

$$\mathbf{K}_k(p) = \left((1 + p^{-1})\mathbf{X}^{\mathrm{p}}_k\mathbf{H}^{\mathrm{T}}_k + S\mathbf{C}^{\mathrm{p}}_k\mathbf{H}^{\mathrm{T}}_k\right) \cdot$$
$$\left((1 + p^{-1})\mathbf{H}_k\mathbf{X}^{\mathrm{p}}_k\mathbf{H}^{\mathrm{T}}_k + (1 + p)\mathbf{X}^{z}_k + S\mathbf{H}_k\mathbf{C}^{\mathrm{p}}_k\mathbf{H}^{\mathrm{T}} + S\mathbf{C}^{z}_k\right)^{-1} ,$$

where simply the covariance matrices are replaced by scaled versions. An alternative way to define the problem is to consider a convex combination of the traces, and thus we are striving for

$$\mathbf{K}_k(p) = \arg\min \left\{ (1 - \alpha)\operatorname{trace}\left(\mathbf{C}^{\mathrm{e}}_k\right) + \alpha\operatorname{trace}\left(\mathbf{X}^{\mathrm{e}}_k(p)\right) \right\}$$

with $\alpha \in [0, 1]$. The result is then given by the gain

$$\mathbf{K}_k(p) = \left(\alpha(1 + p^{-1})\mathbf{X}^{\mathrm{p}}_k\mathbf{H}^{\mathrm{T}}_k + (1 - \alpha)\mathbf{C}^{\mathrm{p}}_k\mathbf{H}^{\mathrm{T}}_k\right) \cdot$$
$$\left(\alpha\left((1 + p^{-1})\mathbf{H}_k\mathbf{X}^{\mathrm{p}}_k\mathbf{H}^{\mathrm{T}}_k + (1 + p)\mathbf{X}^{z}_k\right) + (1 - \alpha)\left(\mathbf{H}_k\mathbf{C}^{\mathrm{p}}_k\mathbf{H}^{\mathrm{T}} + \mathbf{C}^{z}_k\right)\right)^{-1} .$$

$$(2.94)$$

In the following, we prefer the latter definition of the Kalman gain due to the boundedness of the parameter α.

The derived Kalman gains can then be applied within the formulas (2.86), (2.88), and (2.89) in order to compute the estimate $\hat{\underline{x}}_k^{\mathrm{e}}$, the covariance matrix $\mathbf{C}_k^{\mathrm{e}}$, and the shape matrix $\mathbf{X}_k^{\mathrm{e}}$, respectively. Note that also the covariance matrix $\mathbf{C}_k^{\mathrm{e}}$ depends on p due to $\mathbf{K}_k(p)$, which is employed to combine the error characteristics $\mathbf{C}_k^{\mathrm{p}}$ and $\mathbf{C}_k^{z}$. As in Section 2.4.2-B, the optimal p^{opt} is determined to minimize the trace, i.e.,

$$p^{\mathrm{opt}} = \arg\min\left\{(1-\alpha)\,\mathrm{trace}\left(\mathbf{C}_k^{\mathrm{e}}(p)\right) + \alpha\,\mathrm{trace}\left(\mathbf{X}_k^{\mathrm{e}}(p)\right)\right\},$$

which is a one-dimensional convex optimization problem.

The formulas for the prediction step remain for each instance exactly the same, be it the Kalman filter for sets of densities from Section 2.4.1, the LMMSE estimator from Section 2.4.2, or the advanced scheme in this section. However, the filtering step has successively been generalized and the results of this section embrace the previous instances of the filtering step. We do not repeat here the formulas for the prediction and filtering steps but summarize them in Figure 2.8 and Figure 2.9.

2.5.2 Limiting Cases

The approach derived in Section 2.4.2 is directly related to the parameters $\alpha = 0.5$ or $S = 1$, respectively, which necessarily leads to the Kalman gain (2.91). Accordingly, the results depicted in Figure 2.10(a) are the same as in Figure 2.7(a).

From Section 2.4.2-C, we can easily recognize that, for $\alpha = 0$ or, respectively, $S \to \infty$, the gain (2.94) becomes the standard Kalman gain (2.24). Therefore, the blue dashed ellipse in Figure 2.10(b) corresponds to the fusion result in Figure 2.7(b). Furthermore, we have obtained the Kalman filter for sets of Gaussian densities, as it has been derived in Section 2.4.1. It is important to note that p and $\mathbf{K}_k$ are determined independently from each other. However, the advantage that p can, in this case, be computed analytically by means of (2.75) comes at the expense of possibly large set-membership error bounds, which are rather computed as a byproduct.

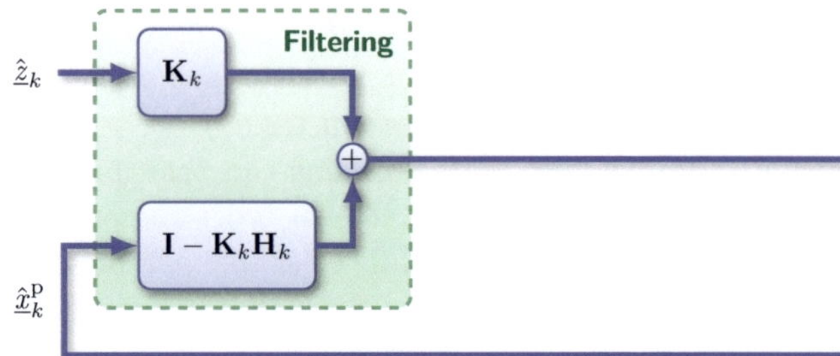

Filtering

Input:

- prior or predicted estimate $\hat{\underline{x}}_k^{\mathrm{p}}$ with error covariance matrix $\mathbf{C}_k^{\mathrm{p}}$ and shape matrix $\mathbf{X}_k^{\mathrm{p}}$

- linear measurement model (2.66)

- observation $\hat{\underline{z}}_k$, sensor noise with stochastic and set-membership error statistics $\mathbf{C}_k^z$ and $\mathbf{X}_k^z$

- weighting parameter α in order to adjust the ratio between both error characteristics

---

Computation of updated estimate and error characteristics

- For given weighting parameter α, the gain $\mathbf{K}_k(p)$ is

$$\mathbf{K}_k(p) \stackrel{(2.94)}{=} \left(\alpha\left(1 + p^{-1}\right)\mathbf{X}_k^{\mathrm{p}}\mathbf{H}_k^{\mathrm{T}} + (1 - \alpha)\mathbf{C}_k^{\mathrm{p}}\mathbf{H}_k^{\mathrm{T}}\right) \cdot$$
$$\left(\alpha\left((1 + p^{-1})\mathbf{H}_k\mathbf{X}_k^{\mathrm{p}}\mathbf{H}_k^{\mathrm{T}} + (1 + p)\mathbf{X}_k^z\right) + (1 - \alpha)\left(\mathbf{H}_k\mathbf{C}_k^{\mathrm{p}}\mathbf{H}^{\mathrm{T}} + \mathbf{C}_k^z\right)\right)^{-1} \tag{2.95}$$

- Computation of updated estimate $\hat{\underline{x}}_k^{\mathrm{e}}$ by means of

$$\hat{\underline{x}}_k^{\mathrm{e}} \stackrel{(2.86)}{=} \left(\mathbf{I} - \mathbf{K}_k(p)\,\mathbf{H}_k\right)\hat{\underline{x}}_k^{\mathrm{p}} + \mathbf{K}_k(p)\,\hat{\underline{z}}_k \tag{2.96}$$

- Computation of updated error covariance matrix $\mathbf{C}_k^{\mathrm{e}}$ according to

$$\mathbf{C}_k^{\mathrm{e}}(p) \stackrel{(2.88)}{=} \left(\mathbf{I} - \mathbf{K}_k(p)\,\mathbf{H}_k\right)\mathbf{C}_k^{\mathrm{p}}\left(\mathbf{I} - \mathbf{K}_k(p)\,\mathbf{H}_k\right)^{\mathrm{T}} + \mathbf{K}_k(p)\,\mathbf{C}_k^z\,\mathbf{K}_k(p)^{\mathrm{T}} \tag{2.97}$$

- The updated shape matrix $\mathbf{X}_k^{\mathrm{e}}$ for the set-membership error is given by

$$\mathbf{X}_k^{\mathrm{e}}(p) \stackrel{(2.89)}{=} \left(1 + p^{-1}\right)\left(\mathbf{I} - \mathbf{K}_k(p)\,\mathbf{H}_k\right)\mathbf{X}_k^{\mathrm{p}}\left(\mathbf{I} - \mathbf{K}_k(p)\,\mathbf{H}_k\right)^{\mathrm{T}} + (1 + p)\,\mathbf{K}_k(p)\,\mathbf{X}_k^z\,\mathbf{K}_k(p)^{\mathrm{T}} \tag{2.98}$$

- The optimal parameter p^{opt} is determined by minimizing

$$p^{\mathrm{opt}} = \arg\min\left\{(1 - \alpha)\,\mathrm{trace}(\mathbf{C}_k^{\mathrm{e}}(p)) + \alpha\,\mathrm{trace}(\mathbf{X}_k^{\mathrm{e}}(p))\right\}$$

The updated point estimate $\hat{\underline{x}}_k^{\mathrm{e}} = \hat{\underline{x}}_k^{\mathrm{e}}(p^{\mathrm{opt}})$ is characterized by the stochastic error characteristics $\mathbf{C}_k^{\mathrm{e}} = \mathbf{C}_k^{\mathrm{e}}(p^{\mathrm{opt}})$ and set-membership error description $\mathbf{X}_k^{\mathrm{e}} = \mathbf{X}_k^{\mathrm{e}}(p^{\mathrm{opt}})$.

Figure 2.8: Summary of the filtering step.

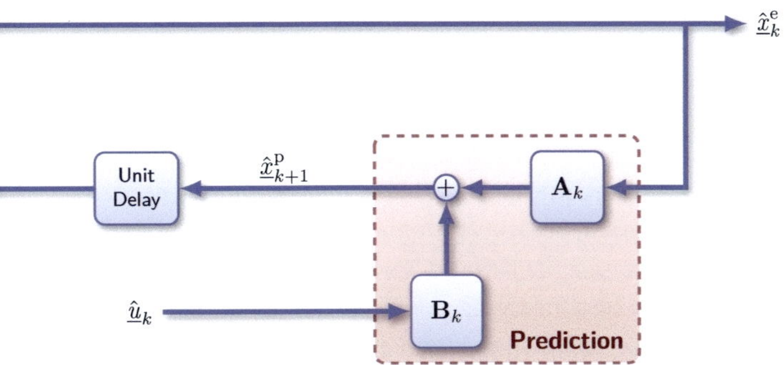

Prediction

Input:

- point estimate $\hat{\underline{x}}_k^{\mathrm{e}}$ with estimated covariance matrix $\mathbf{C}_k^{\mathrm{e}}$ and shape matrix $\mathbf{X}_k^{\mathrm{e}}$

- linear system model (2.65)

- control input $\hat{\underline{u}}_k$, process noise with stochastic and set-membership error statistics $\mathbf{C}_k^u$ and $\mathbf{X}_k^u$

Computation of predicted estimate and error characteristics

- Computation of predicted estimate $\hat{\underline{x}}_{k+1}^{\mathrm{p}}$ by means of

$$\hat{\underline{x}}_{k+1}^{\mathrm{p}} \overset{(2.81)}{=} \mathbf{A}_k\,\hat{\underline{x}}_k^{\mathrm{e}} + \mathbf{B}_k\,\hat{\underline{u}}_k \tag{2.99}$$

- Computation of error covariance matrix $\mathbf{C}_{k+1}^{\mathrm{p}}$ according to

$$\mathbf{C}_{k+1}^{\mathrm{p}} \overset{(2.83)}{=} \mathbf{A}_k\mathbf{C}_k^{\mathrm{e}}\mathbf{A}_k^{\mathrm{T}} + \mathbf{B}_k\mathbf{C}_k^u\mathbf{B}_k^{\mathrm{T}} \tag{2.100}$$

- The family of possible shape matrices $\mathbf{X}_{k+1}^{\mathrm{p}}(p)$ with $p > 0$ is given by

$$\mathbf{X}_{k+1}^{\mathrm{p}}(p) \overset{(2.84)}{=} (1+p^{-1})\,\mathbf{A}_k\mathbf{X}_k^{\mathrm{e}}\mathbf{A}_k^{\mathrm{T}} + (1+p)\,\mathbf{B}_k\mathbf{X}_k^u\mathbf{B}_k^{\mathrm{T}} \tag{2.101}$$

- In order to minimize the MSE (2.85), the parameter p^{opt} is chosen according to

$$p^{\mathrm{opt}} \overset{(2.70)}{=} \mathrm{trace}(\mathbf{A}_k\mathbf{X}_k^{\mathrm{e}}\mathbf{A}_k^{\mathrm{T}})^{\frac{1}{2}} \cdot \mathrm{trace}(\mathbf{B}_k\mathbf{X}_k^u\mathbf{B}_k^{\mathrm{T}})^{-\frac{1}{2}} \tag{2.102}$$

The predicted point estimate $\hat{\underline{x}}_{k+1}^{\mathrm{p}}$ is characterized by the stochastic error characteristic $\mathbf{C}_{k+1}^{\mathrm{p}}$ and the set-membership error description $\mathbf{X}_{k+1}^{\mathrm{p}}$.

Figure 2.9: Summary of the prediction step.

For $\alpha = 1$ or, respectively, $S = 0$, we obtain the formulas of the centralized intersection (2.92). In this case, we encounter a possibly non-decreasing stochastic uncertainty, as illustrated in Figure 2.10(c). The blue dashed covariance ellipsoid is still similar to the green dashed ellipsoid and is in some directions larger than the red one. The gain does not incorporate the covariance matrices.

In Figure 2.10(d), the interesting special case of fusing a purely stochastic and a purely set-membership state estimate is depicted, and it impressively demonstrates how stochastic and set-membership estimation principles are melt together to a unifying concept. As indicated by the gray ellipsoid, for small α, we insist to minimize the stochastic error and therefore tend to primarily trust the set-membership estimate that reports no stochastic error. For $\alpha = 0$, the stochastic estimate is even rejected. Analogously, the opposite applies for large α.

In conclusion, the proposed concept includes the purely stochastic and purely set-membership estimation principles by distrusting one of both error characteristics. The adjustable Kalman gain allows for a smooth transition between stochastic and set-membership estimators, and the generalizations of the standard Kalman filter presented in Section 2.4.1 and Section 2.4.2 are encompassed as special cases.

2.6   Treatment of Nonlinearities in Approximate Kalman Filtering

Although Kalman filtering techniques may severely suffer from the influences of nonlinearities, they are also widely applied to provide estimates on system states that evolve according to nonlinear system dynamics and are observed through nonlinear sensor systems. The simple processing of estimates states the reason why the Kalman filter has also become a well-accepted approach when nonlinearities are encountered. In this case, the process and measurement models are linearized either by a Taylor series expansion, as it is done for the *extended Kalman filter* (EKF) [163], or by a linear regression analysis, for which the *unscented Kalman filter* [95, 97]

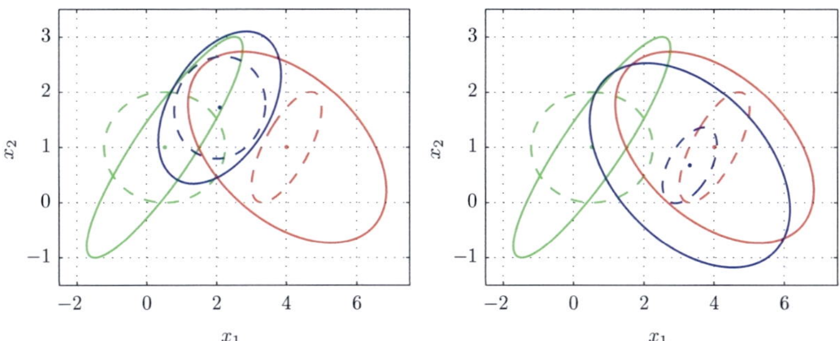

(a) Fusion of two estimates with $\alpha = 0.5$. The result is the same as for the LMMSE estimator in Section 2.4.2.

(b) Fusion of same estimates with $\alpha = 0$. $\mathbf{C}_k^e$ is minimized and the same as for the standard Kalman filter. The set-membership bound increases. The dashed ellipsoids correspond to Fig. 2.7(b).

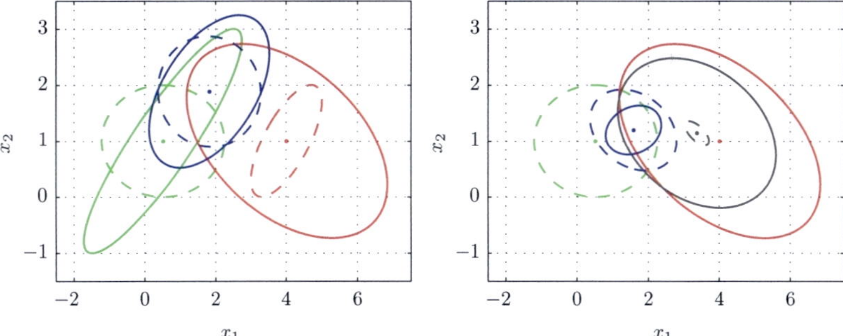

(c) Fusion of same estimates with $\alpha = 1$ gives the centered intersection for the set-membership bound, but the covariance ellipsoid increases. The solid ellipsoids correspond to Fig. 2.7(c).

(d) Fusion of a purely stochastic estimate with a purely set-membership estimate. The result for $\alpha = 0.5$ is blue, the result for $\alpha = 0.1$ is gray.

Figure 2.10: The blue ellipsoids are the uncertainty characteristics for the fused red and green estimates. In contrast to Fig. 2.7, both types of uncertainty are present in each case, but different weighting parameters for the gain (2.94) are chosen. For each estimate, set-membership error bounds are bold and the stochastic error characteristics are drawn in dashed lines.

is a well-known candidate. Especially, the former approach suffers from inconsistent estimates [26]. The regression-based filters implicitly calculate an additional linearization noise and are better able to preserve consistency at the cost of less informative[10] estimation results [114, 115]. In order to enhance the reliability of approximate Kalman filters, the derived estimation methods from the previous sections prove to be a useful tool for the effective treatment of nonlinearities in approximate Kalman filtering. The estimation principles employed in this section have been studied in [183] and [202] and their applicability has been demonstrated against the background of beating heart surgery systems [191] (see Section 2.7.3), where an estimate on the heart surface displacement is to be computed. The approach presented in the following has been published in [202]. We extend this work by a deeper analysis, combine it with the advanced filtering scheme, and discuss an additional way of approximating the nonlinear mappings.

The trouble with nonlinear system and measurement functions is that Gaussian random quantities will not be Gaussian anymore. For instance, quadratic measurement models can result in multi-modal densities whose support is not primarily allocated around the mean. In general, the resulting density can be an arbitrary function for which no finite-dimensional parameterization is obtainable. So, the reason to convert the true density back to a Gaussian one is given by its simple parameterization and processing. For any nonlinear transformation

$$\underline{y} = g(\underline{x}) = g^{\mathrm{Lin}}(\underline{x}) + g^{\mathrm{Nonlin}}(\underline{x}) \tag{2.103}$$

of an uncertain quantity $\underline{x}$, this involves that the nonlinear part $g^{\mathrm{Nonlin}}(\underline{x})$ is omitted such that $\underline{y}$ remains normally distributed. This is done implicitly whenever Kalman filter techniques are used.

As illustrated in Figure 2.11, the key idea of this section consists in replacing $g^{\mathrm{Nonlin}}(\underline{x})$ by a set that is an appropriate bound of the nonlinear part. The unknown but bounded error description, i.e., the set, is hence intended to account for the typically neglected nonlinearities. Of course, linearization errors cannot, in general, be bounded over the entire domain. By employing the available knowledge about the state $\underline{x}_k$, we consider

---

[10]An estimate is *informative* if it is related to a low MSE.

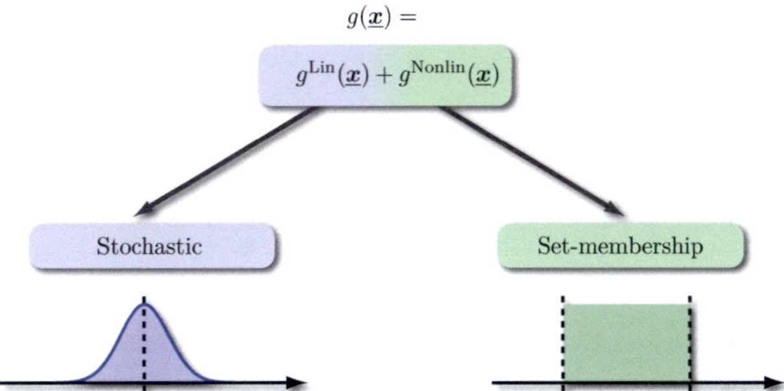

Figure 2.11: Kalman filter methods only consider linear parts of nonlinear system and observation models. The idea here is to incorporate the nonlinearities by means of set-membership characterizations.

these errors only over the most probable region, to which we refer to as *uncertainty region* around the estimate $\hat{\underline{x}}_k^{\mathrm{e}}$. More precisely, we calculate bounds on the linearization errors over a confidence set in which the true state lies with a predefined probability level P. At this point, it is necessary to stress that we essentially refer to the Bayesian viewpoint of confidence sets [86], also called *credible sets* or *Bayesian confidence sets*. The level P then states the probability to which the confidence set covers the true state. For an n_x-dimensional Gaussian random vector $\underline{x} \sim \mathcal{N}(\hat{\underline{x}}, \mathbf{C})$, the confidence set to a certain probability level P is given by an ellipsoid

$$\mathcal{E}(\hat{\underline{x}}, s \cdot \mathbf{C}) = \left\{ \underline{x} \in \mathbb{R}^n \mid (\underline{x} - \hat{\underline{x}})^{\mathrm{T}} \mathbf{C}^{-1} (\underline{x} - \hat{\underline{x}}) \leq s \right\}$$

with the mean as midpoint and the scaled covariance matrix as shape matrix according to definition (2.31). The parameter s can be determined by means of the distribution of the squared Mahalanobis distance, which is that of a chi-squared variate with n_x degrees of freedom [154]. Thus, the scalar s depends on the chosen probability level P and the dimensionality n_x. The confidence set $\mathcal{E}(\hat{\underline{x}}, s \cdot \mathbf{C})$ represents certain sigma bounds on the uncertainty affecting $\hat{\underline{x}}$ and is given by a scaled covariance ellipsoid. Figure 2.12(a) depicts a confidence set for a one-dimensional estimate. Over this set,

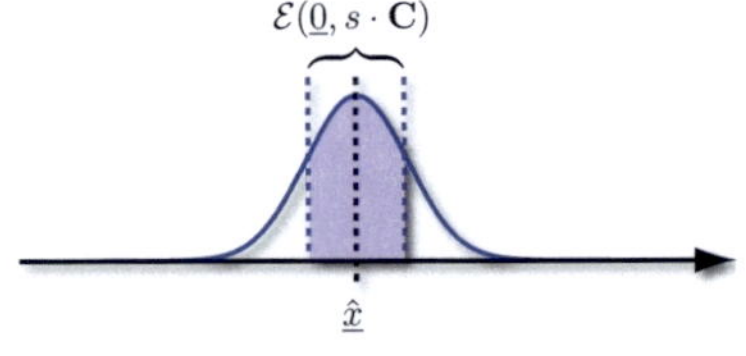

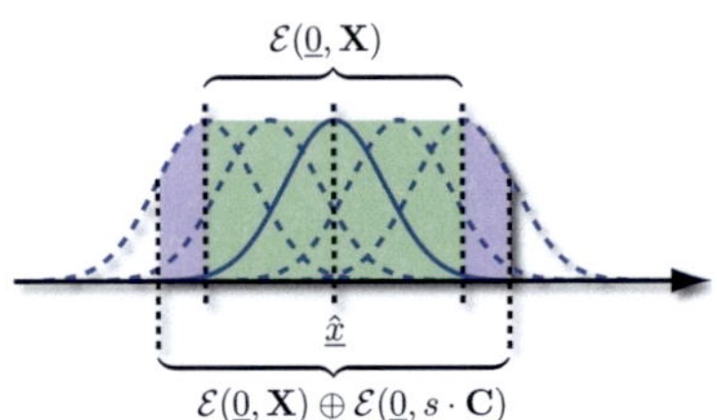

(a) Sigma bound for estimate with stochastic uncertainty.

(b) Upper sigma bound for estimate affected by stochastic and set-membership errors.

Figure 2.12: Confidence sets for an estimate $\hat{\underline{x}}$. The error covariance matrix is represented by a Gaussian density. In the presence of set-membership uncertainty, the confidence set is combined with the bounding set.

linearization errors can be bounded by a set $\mathcal{L}$ according to

$$g^{\mathrm{Nonlin}}\left(\mathcal{E}(\hat{\underline{x}}, s \cdot \mathbf{C})\right) \subseteq \mathcal{L} \; .$$

This set $\mathcal{L}$ can then be conceived as an unknown but bounded perturbation acting upon the transformation, so that (2.103) becomes the uncertain quantity

$$\underline{y} \approx g^{\mathrm{Lin}}(\underline{x}) + \underline{l} \; ,$$

with $\underline{l} \in \mathcal{L}$, which implies that $\underline{y}$ can be modeled by a set of Gaussian densities, as it is done in sections 2.3 and 2.4.1. In line with Section 2.4.2 and as an extension of the work in [202], we can also consider $\underline{y}$ as an uncertain quantity with stochastic and set-membership uncertainty characteristics. In compliance with the preceding sections, we will employ ellipsoidal bounds $\mathcal{L} = \mathcal{E}(\underline{0}, \mathbf{X}^y)$ for the linearization errors. With $\mathcal{E} = \mathcal{E}(\hat{\underline{x}}, s \cdot \mathbf{C}^x)$ being a confidence set to a chosen probability level P, the corresponding confidence set of the nonlinearly transformed state estimate (2.103) is then bounded according to

$$g^{\mathrm{Lin}}(\mathcal{E}) \oplus g^{\mathrm{Nonlin}}(\mathcal{E}) \subseteq g^{\mathrm{Lin}}(\mathcal{E}) \oplus \mathcal{E}(\underline{0}, \mathbf{X}^y) \; .$$

The latter enclosing set is the Minkowski sum of the covariance ellipsoid $\mathcal{E}(\hat{y}, s \cdot \mathbf{C}^y)$ and the error bound $\mathcal{E}(\underline{0}, \mathbf{X}^y)$, where $\hat{y}$ and $\mathbf{C}^y$ are the new mean and covariance matrix after applying the linear mapping g^{Lin} to $\underline{x}$. With $\underline{l} \in \mathcal{E}(\underline{0}, \mathbf{X}^y)$, the set-membership parameter $\mathbf{X}^y$ models the possible impact of the neglected nonlinearities. After the nonlinear transformation, the quantity $\underline{y}$ is hence characterized by the parameters $\hat{y}$, $\mathbf{C}^y$, and $\mathbf{X}^y$. For the one-dimensional case, Figure 2.12(b) shows the overall set bounding the maximum error on the state estimate with the probability P. The computation of the bound and the transformation result are illustrated in Example 2.3. The subsequent subsection provides a more detailed guidance for bounding linearization errors in the prediction and filtering phase of a Kalman filter.

Example 2.3: Bounded linearization errors
We consider a random variable x that is normally distributed with mean 4.5 and standard deviation 0.5. It is transformed by means of the nonlinear mapping

$$g(x) = 0.05 \cdot (x - 1)(x - 3)(x - 4)(x - 5)(x - 8) + 2 \ ,$$

which corresponds to the black curves in Figure 2.13. Figure 2.13(a) shows the resulting density, which is far from being Gaussian and can even become multimodal. In Figure 2.13(b), the nonlinear mapping is linearized, and the 2-sigma bound around the mean is used to compute the set of possible linearization errors. More precisely, these errors take values between the minimum and maximum possible distance between the linear and nonlinear mapping (green area between nonlinear and linearized mapping). The obtained bound is then interpreted as a set of possible means for the set of resulting Gaussian densities.

<h3>2.6.1 Bounding Linearization Errors</h3>

In the following, we will show that the bounds on linearization errors can be propagated through the Kalman prediction and filtering steps and that we are capable of sustaining a certain probability level, to which the maximum

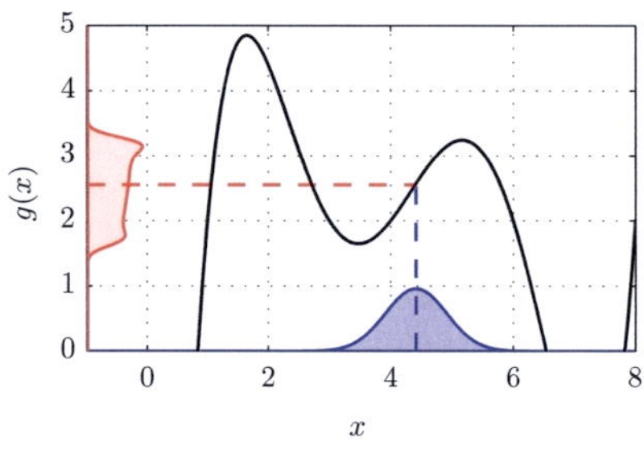

(a) Result of nonlinear transformation is a non-Gaussian density.

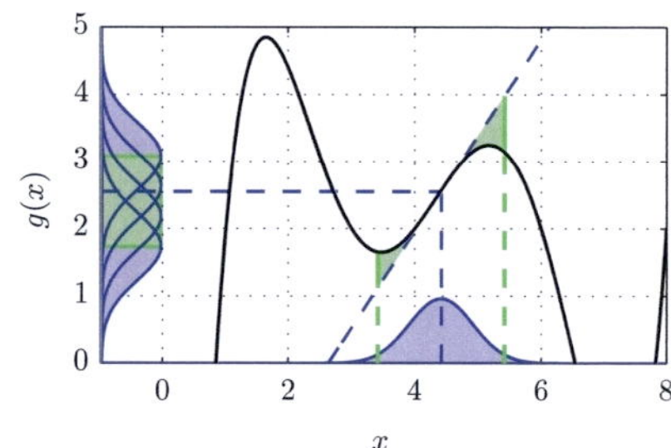

(b) Result is represented by a set of Gaussian densities.

Figure 2.13: Transformations of a Gaussian random variable. The nonlinear result is drawn red in Plot (a). In Plot (b), the nonlinear mapping is linearized and the error is bounded over 2-sigma bounds. The result is a set of Gaussian densities.

linearization error lies within these bounds. In each processing time step, the a priori defined probability level P is used to determine the region over which the linearization errors are to be bounded. For the prediction step, the desired approximations are detailed in Paragraph 2.6.1-A. The derivation of error bounds for the observation model lies in the focus of Paragraph 2.6.1-B.

A   Linearized System Models

Given that the current estimate $\hat{\underline{x}}_k^{\mathrm{e}}$ and its error characteristics $\mathbf{C}_k^{\mathrm{e}}$ and $\mathbf{X}_k^{\mathrm{e}}$ have been computed with respect to a certain probability level P, we aspire to maintain this level when approximating the state transition model

$$\underline{x}_{k+1} = \underline{a}_k(\underline{x}_k, \underline{u}_k)$$

linearly and calculating a predicted state estimate $\hat{\underline{x}}_{k+1}^{\mathrm{p}}$. The shape matrix $\mathbf{X}_k^{\mathrm{e}}$ accounts for the linearization errors made so far and further involved set-membership uncertainties, and the covariance matrix $\mathbf{C}_k^{\mathrm{e}}$ characterizes the linearly processed stochastic uncertainty. The uncertainty region around $\hat{\underline{x}}_k^{\mathrm{e}}$ is then represented by the Minkowski sum

$$\begin{aligned}
\mathcal{C}_k &= \hat{\underline{x}}_k^{\mathrm{e}} + \mathcal{E}(\underline{0}, \mathbf{X}_k^{\mathrm{e}}) \oplus \mathcal{E}(\underline{0}, s \cdot \mathbf{C}_k^{\mathrm{e}}) \\
&= \mathcal{E}(\hat{\underline{x}}_k^{\mathrm{e}}, \mathbf{X}_k^{\mathrm{e}}) \oplus \mathcal{E}(\underline{0}, s \cdot \mathbf{C}_k^{\mathrm{e}}) \;,
\end{aligned}$$

where $\mathcal{E}(\underline{0}, s \cdot \mathbf{C}_k^{\mathrm{e}})$ is the P confidence region for the stochastic uncertainty. In order to simplify matters, we assume that process noise directly affects the control input according to

$$\underline{u}_k = \hat{\underline{u}}_k + \underline{w}_k + \underline{d}_k$$

with $\underline{w}_k \sim \mathcal{N}(\underline{0}, \mathbf{C}_k^u)$ and $\underline{d}_k \in \mathcal{E}(\underline{0}, \mathbf{X}_k^u)$ and that the perturbations can be bounded with respect to the same probability level P by the set

$$\begin{aligned}
\mathcal{U}_k &= \hat{\underline{u}}_k + \mathcal{E}(\underline{0}, \mathbf{X}_k^u) \oplus \mathcal{E}(\underline{0}, \tilde{s} \cdot \mathbf{C}_k^u) \\
&= \mathcal{E}(\hat{\underline{u}}_k, \mathbf{X}_k^u) \oplus \mathcal{E}(\underline{0}, \tilde{s} \cdot \mathbf{C}_k^u)
\end{aligned}$$

The linearization errors of an approximation

$$\underline{a}_k(\underline{x}_k, \underline{u}_k) \approx \mathbf{A}_k \underline{x}_k + \mathbf{B}_k \underline{u}_k + \hat{\underline{a}}_k$$

over the considered regions $\mathcal{C}_k$ and $\mathcal{U}_k$ yield a set

$$\mathcal{R}_k^{\mathrm{err}} = \left\{ \underline{a}_k(\underline{x}_k, \underline{u}_k) - (\mathbf{A}_k \underline{x}_k + \mathbf{B}_k \underline{u}_k + \hat{\underline{a}}_k) \,\middle|\, \underline{x}_k \in \mathcal{C}_k, \underline{u}_k \in \mathcal{U}_k \right\} \;,$$

where $\hat{\underline{a}}_k$ is the translation vector[11] of the affine mapping. The set $\mathcal{R}_k^{\mathrm{err}}$ of linearization errors can, for instance, be bounded by componentwise taking the supremum

$$\underline{r}_k^{\sup} = \sup_{\underline{r} \in \mathcal{R}_k^{\mathrm{err}}} \underline{r} = \sup_{\substack{\underline{x}_k \in \mathcal{C}_k \\ \underline{u}_k \in \mathcal{U}_k}} \left[\underline{a}_k(\underline{x}_k, \underline{u}_k) - (\mathbf{A}_k \underline{x}_k + \mathbf{B}_k \underline{u}_k + \hat{\underline{a}}_k) \right]$$

and the infimum $\underline{r}_k^{\inf}$ correspondingly. In doing so, we obtain an n_x-dimensional rectangle, which itself can be interpreted as the Minkowski sum of n_x one-dimensional ellipsoids, i.e., intervals

$$\left\{ \underline{x} \,\middle|\, x_i \in [(\underline{r}_k^{\inf})_i, (\underline{r}_k^{\sup})_i], x_j = 0 \text{ for } j \neq i \right\} \subset \mathbb{R}^{n_x} \;.$$

---

[11]Note that generally linear systems, as in (2.65), are considered. $\hat{\underline{a}}_k$ can be simply seen as an additional input vector.

So, an enclosing ellipsoid $\mathcal{E}(\hat{\underline{r}}_k^{\mathrm{err}}, \mathbf{R}_k^{\mathrm{err}})$ is obtained through an outer approximation (2.36). A simpler but more conservative bound is a ball $\mathcal{E}(\underline{0}, r^{\mathrm{sup}} \cdot \mathbf{I})$ with scaling factor

$$r^{\mathrm{sup}} = \sup_{\substack{\underline{x}_k \in \mathcal{C}_k \\ \underline{u}_k \in \mathcal{U}_k}} \left\| \underline{a}_k(\underline{x}_k, \underline{u}_k) - \left(\mathbf{A}_k \underline{x}_k + \mathbf{B}_k \underline{u}_k + \hat{\underline{a}}_k\right) \right\|_2^2 ,$$

where the maximum possible error is used as a bound in every direction. In Figure 2.14, linearization errors for a two-dimensional function that maps the components x_1 and x_2 of the state vector to a component y are drawn, where a bound for the maximum error in y is an interval.

The Kalman prediction step for the set-membership uncertainties now yields

$$\mathcal{X}_{k+1}^{\mathrm{p}} = \left(\mathbf{A}_k \mathcal{E}(\hat{\underline{x}}_k^{\mathrm{e}}, \mathbf{X}_k^{\mathrm{e}}) \oplus \mathbf{B}_k \mathcal{E}(\hat{\underline{u}}_k, \mathbf{X}_k^{u}) + \hat{\underline{a}}_k\right) \oplus \mathcal{E}(\hat{\underline{r}}_k^{\mathrm{err}}, \mathbf{R}_k^{\mathrm{err}}) ,$$

which directly corresponds to the prediction step derived in sections 2.4.1 and 2.4.2 with the additional set-membership error $\mathcal{E}(\hat{\underline{r}}_k^{\mathrm{err}}, \mathbf{R}_k^{\mathrm{err}})$. For $\mathcal{X}_{k+1}^{\mathrm{p}}$, an ellipsoidal outer approximation with shape matrix $\mathbf{X}_{k+1}^{\mathrm{p}}$ has then to be computed. The formulas (2.99), (2.100), and (2.101) can hence be employed to compute the estimate $\hat{\underline{x}}_{k+1}^{\mathrm{p}}$, the error covariance matrix $\mathbf{C}_{k+1}^{\mathrm{p}}$, and the set-membership error matrix $\mathbf{X}_{k+1}^{\mathrm{p}}$, respectively. The covariance matrix characterizing the stochastic uncertainty does not depend on the bound for the linearization errors.

B   Linearized Sensor Models

In accordance with the preceding paragraph, linearization errors for the measurement mapping will be taken into account by the error characteristics of the filtered state estimate $\hat{\underline{x}}_k^{\mathrm{e}}$. The sensor model is approximated by a linear mapping

$$\hat{\underline{z}}_k = \underline{h}_k(\underline{x}_k) + \underline{v}_k + \underline{e}_k \approx \mathbf{H}_k \underline{x}_k + \hat{\underline{h}}_k + \underline{v}_k + \underline{e}_k ,$$

with $\underline{v}_k \sim \mathcal{N}(\underline{0}, \mathbf{C}_k^z)$ and $\underline{e}_k \in \mathcal{E}(\underline{0}, \mathbf{X}_k^z)$. We again aspire to model linearization errors as an additional set-membership disturbance, and therefore we consider a P confidence region around the observation $\hat{\underline{z}}_k$, i.e.,

$$\mathcal{V}_k = \mathcal{E}(\underline{0}, \tilde{s} \cdot \mathbf{C}_k^z) \oplus \mathcal{E}(\underline{0}, \mathbf{X}_k^z) ,$$

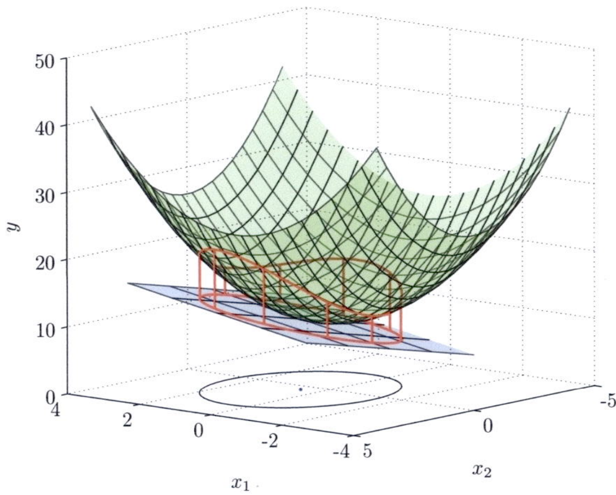

Figure 2.14: Components x_1 and x_2 are mapped to y, according to a parabolic function. The ellipsoid for bounding linearization errors becomes the black ellipse in the x_1-x_2-plane. The linearized mapping is depicted by the blue plane. All linearization errors are positive and are drawn red.

so, with probability P, there exists a vector $\underline{v}_k \in \mathcal{V}_k$ with

$$\hat{\underline{z}}_k = \underline{h}_k(\underline{x}_k) + \underline{v}_k \ ,$$

which entails the inclusion

$$\hat{\underline{z}}_k - \underline{h}_k(\underline{x}_k) \in \mathcal{V}_k \ . \tag{2.104}$$

By $\mathcal{X}_k^h$, we denote the set of all possible $\underline{x}_k$ that fulfill this inclusion. The maximum linearization error can then, as in the previous paragraph, be bounded by the componentwise supremum

$$\underline{R}_k^{\mathrm{sup}} := \sup_{\underline{x}^h \in \mathcal{X}_k^h} \left[\underline{h}_k(\underline{x}_k^h) - \left(\mathbf{H}_k \, \underline{x}_k^h + \hat{\underline{h}}_k \right) \right]$$

and infimum $\underline{R}_k^{\mathrm{inf}}$, respectively. Again, this rectangle can be enclosed by a bounding ellipsoid $\mathcal{E}(\hat{\underline{r}}_k^{\mathrm{err}}, \mathbf{R}_k^{\mathrm{err}})$.

The main difficulty lies in determining the set $\mathcal{X}_k^h$ for the inclusion (2.104). A common approach, which is especially used in the context of set-membership state estimation [40, 111], consists of defining auxiliary linear equations. Such an auxiliary mapping $\underline{x}_k \mapsto \mathbf{H}_k^{\mathrm{aux}}\underline{x}_k + \hat{\underline{h}}_k^{\mathrm{aux}}$ fulfills the relation

$$\left\{\underline{h}_k(\underline{x}_k) + \underline{v}_k \mid \underline{v}_k \in \mathcal{V}_k\right\} \subseteq \mathcal{E}(\mathbf{H}_k^{\mathrm{aux}}\,\underline{x}_k + \hat{\underline{h}}_k^{\mathrm{aux}}, \mathbf{L})$$

for every $\underline{x}_k \in \mathbb{R}^{n_x}$ and a nonnegative definite matrix $\mathbf{L} \in \mathbb{R}^{n_x \times n_x}$. In particular for a specific measurement $\hat{\underline{z}}_k$, we can deduce the implication

$$\hat{\underline{z}}_k - \underline{h}_k(\underline{x}_k) \in \mathcal{V}_k \implies \hat{\underline{z}}_k - (\mathbf{H}_k^{\mathrm{aux}}\,\underline{x}_k + \hat{\underline{h}}_k^{\mathrm{aux}}) \in \mathcal{E}(0, \mathbf{L}_k)$$

and hence the inclusion

$$\left\{\underline{x}_k \in \mathbb{R}^{n_x} \mid \hat{\underline{z}}_k - \underline{h}_k(\underline{x}_k) \in \mathcal{V}_k\right\}$$
$$\subseteq \left\{\underline{x}_k \in \mathbb{R}^{n_x} \mid \hat{\underline{z}}_k - (\mathbf{H}_k^{\mathrm{aux}}\,\underline{x}_k + \hat{\underline{h}}^{\mathrm{aux}}) \in \mathcal{E}(0, \mathbf{L}_k)\right\} =: \mathcal{X}_k^h \,,$$

where the latter set is employed to define $\mathcal{X}_k^h$, over which the linearization errors are bounded.

The auxiliary function can also be utilized to compute a conservative bound $\mathcal{E}(\hat{\underline{r}}_k^{\mathrm{err}}, \mathbf{R}_k^{\mathrm{err}})$ for the set of possible linearization errors, i.e.,

$$\left[(\mathbf{H}_k^{\mathrm{aux}}\,\underline{x}_k + \hat{\underline{h}}^{\mathrm{aux}}) - (\mathbf{H}_k\,\underline{x}_k + \hat{\underline{h}}_k)\right] \in \mathcal{E}(\hat{\underline{r}}_k^{\mathrm{err}}, \mathbf{R}_k^{\mathrm{err}})$$

for all $\underline{x}_k \in \mathcal{X}_k^h$, instead of computing a bound for the errors $[\underline{h}_k(\underline{x}_k) - (\mathbf{H}_k\,\underline{x}_k + \hat{\underline{h}}_k)]$. The estimate is then specified through

$$\hat{\underline{x}}_k^{\mathrm{e}} = (\mathbf{I} - \mathbf{K}_k\mathbf{H}_k)\hat{\underline{x}}_k^{\mathrm{p}} + \mathbf{K}_k\left(\hat{\underline{z}}_k - (\hat{\underline{h}}_k + \hat{\underline{r}}_k^{\mathrm{err}})\right)$$

according to (2.96), where the derived error bound has the generally nonzero midpoint $\hat{\underline{r}}_k^{\mathrm{err}}$. Formerly made linearization errors are bounded by $\mathcal{E}(\underline{0}, \mathbf{X}_k^{\mathrm{p}})$. The associated set-membership error description of the updated estimate is related to

$$\mathcal{X}_k^{\mathrm{e}} = (\mathbf{I} - \mathbf{K}_k\mathbf{H}_k)\mathcal{E}(\underline{0}, \mathbf{X}_k^{\mathrm{p}}) \oplus \mathbf{K}_k\left(\mathcal{E}(\underline{0}, \mathbf{X}_k^{z}) \oplus \mathcal{E}(\underline{0}, \mathbf{X}_k^{\mathrm{err}})\right) \,,$$

for which an ellipsoidal bound with shape matrix $\mathbf{X}_k^{\mathrm{e}}$ is obtained analogously to (2.98). For the covariance matrix, the formula (2.97) is applied. The Kalman gain can be determined to minimize the covariance matrix by employing (2.24), to minimize both error characteristics equally by employing (2.91), or to minimize a weighted version of the gain by using (2.95) in its most general and flexible instance.

2.6.2 Extended and Linear Regression Kalman Filtering

This subsection provides a brief guide for the purpose of identifying linearization errors and combines approaches that are used in [199] and [202]. More precisely, it is a short survey of well-known Kalman filter implementations for nonlinear estimation problems and particularly focuses on unveiling hidden linearization errors.

A Extended Kalman Filtering

The extended Kalman filter [163] has to be considered as the most widely used method for applying Kalman filtering techniques to nonlinear state estimation problems. For that purpose, differentiable system and measurement functions are approximated by first-order Taylor series expansions. Such an expansion of a nonlinear mapping $\underline{x} \mapsto \underline{g}(\underline{x})$ is evaluated at the current state estimate $\hat{\underline{x}}_k$, i.e.,

$$
\begin{aligned}
\underline{g}_l(\underline{x}) &= \underline{g}_l(\hat{\underline{x}}) + \sum_{i=1}^{n_x} \left(\left. \frac{\partial \underline{g}_l}{\partial \underline{x}_i} \right|_{\underline{x}=\hat{\underline{x}}} \right) (\Delta\underline{x})_i \\
&\quad + \sum_{i=1,j=1}^{n_x} \left(\left. \frac{\partial \underline{g}_l}{\partial \underline{x}_i \partial \underline{x}_j} \right|_{\underline{x}=\hat{\underline{x}}} \right) \frac{1}{2!} (\Delta\underline{x})_i (\Delta\underline{x})_j + \ldots \\
&= \underline{g}_l(\hat{\underline{x}}) + \underbrace{\left(\left. \frac{\partial \underline{g}_l}{\partial \underline{x}} \right|_{\underline{x}=\hat{\underline{x}}} \right)}_{=:(\mathbf{A}_k)_l} \Delta\underline{x} \\
&\quad + \sum_{i=1,j=1}^{n_x} \left(\left. \frac{\partial \underline{g}_l}{\partial \underline{x}_i \partial \underline{x}_j} \right|_{\underline{x}=\hat{\underline{x}}} \right) \frac{1}{2} (\Delta\underline{x})_i (\Delta\underline{x})_j + \ldots
\end{aligned}
$$

with $\Delta \underline{x}_k = (\underline{x} - \hat{\underline{x}}_k)$, where $\mathbf{A}_k$ is the sought system or measurement matrix and the subscript i in a vector/matrix x_i denotes the ith row of $\underline{x}$. The linear approximation of $\underline{g}$ then yields

$$\underline{g}(\underline{x}) \approx \mathbf{A}_k \, \underline{x} + \underbrace{\underline{g}(\hat{\underline{x}}) - \mathbf{A}_k \, \hat{\underline{x}}}_{=: \hat{\underline{a}}} \ .$$

Consequently, a system or measurement mapping is approximated by its Jacobian matrix at $\hat{\underline{x}}_k^{\mathrm{e}}$ or $\hat{\underline{x}}_k^{\mathrm{p}}$, respectively, and all higher-order terms are neglected. In order to assess the influence of the higher-order terms, the Cauchy or Lagrange form of the remainder can, for instance, be examined. With them, the nonlinear part can be bounded over the considered uncertainty region around the current state estimate and the input or measurement vector. As an example, the Hessian matrices can be employed to compute the radius

$$r = \sup_{\underline{x} \in \mathcal{X}} \frac{1}{2} T \left\| \underline{x} - \hat{\underline{x}} \right\|_\infty$$

of a bounding ball $\mathcal{E}(\underline{0}, r^2 \cdot \mathbf{I})$, where T is given by

$$T = \sum_{i,j=1}^{N} \sup_{\underline{x} \in \mathcal{X}, l} \left\| \frac{\partial^2 \underline{g}_l}{\partial \underline{x}_i \partial \underline{x}_j}(\underline{x}) \right\| \ .$$

The concept derived in Subsection 2.6.1 then enables us to take into account the bounds for the remainder during prediction and filtering steps.

B   Linearization Around Set

Especially when adopting the viewpoint of Section 2.4.1, where an estimate is characterized by an ellipsoid of conditional means, we can easily accept that the nonlinear mapping should not be linearized about a single operating point, but about the entire set of possible means. In [132] and [199], an approach for approximating a nonlinear mapping $\underline{g}$ around a set of means has been suggested. It is explained here on the basis of the system model, which must be linearized not only around the state variable but also around the input variable. The system model will be approximated by an affine mapping, i.e.,

$$\underline{x}_{k+1} = \underline{a}_k(\underline{x}_k, \underline{u}_k) \approx \mathbf{A}_k\,\underline{x}_k + \mathbf{B}_k\,\underline{u}_k + \hat{\underline{a}}_k \ . \tag{2.105}$$

The idea is to choose approximation points for each ellipsoid, which are commonly assumed to lie equidistantly on the principal axes[12] of the ellipsoids. Let $\underline{x}_k^{\mathrm{e},(1)}, \ldots, \underline{x}_k^{\mathrm{e},(N)}$ denote the approximation points of the ellipsoid $\mathcal{E}(\hat{\underline{x}}_k^{\mathrm{e}}, \mathbf{X}_k^{\mathrm{e}})$ of means and $\underline{u}_k^{(1)}, \ldots, \underline{u}_k^{(M)}$ the approximation points of the ellipsoidal set $\mathcal{E}(\hat{\underline{u}}_k, \mathbf{X}_k^u)$ of possible inputs. Then,

$$\underline{d}_k^{(i,j)} := \underline{a}_k\left(\underline{x}_k^{\mathrm{e},(i)}, \underline{u}_k^{(j)}\right) - \mathbf{A}_k\,\underline{x}_k^{\mathrm{e},(i)} - \mathbf{B}_k\,\underline{u}_k^{(j)} - \hat{\underline{a}}_k \ ,$$

for $i = 1, \ldots, N$, $j = 1, \ldots, M$, denote the errors between the nonlinear mapping and its linear approximation at these points. The mappings $\mathbf{A}_k$, $\mathbf{B}_k$ and the base point $\hat{\underline{a}}_k$ are computed to satisfy

$$\{\mathbf{A}_k, \mathbf{B}_k, \hat{\underline{a}}_k\} = \underset{\mathbf{A}_k, \mathbf{B}_k, \hat{\underline{a}}_k}{\arg\min} \ \sum_{i,j=1}^{N,M} \omega_{i,j}^{-1} \left[\underline{d}_k^{(i,j)}\right]^{\mathrm{T}} \left[\underline{d}_k^{(i,j)}\right] \ , \tag{2.106}$$

with weighting factors $\omega_{i,j}$. The solution of this weighted least squares problem [113] is given by

$$\begin{bmatrix} \mathbf{A}_k^{\mathrm{T}} \\ \mathbf{B}_k^{\mathrm{T}} \\ \hat{\underline{a}}_k^{\mathrm{T}} \end{bmatrix} = \left(\mathbf{F}_k \mathbf{Q}_k^{-1} \mathbf{F}_k^{\mathrm{T}}\right)^{-1} \mathbf{F}_k \mathbf{Q}_k^{-1} \mathbf{f}_k \ ,$$

where $\mathbf{F}_k$, $\mathbf{f}_k$, and $\mathbf{Q}$ are defined by

$$\mathbf{F}_k = \begin{bmatrix} \underline{x}_k^{\mathrm{e},(1)} & \cdots & \underline{x}_k^{\mathrm{e},(1)} & \underline{x}_k^{\mathrm{e},(2)} & \cdots & \underline{x}_k^{\mathrm{e},(2)} & \underline{x}_k^{\mathrm{e},(3)} & \cdots & \underline{x}_k^{\mathrm{e},(N)} \\ \underline{u}_k^{(1)} & \cdots & \underline{u}_k^{(M)} & \underline{u}_k^{(1)} & \cdots & \underline{u}_k^{(M)} & \underline{u}_k^{(1)} & \cdots & \underline{u}_k^{(M)} \\ 1 & \cdots & 1 & 1 & \cdots & 1 & 1 & \cdots & 1 \end{bmatrix} \ ,$$

$$\mathbf{f}_k = \begin{bmatrix} \underline{a}_k\big(\underline{x}_k^{\mathrm{e},(1)}, \underline{u}_k^{(1)}\big) & \cdots & \underline{a}_k\big(\underline{x}_k^{\mathrm{e},(N)}, \underline{u}_k^{(M)}\big) \end{bmatrix}^{\mathrm{T}} \ ,$$

and $\mathbf{Q}_k = \mathrm{diag}(\omega_{1,1}, \ldots, \omega_{N,M})$, respectively.

---

[12]They can be determined in the same way as for the unscented Kalman filter [97].

The result can now be used to set up the linear approximation in (2.105), and the procedure in the preceding Paragraph 2.6.1-A leads to the bounding sets for the linearization errors. The same linearization technique can be performed for nonlinear measurement models $\underline{h}_k$ if set-membership uncertainty on the prior estimate is present. Of course, it is also reasonable to linearize around the complete uncertainty region of an estimate, i.e., to use the set-membership error bound combined with the P confidence region in order to determine the approximation points for the computation of (2.105).

C Linear Regression Kalman Filtering

Well-known examples for linear regression Kalman filters [114] are the unscented Kalman filter [95, 97, 100] or Gaussian filters like [25, 82]. These filters determine statistical linearizations of nonlinear system and measurement models by means of a certain number L of regression points $\{\underline{x}_i\}_{i=1,\dots,L}$, which are usually taken around the mean of the current state estimate. Thus, this is approach is similar to the approximation technique of the previous Paragraph 2.6.2-B. For the regression points, the function values of the considered nonlinearity $\underline{y}_i = g(\underline{x}_i)$, be it the system function or the measurement function, are calculated in order to obtain the set $\{\omega_i, \underline{x}_i, \underline{y}_i\}_{i=1,\dots,N}$, where ω_i are chosen weights. In general, prediction and filtering results are directly computed on the basis of these regression points, so that the underlying linear mapping remains hidden to the user. As explicated in [114] and [202], this linear mapping can be determined as the solution of

$$\{\mathbf{A}, \underline{\hat{a}}\} = \arg \min_{\mathbf{A}, \underline{b}} \sum_{i=1}^{L} \omega_i \cdot \underline{e}_i^{\mathrm{T}} \cdot \underline{e}_i \, ,$$

where the sum of weighted squared errors with $\underline{e}_i = \underline{y}_i - (\mathbf{A}\underline{x}_i + \underline{b})$ is minimized. This least-squares problem resembles (2.106) in the previous paragraph and has the solution

$$\mathbf{A} = \mathbf{C}_{xz}^{\mathrm{T}} \mathbf{C}_{xx}^{-1} \quad \text{and} \quad \underline{\hat{a}} = \underline{\hat{y}} - \mathbf{A} \cdot \underline{\hat{x}}$$

with $\hat{\underline{x}} = \sum_{i=1}^{L} \omega_i \cdot \underline{x}_i$, $\hat{\underline{y}} = \sum_{i=1}^{L} \omega_i \cdot \underline{y}_i$, $\mathbf{C}_{xx} = \sum_{i=1}^{L} \omega_i \cdot (\underline{x}_i - \hat{\underline{x}}) \cdot (\underline{x}_i - \hat{\underline{x}})^{\mathrm{T}}$, and $\mathbf{C}_{xy} = \sum_{i=1}^{L} \omega_i \cdot (\underline{x}_i - \hat{\underline{x}}) \cdot (\underline{y}_i - \hat{\underline{y}})^{\mathrm{T}}$. The resulting affine mapping $\underline{x} \mapsto \mathbf{A}\,\underline{x} + \hat{\underline{a}}$ can be used to compute a state estimate by means of the standard Kalman filter equations. This estimate is the same as that of the linear regression Kalman filter.

The influences of the linearization errors can statistically be examined by means of the error covariance matrix

$$\mathbf{C}_k^* = \sum_{i=1}^{L} \omega_i \cdot \underline{e}_i \cdot \underline{e}_i^{\mathrm{T}} \ .$$

However, instead of considering this statistical characterization of linearization errors, which may act as an additional linearization noise but is only based on a few sample points, one can now employ the systematic bounds from Subsection 2.6.1.

2.7 Applied Simultaneous Stochastic and Set-membership Estimation

As a proof of concept, we consider different applications of simultaneous stochastic and set-membership state estimation. Of course, in several cases, non-stochastic and systematic errors affecting the state can be estimated by standard bias-aware Kalman filtering techniques and canceled out, e.g., by following the direction of [60] and subsequent approaches [50]. However, by contenting ourselves with error bounds, we bypass the need to specify a certain error characteristic, such as constancy. For instance, the unknown but bounded error term can even be a second stochastic disturbance with a compactly supported probability density and therefore may not behave systematically. In particular, neglected nonlinearities cannot be subjected to a systematic error behavior, but also simple calibration errors can be difficult to characterize in terms of a random variable or bias, as revealed by the experiments explained in Subsection 2.7.2. Similarly, for the estimation of the heart surface displacement, the proposed approach has successfully been applied to account for stochastic as well as systematic errors, as pointed out in Subsection 2.7.3. Particular attention has to be directed to sensor networks where various sources of uncertainties can be identified that require a methodical treatment of stochastic and set-membership errors. As examples, quantization of sensor data and event-based state estimation are discussed in Subsection 2.7.4. The first subsection shows how the methods from Section 2.6 are employed to bound linearization errors.

2.7.1 In Nonlinear Systems

In this first scenario, we employ the approach proposed in Section 2.6 in order to bound linearization errors when applying the extended Kalman filter. The dynamics of an unforced van-der-Pol oscillator [183] are considered that are governed by the second-order differential equation

$$\frac{d^2x}{dt^2} - \mu(1 - x^2)\frac{dx}{dt} + x = 0 \; ,$$

which is damped by a nonlinear coefficient when x approaches high values. This equation can be rewritten as a system

$$\begin{bmatrix} \dot{x}^{(1)}(t) \\ \dot{x}^{(2)}(t) \end{bmatrix} = \begin{bmatrix} x^{(2)}(t) \\ \mu\big(1 - (x^{(1)}(t))^2\big)\, x^{(2)}(t) - x^{(1)}(t) \end{bmatrix}$$

of first-order differential equations. The Euler method with step size Δt can be employed to obtain the discrete-time system model

$$\begin{bmatrix} x_{k+1}^{(1)} \\ x_{k+1}^{(2)} \end{bmatrix} = f(\underline{x}_k, \Delta t) = \begin{bmatrix} x_k^{(1)} \\ x_k^{(2)} \end{bmatrix} + \Delta t \begin{bmatrix} x_k^{(2)} \\ \mu\big(1 - (x_k^{(1)})^2\big)\, x_k^{(2)} - x_k^{(1)} \end{bmatrix} \; .$$

In the estimation algorithm, it is assumed that the nonlinear model

$$\underline{x}_{k+1} = f(\underline{x}_k, \Delta t) + \underline{w}_k$$

is perturbed by the process noise $\underline{w}_k \sim \mathcal{N}(\underline{0}, 0.02\mathbf{I})$. The system matrix for the linearized model is finally given by the Jacobian

$$\mathbf{A}_k(\underline{x}) = \begin{bmatrix} 1 & \Delta t \\ -2\Delta t\, \mu\, x^{(1)}\, x^{(2)} - 1 & 1 + \Delta t\, \mu(1 - (x^{(1)})^2) \end{bmatrix} \; ,$$

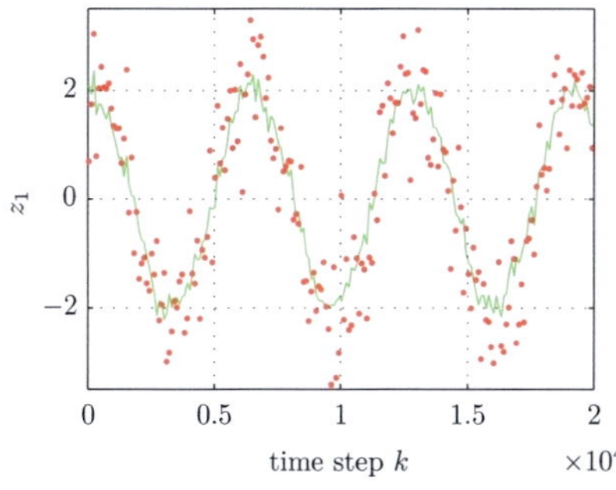

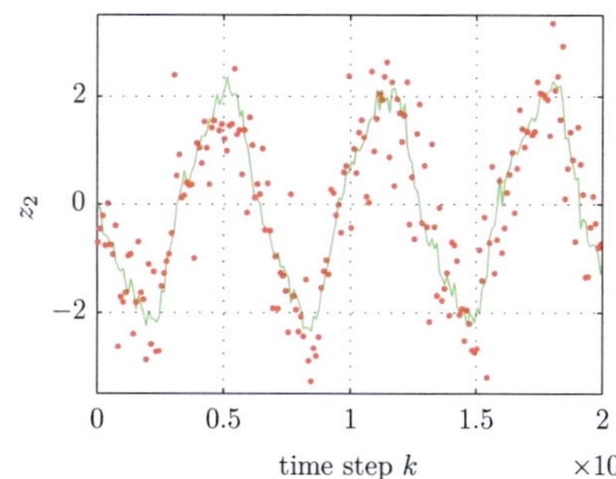

(a) Red dots: Component z_1 of measurements.

(b) Red dots: Component z_2 of measurements.

Figure 2.15: Output signal and measurements for the simulated van der Pol oscillator.

which is employed by the EKF to compute the predicted covariance matrix (2.20). The state is directly observed according to

$$\hat{\underline{z}}_k = \underline{x}_k + \underline{v}_k$$

with the measurement noise $\underline{v}_k \sim \mathcal{N}(\underline{0}, 0.5\mathbf{I})$.

For the simulation, we model the true signal, i.e., the ground truth, by setting $\Delta t_{\text{true}} = 0.001$ and compute $20\,000$ time steps. The noisy signal and 200 obtained measurements are depicted in Figure 2.15. For the discrete-time EKF, the step size $\Delta t_{\text{EKF}} = 0.1$ is chosen. Hence, 200 filtering steps are performed. The damping coefficient is, in both cases, $\mu = 0.4$. The initial estimate is $\hat{\underline{x}}_0^{\text{P}} = [2, 0]^{\text{T}}$ with error covariance matrix $\mathbf{C}_0^{\text{P}} = \mathbf{I}$ and zero shape matrix $\mathbf{X}_0^{\text{P}} = \mathbf{0}$, i.e., no unknown but bounded errors are assumed to be present at the beginning. The estimated state is shown in Figure 2.16, where the advanced filtering step from Section 2.5 has been employed. The weighting parameter is set to $\alpha = 0.1$ in order to favor a minimal covariance matrix. The choice of different weighting parameters, i.e. the choice of different gains, is considered more detailed in Subsection 2.7.4. Figure 2.17 shows an particularly interesting feature: The set-membership error bounds

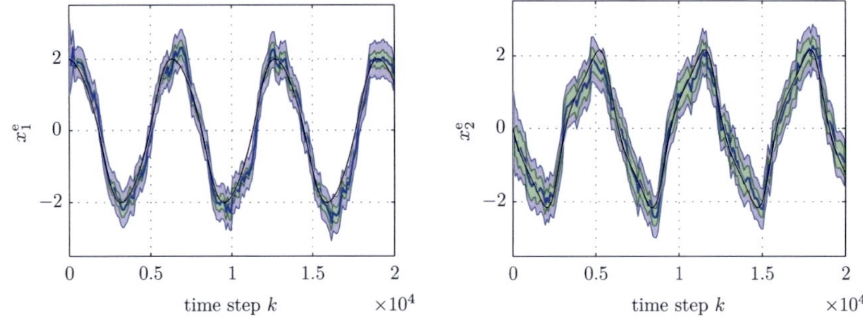

(a) Estimate of first component.

(b) Estimate of second component.

Figure 2.16: Estimated system state with error bounds. The set-membership error bound is shown in green. Additional 3-sigma bounds for the stochastic errors are drawn blue.

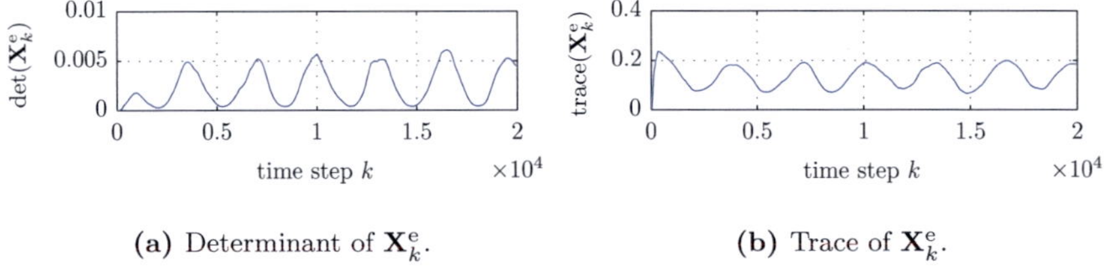

(a) Determinant of $\mathbf{X}_k^e$.

(b) Trace of $\mathbf{X}_k^e$.

Figure 2.17: Determinant and trace of the computed shape matrices.

become small when the linearization errors are small, and they increase when the nonlinear damping factor is more dominant. This is compliant with the property of the van der Pol oscillator that it behaves linearly near the origin so that only small linearization errors are caused.

2.7.2 In Model Predictive Control

The proposed simultaneous stochastic and set-membership estimation principles have been combined with model predictive control methods in [192]. An application-specific reward function is to be calculated that assesses the anticipated impact of future control inputs and measurements. With stochastic and set-membership errors being present, a rigorous generalization of deterministic reward functions has to be derived that maps random and set-valued variables to scalar well-defined values. One can commence with the consideration of a deterministic reward function, which becomes both a random and a set-valued function if the arguments consists of stochastic and unknown but bounded quantities. For instance, the expectation value and infimum functionals can then be utilized to condense the result into a certain reward value.

Experiments with a walking robot that particularly suffers from calibration errors have been conducted in [192]. The controller cannot be sure that the correct steering angles are applied. The unknown calibration error is bounded by a set, and further process and sensor noise is assumed to affect the system randomly. It turned out that the combined stochastic and set-membership approach is more cautious and reliable than an extended bias-aware Kalman filter that attempts to compute an estimate on the calibration error.

For specific systematic errors, bias-aware Kalman filtering techniques like [50] and [60] can provide promising estimation results, but the conducted experiments have demonstrated that the treatment of unknown systematic errors by stochastic methods can be by far more difficult and less reliable. On the other hand, bounding an error by a set implies that the error remains unknown within the bounds, so no insight to the actual error behavior is gained, which might be desired in some applications.

2.7.3 In Beating Heart Surgery

The use of simultaneous stochastic and set-membership state estimation techniques has particularly been investigated in robotic beating heart surgery [191]. The general idea is to synchronize surgical instruments with the heart surface motion so as to perform the surgery "off pump" and to eliminate the need to artificially stop the heart. Furthermore, there is a clear tendency towards minimally invasive heart surgery. The estimator shall continually provide the surgeon with an impression of the quality and reliability of the estimated heart displacement. The state estimate on the heart displacement is supplemented by an uncertainty region, as illustrated in Figure 2.18, that is computed from the set-membership error bound and a confidence region around random errors. This region can then be employed to distinguish between safe and unsafe operating conditions and to prevent the robot from accidentally harming healthy tissue. An important further application is to equip the surgical system with a haptic guidance in form of safety critical limits or soft virtual fixtures.

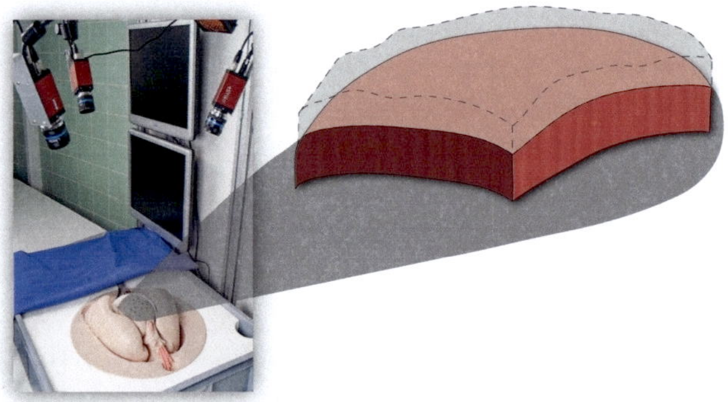

Figure 2.18: Experimental setup with pressure-regulated artificial heart. Heart surface with computed uncertainty region that embraces stochastic and set-membership errors.

The uncertainty region is derived by systematically identifying and describing sources of stochastic and unknown but bounded errors. The concept has been evaluated by means of a pressure-regulated artificial heart that is observed by three camera systems. For instance, specified standard deviations of the pressure regulator serve as a stochastic uncertainty description, and the limited resolution of the camera system is related to an unknown but bounded error affecting each reconstructed landmark position. A main advantage of the proposed concept is that the computed region dynamically adapts to any changes such as occlusions. A detailed analysis and simulation results can be found in [191].

2.7.4 In Sensor Networks

In sensor networks, it is often inevitable to keep the volume and frequency of data transfers as low as possible. Quantization of measurements contributes to reducing the data volume to be transferred, and, with the help of event-based estimation, data needs only to be transmitted at predefined instances of time. Although system and measurement noise terms in this section are purely stochastic, quantization and event-based data processing append set-membership uncertainties to the estimation problem. The considered system models are simplified versions of the discretized Euler equations in [183]. For the first scenario, we revisit the example discussed in [205].

Quantized Data In order to downsize the data to be transmitted in a sensor network, the measurement data can be quantized, which introduces a set-membership error to the estimation problem. We analyze the effect of quantization errors by means of a three-dimensional state that is related to the discrete-time state transition model

$$\underline{x}_{k+1} = \mathbf{A}\,\underline{x}_k + \underline{w}_k + \underline{d}_k$$

with

$$\mathbf{A} = \begin{bmatrix} 1 & h & h \\ -h & 1 & h \\ -0.51h & -0.51h & 1 \end{bmatrix} .$$

The step size $h = \Delta t$ determines how well this system approximates the underlying differential equation. The ground truth is generated by setting $h = 0.0001$, which is depicted as the black dashed lines in Figure 2.19. In the implementation of the estimator, $h = 0.1$ is used, and zero-mean process noises $\underline{w}_k$ and $\underline{d}_k$ with stochastic and set-membership error characteristics $\mathbf{C}_k^w = \mathrm{diag}([0.2, 0.15, 0.1])$ and $\mathbf{X}_k^d = \mathrm{diag}([0.1, 0.1, 0.1])$ are assumed. At each time step k, 10 measurements are received that are each related to the state by the sensor model

$$\underline{z}_k = \underline{x}_k + \underline{v}_k \ ,$$

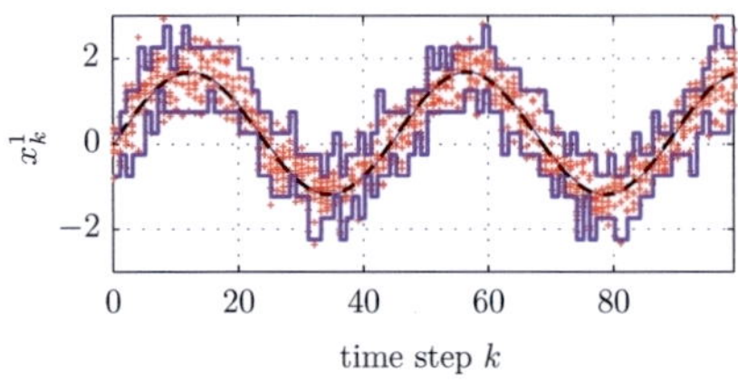

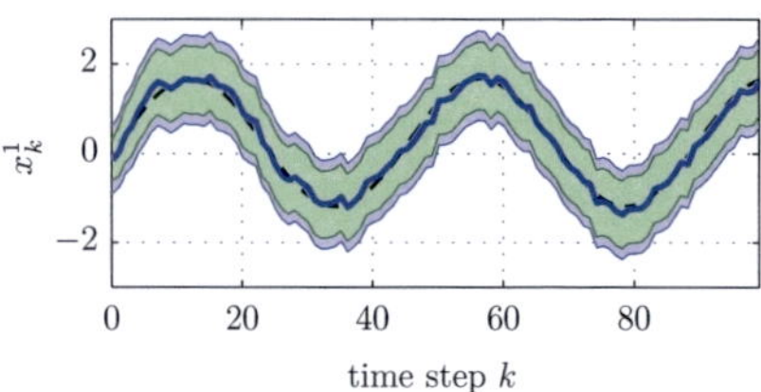

(a) Simulated measurements. The lowest and highest value after discretization are shown in purple.

(b) Results of the filtering algorithm that is optimized for $\mathbf{C}_k^{\mathrm{e}}$, i.e., $\alpha = 0$.

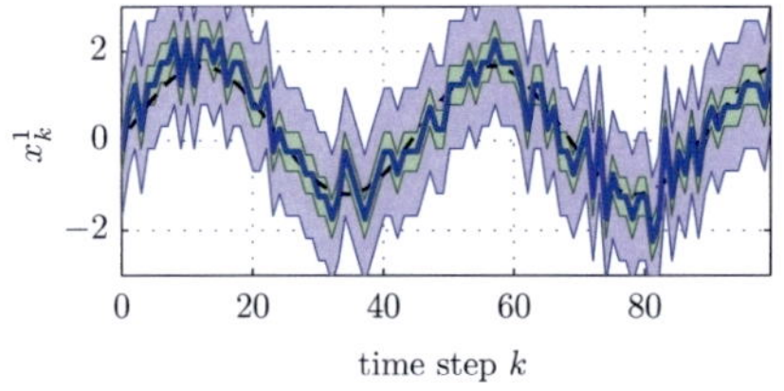

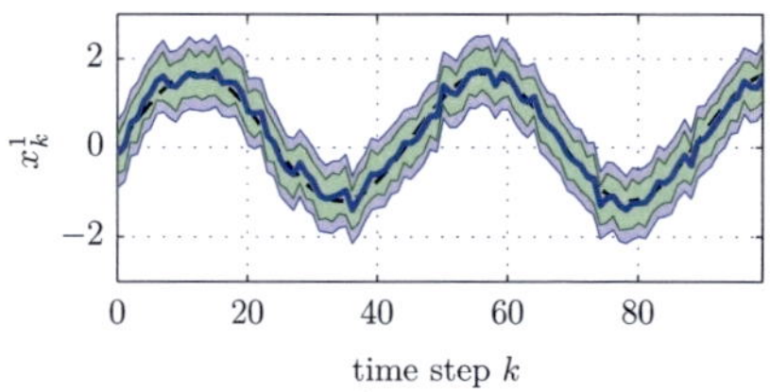

(c) Results of the filtering algorithm that is optimized for $\mathbf{X}_k^{\mathrm{e}}$, i.e., $\alpha = 1$.

(d) Results of the filtering algorithm with equilibrium weighting, i.e., $\alpha = 0.5$.

Figure 2.19: Estimation results in case of quantized measurements. Only the first component of the state vector is shown. The thick blue line is the estimation result. The green area bounds the set-membership error. The light blue 3-sigma bounds of the stochastic error characteristics are added to the green bound.

which is altered by the measurement noise $\underline{\boldsymbol{v}}_k \sim \mathcal{N}(\underline{0}, \mathbf{C}_k^v)$ with $\mathbf{C}_k^v = \mathrm{diag}([0.25, 0.5, 0.75])$. Set-membership errors during the filtering step come into play due to the quantization

$$q(z_k^{(i)}) = 2 \left\lfloor \frac{z_k^{(i)}}{2} \right\rfloor + 0.25$$

for each component $z^{(i)}, i = 1, 2, 3$, of $\hat{\underline{z}}_k$. Hence, the measurement space is subdivided into $[0.5)^3$-cubes. Each cube can be bounded by a ball with shape matrix $\mathbf{X}_k^z = \frac{3}{16}\mathbf{I}$, which characterizes the unknown but bounded quantization error. Figure 2.19(a) illustrates the measurement process. The filtering results are depicted in figures 2.19(b)–(d) for different choices of α for the gain (2.95). Figure 2.19(b) shows the result where the stochastic uncertainty is minimized and corresponds to the approach from Section 2.4.1. In Figure 2.19(c), the set-membership uncertainty is minimal by setting $\alpha = 1$. With $\alpha = 0.5$, the result in Figure 2.19(d) is attained. The estimated signal in Figure 2.19(c) is not smooth and apparently suffers from the stochastic noise. This implies, with a gain being optimized for set-membership errors, the high-frequency measurement noise cannot be filtered out properly. The signal in Figure 2.19(b) appears to be very smooth while the result Figure 2.19(d) has the tightest error bounds.

Event-Based Communication   We consider a setup that is similar to the preceding example. The state transition model

$$\underline{x}_{k+1} = \mathbf{A}\,\underline{x}_k + \underline{w}_k + \underline{d}_k$$

has the different system matrix

$$\mathbf{A} = \begin{bmatrix} 1 & 2h & 2h \\ -2h & 1 & 2h \\ -0.51h & -0.51h & 1 \end{bmatrix} .$$

The ground truth is again generated with $h = 0.0001$ depicted as the black dashed lines in figures 2.20(a)–(d). For the state estimator, we consider the step size $h = 0.1$. The zero-mean noise terms $\underline{w}_k$ and $\underline{d}_k$ have the covariance and shape matrices $\mathbf{C}_k^w = \mathrm{diag}([0.2, 0.15, 0.1])$ and $\mathbf{X}_k^d = \mathrm{diag}([0.1, 0.1, 0.1])$, respectively. The state is again directly observed according to

$$\underline{z}_k = \underline{x}_k + \underline{v}_k$$

with $\underline{v}_k \sim \mathcal{N}(\underline{0}, \mathbf{C}_k^v)$ with $\mathbf{C}_k^v = \mathrm{diag}([0.25, 0.5, 0.75])$. Hence, the sensor itself is only affected by a random noise. In contrast to the previous example, the sensor data is only communicated to the estimator when the difference of the current measurement to the last reported measurement is larger than 0.7, i.e., when

$$\|\underline{\hat{z}}_{\mathrm{last}} - \underline{\hat{z}}_k\| > 0.7 .$$

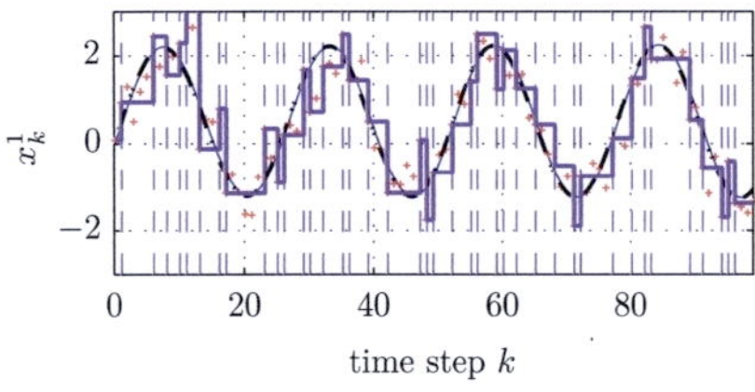

(a) Simulated measurements. Time steps of transmission are marked by purple dashed lines. Sensor observations are red. The solid purple line represents the measurements used by the estimator.

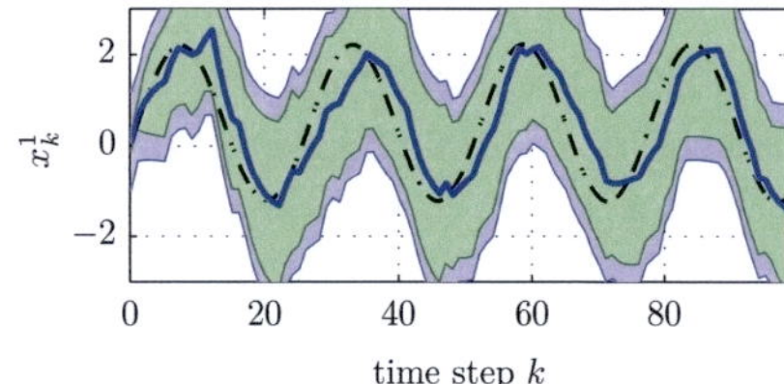

(b) Results of the filtering algorithm that is optimized for $\mathbf{C}_k^{\mathrm{e}}$, i.e., $\alpha = 0$.

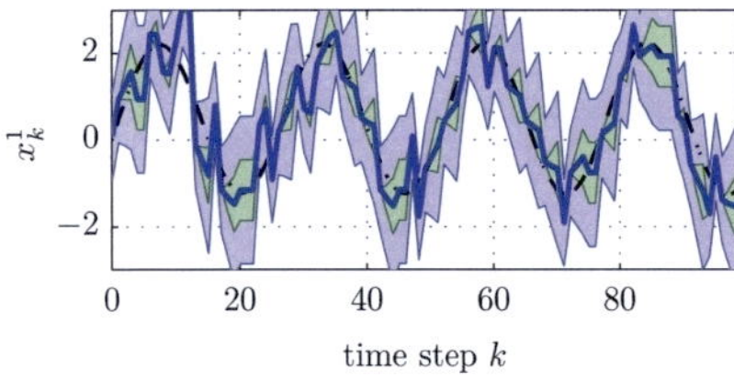

(c) Results of the filtering algorithm that is optimized for $\mathbf{X}_k^{\mathrm{e}}$, i.e., $\alpha = 1$.

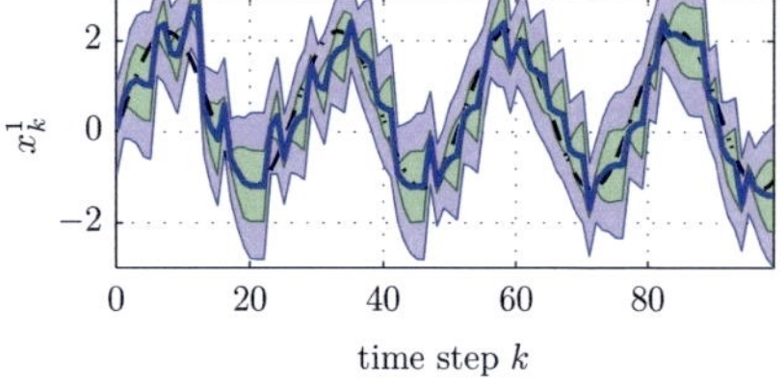

(d) Results of the filtering algorithm with equilibrium weighting, i.e., $\alpha = 0.5$.

Figure 2.20: Estimation results in case of event-based filtering. Only the first component of the state vector is shown. The thick blue line is the estimation result. The green area bounds the set-membership error. The light blue 3-sigma bounds of the stochastic error characteristics are added to the green bound.

At each time step, only one measurement is observed. The events of communication are marked by dashed purple lines in Figure 2.20(a). The purple stairs function represents the measurements that are employed by the estimator as long as no change is reported by the sensor. Until a new measurement arrives, the estimator assumes that the observations are contained in a ball with radius $r = 0.7$ around the last measurement $\hat{\underline{z}}_{\mathrm{last}}$. Hence, this ball represents a set-membership measurement noise. When a new measurement arrives, this measurement $\hat{\underline{z}}_k$ is used to update the state estimate. Also, $\hat{\underline{z}}_{\mathrm{last}}$ is set to $\hat{\underline{z}}_k$ until the next event is reported. Of course, for a new arriving measurement, no set-membership error is present. This states the reason why the green bound in Figure 2.20(c), where the set-membership uncertainty is minimized, vanishes at the purple time steps. The worst result is shown in Figure 2.20(b), where the set-membership error bound is only a byproduct, and the stochastic error is minimized. Here, the estimate is less informative due to large error bounds and suffers from a lag, which is caused by placing too much trust in the inter-event measurements $\hat{\underline{z}}_{\text{last}}$. Again, Figure 2.20(d) shows the minimum total error bound, although the difference to Figure 2.20(c) is not significant. By employing this event-based filtering scheme, only 41 instead of 100 data transfers from the sensor to the estimator are required.

2.8 Conclusions from Chapter 2

In many situations, it can be a tough decision to commit oneself to a certain uncertainty model. Although either purely stochastic or purely set-membership models are most widely employed in state estimation systems, they only allow for an incomplete picture of the imperfect knowledge about the real system: A set can hardly account for outliers of a normally distributed error, and characterizing an unknown error behavior by a specific probability density is false and misdirecting. A combined stochastic and set-membership uncertainty model is possibly able to encompass almost every type of errors that are encountered in state estimation theory. Set-membership descriptions are most appropriate to account for ignorance. Of course, this type of uncertainty must be bounded since the presence of a completely unknown and unbounded error would not allow drawing any conclusions. With stochastic quantities, knowledge about possible realizations can be incorporated, and it can be assessed how likely they are to occur.

The major objective pursued by this chapter has been to derive an estimation concept that is a generalization of purely stochastic and purely set-membership estimation principles but includes them as special cases. Sets of translated Gaussian densities are a direct consequence of the Bayesian perspective on the Kalman filter algorithm. The presence of additive unknown but bounded errors only affects the means of the underlying Gaussian densities, and the computation of the Kalman gain only involves the corresponding covariance matrices. Here, the estimated set-membership error bounds can be considered rather as a byproduct, since the stochastic error is minimized while the extent of set-membership error bounds is ignored. Therefore, we consider the total mean squared error (MSE) that comprises the variance of the stochastic errors as well as a bound on the maximum bias. In doing so, a Kalman gain is derived that minimizes the MSE in the presence of both stochastic and additional unknown but bounded uncertainties, which are represented by Gaussian random variables and ellipsoidal sets, respectively. As a result, a generalization of the well-known Kalman filtering scheme is attained that reduces to the standard Kalman filter in the absence of set-membership uncertainty and that otherwise becomes the intersection of sets in case of vanishing stochastic uncertainty. Considering the total MSE is related to the underlying conception that a covariance ellipsoid and a rigid ellipsoidal bound are interpreted in the same manner by the Kalman gain. Of course, covariance ellipsoids are far from being interpretable as rigid bounds. Therefore, we have proposed an extension to the derived Kalman gain that leaves it to the user to favor either the minimization of the stochastic uncertainty or the minimization of the set-membership uncertainty. The gain is adjusted by a simple scalar parameter. The extreme cases consists of a standard Kalman filter that computes consistent shape matrices for unknown but bounded errors as a sideline and a set-membership estimator that produces consistent error covariance matrices as a byproduct. As a consequence, a simple estimation concept has been derived whose implementation is not more complicated than for a standard Kalman filter or an ellipsoidal estimator. This result proves that a simultaneous incorporation of stochastic and set-membership errors does entail neither elaborate algorithms nor diffuse approximations.

Even if only stochastic errors are present, the proposed concept can be employed to significantly increase the reliability of extended or linear regression Kalman filters. Furthermore, several scenarios have been studied where an explicit distinction between stochastic and set-membership errors appears promising. The results of this chapter lays the spadework for the following chapter that studies how to implement these estimation concepts in a distributed or decentralized fashion.

CHAPTER
3

Distributed and Decentralized Kalman Filtering: Challenges and Solutions

Overview of Chapter 3

State estimation techniques like the concepts studied in the previous chapter provide the means to gain insight to an uncertain state from noisy measurements and input signals. The rapid advances in sensor and communication technology entail an increasing demand for implementing these estimation algorithms in distributed networked systems [71,118,119]. The general idea is that data is collected and processed locally on different sensor nodes, with the aim of monitoring large-scale phenomena, distributing computational resources, and increasing robustness to failures. With a wide scope of applications such as, inter alia, monitoring volcanic eruptions [176], detection and forecasting of hazards [104], indoor person localization [211], or cooperative multi-platform estimation problems [81], sensor network designs have to address a vast variety of different requirements and limitations. The advancements in sensor network technologies are accompanied by a paradigm shift towards a large number of miniaturized, low power-consuming sensor devices instead of few powerful, high-resolution systems. As a consequence of this trend, a single node can in general provide neither informative nor precise data, but the entire network can in return cover large areas and increase the spatial resolution considerably. A sensor network inherently displays an outstanding potential to solve problems in a cooperative fashion and is hence, in many settings, superior to a single high-performance central processing unit.

Even if a central node is present, the requirements regarding communication bandwidth and processing power would be extensive in a large-scale network without a local preprocessing of sensor data. Due to local processing of accruing data, assumptions made in standard formulations of information processing and state estimation algorithms may not apply anymore. In particular, the estimation principles studied in Chapter 2 essentially necessitate a central component that handles the required processing steps. In contrast, this chapter is devoted to the task of distributing the estimation techniques from Chapter 2 within a network of sensor nodes. It commences, in Section 3.1, with a general discussion of the challenges related to network-centric systems and introduces the three general types of estimation architectures. Besides stochastic and unknown but bounded process and measurement noise, the approach of a local data processing and subsequent fusion poses an additional source of uncertainty: Dependencies among local data sets are either too expensive to keep track of or simply remain hidden to the fusion nodes. Hence, they can be viewed as another uncertainty affecting the state estimation process. Section 3.2 discusses the causes of interdependencies. The most evident solution is to maintain a central instance within the network but also to intelligently preprocess sensor data along any communication path in order to assure a scalable processing of information. Section 3.3 gives special attention to such centralized schemes, where the information form of estimates reveals itself to be especially well suited. The simultaneous treatment of stochastic and set-membership uncertainties inside the information filter is considered in Section 3.4. In a fully decentralized network, each node is capable of operating autonomously. Dependencies therefore cannot be tracked down and have to be bounded conservatively. Section 3.5 introduces basic approaches, which are further developed in Section 3.6 in order to bound unknown dependencies more tightly.

3.1 State Estimation in Networked Systems

Networks of sensors and actuators conceptually provide the means to cooperatively manage vast amounts of data, to monitor large-scale phenomena from different perspectives, and to be adaptable to rapidly and unpredictably changing environments. Along with these aspects, a networked system entails numerous advantages but also imposes many additional challenges on an information processing system. Critical limitations in communication and computational resources have to be taken into account in order to properly adapt existing and design new estimation algorithms. This section introduces the general types of estimation architectures and states the challenges of networked state estimation more precisely.

3.1.1 Centralized, Distributed, and Decentralized State Estimation Architectures

The terms for the three general types of estimation architectures—centralized, distributed, and decentralized—are often used in an ambiguous fashion. In a *centralized* system, as indicated by Figure 3.1, a single computer system is charged with acquiring sensor data from the devices, computing an estimate of the state, and making decisions. So, all measurement data must be send to the central node, which has sole responsibility for computing an informative estimate. A *distributed* estimation architecture refers to a collection of cooperatively operating devices that, viewed from the outside, still appears as a single system. The observed data can be processed on different nodes, but for computing a final estimate, a central system may be required. An example is a hierarchical structure, as illustrated in Figure 3.2, where local estimates are computed on intermediate nodes that undertake some tasks of the central unit. In Section 3.3.3, an optimally distributed Kalman filtering scheme is considered, where each node performs prediction and filtering steps locally. However, the local quantities taken alone have no meaning but, in conjunction with each other, a globally optimal estimate can be computed. This scheme even completely fails when only a single node drops out of the network, unlike a *decentralized* estimation architecture, as depicted in Figure 3.3. Here, the estimation problem is solved locally on each node using local data. The nodes are intended to operate independently and can share their information with each other for the purpose of solving the higher-level estimation problem. In general, a decentralized sensor network cannot achieve the estimation quality of a centralized system but is inherently more flexible and robust to failures. From the perspective of applications, a decentralized structure therefore often has clear advantages.

Within a fully centralized architecture, the estimation algorithms from Chapter 2 can directly be employed. For distributing the workload to the sensor nodes, at least an efficient acquisition and (pre-)processing of multisensor data is necessary. Centralized Kalman filters can generally exploit the conditional independence[1] of received measurements, whereas a distributed and decentralized estimation scheme encounters the additional challenge that even independently processed estimates are not necessarily independent. The reasons for dependencies between locally processed data are manifold: The nodes may share common prior information, the same process noise is exploited by different local state transition models, sensor noises may be correlated, etc. The central problem is, in general, that communication of locally processed estimate can cause "data incest" [59]— double-counting of information. The challenges concerning state estimation in sensor networks are summarized in the following subsection.

---

[1]The Kalman filter can also be extended to correlated and colored process and measurement noise [161].

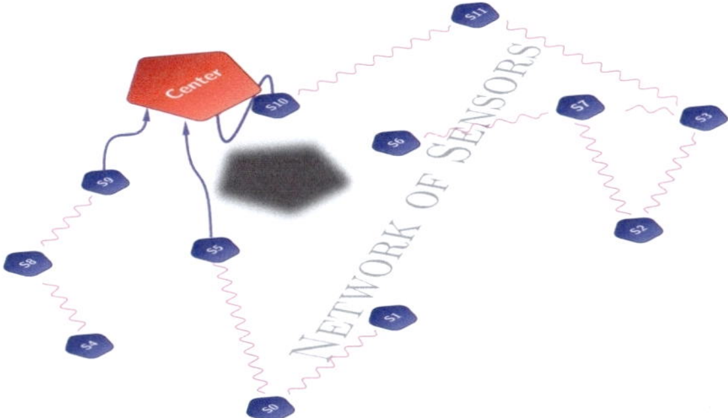

Figure 3.1: Centralized fusion architecture.

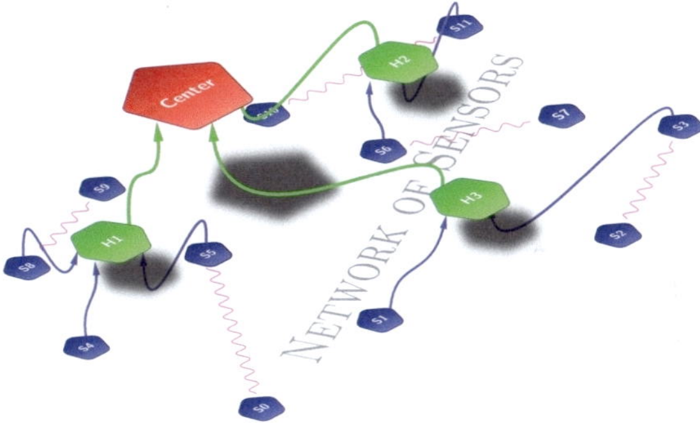

Figure 3.2: Hierarchical two-hop network with one intermediate layer.

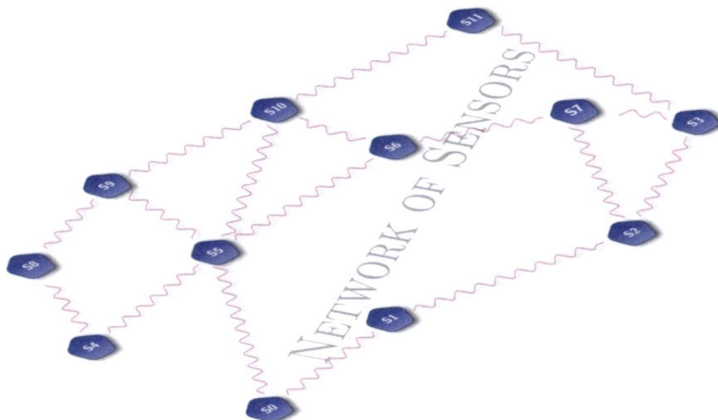

Figure 3.3: Fully decentralized estimation network without central node.

3.1.2   Challenges

The aforementioned advantages of a networked system are opposed to certain limitations and constraints that must be considered during the development of state estimation algorithms. The design of state estimation concepts therefore cannot be detached from the technical aspects but can be abstracted from concrete technical realizations of a network. For the purpose of observing a spatially distributed phenomenon, for instance, volcanic activity [176] and tsunami waves [36], a network of sensor systems is often inevitable, but it can still be efficient to send the measurements to a data sink or gateway node such that a state estimate can be computed on a single high-performance computer system. A centralized network architecture, as depicted in Figure 3.1 inherits the advantage that standard estimation algorithms can basically be employed. Of course, an intelligent preprocessing of sensor data can significantly contribute to reducing the communication load and to relieving workload of the center node. However, the necessity of continuously transmitting data to a central system can also be very unfavorable when sensor nodes are designed to operate autonomously, communication bandwidth and processing power are limited, and a continual connectivity cannot be guaranteed. For instance, wireless mobile ad-hoc networks like car-2-X networks [56] consist of numerous independent nodes and communication links may change frequently. In contrast to a centralized processing, each node needs to store and process a local copy of the state estimate or at least parts of it, as illustrated in Figure 3.3. A certain downside of such a decentralized fusion architectures is the difficulty to extract meaningful information from interchanged estimates without endangering reliability and running the risk of being overconfident. Hierarchical networks, shown in Figure 3.2, fall in between centralized and decentralized structures and can in some cases ease the information processing.

This chapter is devoted to the processing of sensor data in a distributed fashion, while general requirements on a state estimation system and solutions have been discussed in Chapter 2. However, it is worth to emphasize that, for a sensor network, a systematic and mathematically sound treatment of noisy data becomes even more relevant than in single-sensor systems. Although modern highly precise sensor technology alleviates the need for elaborated estimation algorithms and enhances the confidence in the observed data, the scientific endeavor with respect to sensor networks, however, points in the reverse direction. The technological advancements towards miniaturized sensor systems allows considering the usage of a massive amount of low-cost sensor nodes, for instance, for person localization systems [211], and ultimately up to smart dust [144,172] consisting of nodes with a volume of about one cubic millimeter. Low-powered and fault-prone systems again place tremendous demands on the implemented estimation methods. Simultaneous stochastic and set-membership state estimation here becomes particularly useful: Quantization of estimates and measurements can contribute to downsizing the data volume and event-based state estimation to reducing the data rate.

3.2 An Additional Source of Uncertainty: Unknown Dependencies

Fusion of estimates might appear to be an easy task—at first sight. At a certain time step k, two estimates $\hat{\underline{x}}_A$ and $\hat{\underline{x}}_B$ on the same state can be viewed as observations of selfsame state according to

$$\begin{bmatrix} \hat{\underline{x}}_A \\ \hat{\underline{x}}_B \end{bmatrix} = \mathbf{H}\,\underline{x} = \begin{bmatrix} \mathbf{I} \\ \mathbf{I} \end{bmatrix} \underline{x} \;.$$

The Kalman filtering step can then be exploited to derive a fused estimate, which embodies, more precisely, a maximum likelihood fusion since no prior is available. The estimates can also directly be fused by

$$\hat{\underline{x}}_{\mathrm{fus}} = \mathbf{K}_A\,\hat{\underline{x}}_A + \mathbf{K}_B\,\hat{\underline{x}}_B$$

and determining appropriate gains $\mathbf{K}_A$ and $\mathbf{K}_B$. The fusion of estimates is discussed more detailed in Section 3.5.1. However, we can only attribute a meaning to $\hat{\underline{x}}_{\mathrm{fus}}$ if we are in the position to also compute corresponding error matrices $\mathbf{C}_{\mathrm{fus}}$ and $\mathbf{X}_{\mathrm{fus}}$ that do not understate the true errors. At this point, fusion of estimates becomes a rather complicated task, and we aspire to identify the reasons for this in the following.

The previous section singles out dependencies between estimates as a major challenge for state estimation in networked systems. In simple terms, the same errors are modeled on different sensor nodes, but especially autonomously operating nodes are unable to assess the information they share in common with each other. For stochastic uncertainty models, this issue is directly related to stochastic dependencies: Two random variables $\boldsymbol{x}$ and $\boldsymbol{y}$ are dependent if the inequation $f(x,y) \neq f(x) \cdot f(y)$ holds for the corresponding probability densities. Since this chapter is dedicated to state estimation for linear models and Gaussian noise variables, dependencies are unambiguously characterized by cross-correlations. Hence, the dependencies between two estimates $\hat{\underline{x}}_A$ and $\hat{\underline{x}}_A$ are defined by the cross-covariance matrix $\mathbf{C}_{AB} = \mathrm{E}[(\hat{\underline{x}}_A - \boldsymbol{x})(\hat{\underline{x}}_B - \boldsymbol{x})^{\mathrm{T}}]$ of the estimation errors. As aforementioned, it can prove difficult to keep track of dependencies, but when ignoring them, we possibly have to reckon with biased fusion results and severely underestimated errors.

The good news is that interdependencies along set-membership errors do not require any attention, since employing rigid error bounds bypasses the need to consider a certain error behavior within these bounds. For instance, if two sets enclose exactly the same unknown but bounded error, their intersection is still a valid bound. Hence, this chapter is dedicated to the task of tackling the effects of cross-correlations, especially when they are unknown. The reasons for interdependencies between estimation errors can essentially be subdivided into two types, as explained below.

3.2.1 Common Prior and Sensor Information

The most apparent reason for dependencies between local estimates is that data is simply double-counted. Two sensor nodes A and B that are initialized with the same prior estimate $\hat{\underline{x}}_k^{\mathrm{p}}$ on the state, i.e., $\hat{\underline{x}}_A := \hat{\underline{x}}_k^{\mathrm{p}}$ and $\hat{\underline{x}}_B := \hat{\underline{x}}_k^{\mathrm{p}}$, have fully correlated errors, i.e., $\mathrm{E}[(\hat{\underline{x}}_A - \boldsymbol{x})(\hat{\underline{x}}_B - \boldsymbol{x})^{\mathrm{T}}] = \mathbf{C}_A = \mathbf{C}_B = \mathbf{C}_k^{\mathrm{p}}$. State estimation becomes in particular difficult when common data cannot be backtracked to its source. Data are, in general, further processed along any communication path and cycles in the network topology can then, for instance, prevent common sensor data to be separated from the independent information. Such a situation is a transitive data exchange: A prior node P transmits its local estimate to two nodes A and B. After a while, nodes A and B send their local data to a node E, which must be aware of the information of source P in order to prevent double-counting. This situation is studied in the following example, which also conveys an impression of the effect on the fusion results.

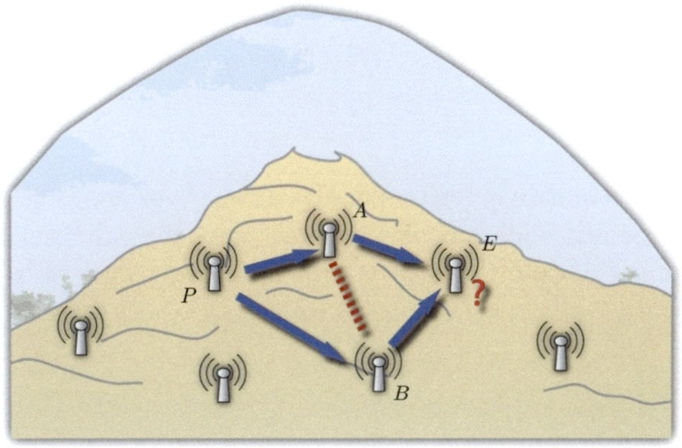

Figure 3.4: The final fusion node E is not aware of common information that is shared by the two transmitting nodes A and B. These have before received data from the same node P.

Example 3.1: Effect of common sensor data

In the simplified situation that measurements are directly related to the state, i.e., by the unit matrix, we can fuse these measurements according to the Kalman filtering step in Section 2.1.1-B by considering one measurement as the prior estimate for the other[2]. A first sensor node P reports a prior estimate with variance $C_P^{\mathrm{p}} = 8$, which is send to two sensor nodes A and B with

---

[2]The combination of local estimates is rather a maximum likelihood fusion than a Bayesian filtering step (see Section 3.5.1).

measurement noise variances $C_A^z = 8$ and $C_B^z = 4$. The corresponding Kalman gains are

$$K_A = C_P^{\mathrm{p}}(C_P^{\mathrm{p}} + C_A^z) = \frac{1}{2} \quad \text{and} \quad K_B = C_P^{\mathrm{p}}(C_P^{\mathrm{p}} + C_B^z) = \frac{2}{3} \, ,$$

and thus the local fusion results have the covariances

$$C_A^{\mathrm{e}} = C_P^{\mathrm{p}} - K_A C_P^{\mathrm{p}} = 4 \quad \text{and} \quad C_B^{\mathrm{e}} = C_P^{\mathrm{p}} - K_B C_P^{\mathrm{p}} = 2\frac{2}{3} \, .$$

Both local estimates are now send to a further node E that combines them according to

$$C_{\mathrm{fus}} = C_A^{\mathrm{e}} - C_A^{\mathrm{e}}(C_A^{\mathrm{e}} + C_B^{\mathrm{e}})^{-1} C_A^{\mathrm{e}}$$
$$= 4 - \frac{16}{4 + 2\frac{2}{3}} = 1\frac{3}{5} \, .$$

The communication path is depicted in Figure 3.4. Apparently, the information provided by the first node is double-counted, since the optimal solution is to fuse the estimate of sensor A only with the measurement of sensor B and not with its local estimate. Then, the error variance is

$$C_{\mathrm{fus}}^{\mathrm{opt}} = C_A^{\mathrm{e}} - C_A^{\mathrm{e}}(C_A^{\mathrm{e}} + C_B^z)^{-1} C_A^{\mathrm{e}}$$
$$= 4 - \frac{16}{4 + 4} = 2 \, ,$$

where C_B^z instead of C_B^{e} has been used. By fusing the local estimates without taking common information into account, an error variance is reported that spuriously underestimates the true variance since only $C_{\mathrm{fus}}^{\mathrm{opt}}$ is attainable at best.

The overall problem is that two estimates $\hat{\underline{x}}_A = \mathrm{E}[\underline{x} \,|\, \mathcal{Z}_A]$ and $\hat{\underline{x}}_B = \mathrm{E}[\underline{x} \,|\, \mathcal{Z}_B]$ are compiled from two measurement sequences $\mathcal{Z}_A$ and $\mathcal{Z}_B$ that possibly share common data $\mathcal{Z}_A \cap \mathcal{Z}_B \neq \emptyset$. The measurements in the intersection and specifically the according measurement errors are double-counted when the estimates are naively fused, and we have to reckon with a biased estimation result.

A general strategy to prevent biased fusion results consists of storing common information separately and removing it from the fusion result.

Each node can, for instance, be equipped with a *channel filter* [27, 71, 123] that stores the information that is shared with neighboring nodes. In a hierarchical networks, this task can be passed to higher-level fusion nodes. In the absence of process noise—the second source of dependencies—even the estimation quality of a centralized estimation scheme can be achieved. The channel filter can be established by means of the information form of the Kalman filter, as explained in Section 3.5.2.

3.2.2 Common Process Noise

The second reason of dependencies can be discovered in the parallel prediction steps among the sensor nodes. In each local state transition model, the same process noise is employed. So the same error is incorporated multiple times and cannot be filtered out. The example below illustrates the effect of dependent errors due to common process noise.

Example 3.2: Effect of common process noise

We consider the situation of two nodes equipped with equal sensors. Both nodes compute local estimates and are each initialized with the first measurement. Let the sensor noises have the variances $C_A = 4$ and $C_B = 4$. First, we consider an optimal processing: The observation in sensor A serves as an prior and is fused according to the Kalman filtering step with the observation of sensor B. The error variance associated with the fused estimate is then

$$C_{\text{fus}}^{\text{e}} = C_A + C_A(C_A + C_B)^{-1}C_A = 2 \,,$$

where the measurement mapping H is the identity. The estimate is predicted according to the simple process model

$$\boldsymbol{x}_{k+1} = \boldsymbol{x}_k + \boldsymbol{w}_k$$

with $\boldsymbol{w}_k \sim \mathcal{N}(0,1)$. The predicted error variance is then $C_{\text{fus}}^{\text{p}} = 3$.

Now, we consider the case that the initial estimates are both predicted locally with the same process model. The predicted local variances yield $C_A^{\text{p}} = C_B^{\text{p}} = 5$. After that, we fuse the local estimates again by means of a Kalman filtering step and obtain the variance

$$C_{\text{fus}}^{\text{p}} = C_A^{\text{p}} + C_A^{\text{p}}(C_A^{\text{p}} + C_B^{\text{p}})^{-1}C_A^{\text{p}} = 5 + 25/10 = 2.5$$

that apparently differs from the optimal result. More precisely, the actual variance is noticeably underestimated, and a naive fusion after prediction yields a overconfident estimate.

The effect of common process noise has essentially been studied against the background of target tracking applications [8,9]. In track-to-track fusion problems, local estimates of the same object's state, particularly its position, are computed on different nodes and are to be fused. In the absence of any other dependencies, it is possible to also keep track of the cross-covariance matrices of the errors, which can then be employed for a consistent fusion of the tracks. The corresponding fusion algorithm is discussed in Section 3.5.1.

3.3 Multisensor Kalman Filtering

In this section, we discuss how to extend the standard Kalman filtering scheme from Section 2.1.1-B in order to efficiently process multiple sensor data. This means that still a central processing node is required that computes the state estimates, as illustrated in figures 3.1 and 3.2. However, by means of the methods explained in subsections 3.3.2 and 3.3.3, sensor data can be preprocessed locally so that the communication load as well as the communication rate can be reduced. This section begins by describing how multiple measurements can be processed within the standard Kalman filter. In each subsection, the linear system model

$$\underline{x}_{k+1} = \mathbf{A}_k \, \underline{x}_k + \mathbf{B}_k (\underline{\hat{u}}_k + \underline{w}_k) \tag{3.1}$$

is considered, with control input $\underline{\hat{u}}_k$ being affected by $\underline{w}_k \sim \mathcal{N}(\underline{0}, \mathbf{C}_k^u)$. The linear measurement model of sensor node $i \in \{1, \dots, N\}$ is defined by

$$\underline{z}_k^i = \mathbf{H}_k^i \, \underline{x}_k + \underline{v}_k^i \;, \tag{3.2}$$

where $\underline{v}_k^i \sim \mathcal{N}(\underline{0}, \mathbf{C}_k^{z,i})$ denotes the measurement noise. Hence, this section only considers stochastic uncertainties, since dependencies between set-membership errors have no effect. The treatment of set-membership uncertainties withing the information filter lies then in the focus of Section 3.4.

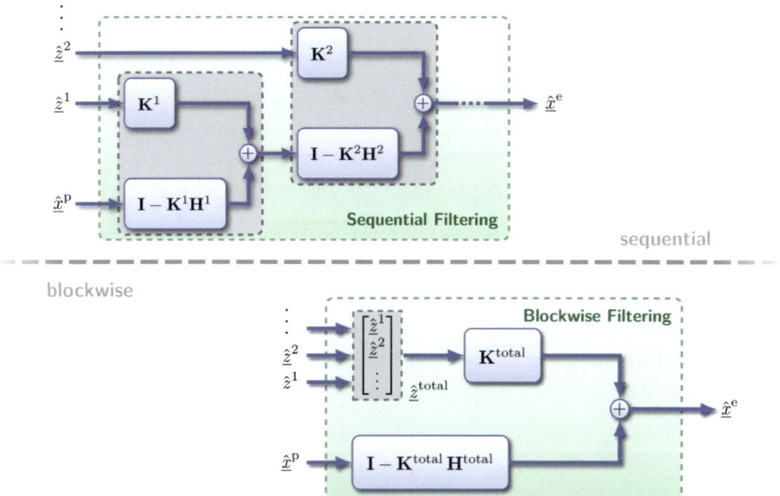

Figure 3.5: Sequential and blockwise processing of multiple measurements.

3.3.1 Sequential or Blockwise State Estimation

For the implementation of the Kalman filtering step for multiple measurement data, principally two possibilities can be named—*sequential*[3] and *blockwise* filtering [9]. Both are related to the right- and left-hand side of equation (2.7). For the first possibility, the conditional independence of the observations given the current state can be exploited such that a sequential incorporation of measurement data can be established. The multisensor measurement data $\mathcal{Z}_k = \{\hat{\underline{z}}_k^1, \ldots, \hat{\underline{z}}_k^N\}$ related to the models (3.2) can be processed through the Kalman filtering step (2.25) element-by-element, which corresponds to applying Bayes' rule to each likelihood (2.7). The

---

[3]This notation is commonly used for the special case of multiple scalar measurements with (fully) diagonal joint covariance matrix. A single matrix inversion can thereby be replaced by multiple scalar divisions [161].

posterior estimate yields

$$\hat{\underline{x}}_k^{\mathrm{e}} = (\mathbf{I} - \mathbf{K}_k^N \mathbf{H}_k^N) \cdot$$

$$\left(\dots \left((\mathbf{I} - \mathbf{K}_k^2 \mathbf{H}_k^2) \overbrace{\left((\mathbf{I} - \mathbf{K}_k^1 \mathbf{H}_k^1) \hat{\underline{x}}_k^{\mathrm{p}} + \mathbf{K}_k^1 \hat{\underline{z}}_k^1 \right)}^{=\hat{\underline{x}}_k^{\mathrm{e},1}} + \mathbf{K}_k^2 \hat{\underline{z}}_k^2 \right) + \dots \right) + \mathbf{K}_k^N \hat{\underline{z}}_k^N$$
$$\underbrace{\phantom{(\mathbf{I} - \mathbf{K}_k^2 \mathbf{H}_k^2)((\mathbf{I} - \mathbf{K}_k^1 \mathbf{H}_k^1)\hat{\underline{x}}_k^{\mathrm{p}} + \mathbf{K}_k^1 \hat{\underline{z}}_k^1) + \mathbf{K}_k^2 \hat{\underline{z}}_k^2}}_{=\hat{\underline{x}}_k^{\mathrm{e},2}}$$

$$(3.3)$$

after N sequential Kalman filtering steps. The second possibility is to perform the filtering step en bloc and to collect all measurements in a single vector of dimension $(N \cdot n_z)$. Therefore, only a single update

$$\hat{\underline{x}}_k^{\mathrm{e}} = (\mathbf{I} - \mathbf{K}_k^{\mathrm{total}} \mathbf{H}_k^{\mathrm{total}}) \, \hat{\underline{x}}_k^{\mathrm{p}} + \mathbf{K}_k^{\mathrm{total}} \begin{bmatrix} \hat{\underline{z}}_k^1 \\ \vdots \\ \hat{\underline{z}}_k^N \end{bmatrix}$$

is required. Figure 3.5 provides a diagram of both schemes. In compliance with the estimate, the posterior covariance matrix can also be computed either sequentially or blockwise.

In general, either way of multisensor data processing renders any other fusion architecture than a centralized one impractical. For a blockwise processing, it is easy to admit that all data must be available at the center node and, for instance by referring back to Figure 3.2, the (green) intermediate nodes have to pass the data received from the subjacent nodes on to their superordinated parent nodes. Also, a sequential processing does not allow for an intelligent preprocessing of sensor data in order to reduce the communication load. Before a sequential incorporation of sensor data is possible, each single measurement together with the corresponding error matrix must be accessible by the center node so that it is capable of computing the required Kalman gains for (3.3). These limitations can be overcome by employing algebraical reformulations of the Kalman filter that will be discussed in the following two subsections.

3.3.2 Information Filtering: An Inverse Covariance Matrix Formulation

The previous subsection unveiled that the standard Kalman filtering step, i.e., equations (2.25) and (2.26), cannot easily be extended to a distributable treatment of multiple observations $\mathcal{Z}_k = \{\hat{\underline{z}}_k^1, \ldots, \hat{\underline{z}}_k^N\}$, as it would be useful in most multisensor data fusion problems. More specifically, the fusion step cannot simply be expressed in terms of the individual Kalman gains $\mathbf{K}_k^i$ and innovations $(\hat{\underline{z}}_k^i - \mathbf{H}_k^i \, \hat{\underline{x}}_k^{\mathrm{P}})$, i.e.,

$$\hat{\underline{x}}_k^{\mathrm{e}} \neq \hat{\underline{x}}_k^{\mathrm{P}} + \sum_{i=1}^{N} \mathbf{K}_k^i \left(\hat{\underline{z}}_k^i - \mathbf{H}_k^i \, \hat{\underline{x}}_k^{\mathrm{P}} \right) , \tag{3.4}$$

since the individual innovations are affected by a common process noise, which causes them to be correlated with each other. The sequential processing (3.3) consists of nested filtering steps and the required gains therefore cannot be computed at once.

A distributable form of the Kalman filtering step is constituted by the *information filter* that essentially embodies an algebraic reformulation of the Kalman filter formulas and provides estimates on the information about an uncertain state rather than on the state itself [133]. Instead of an immediate estimate on the state, the information vector

$$\hat{\underline{y}}_k := \mathbf{C}_k^{-1} \hat{\underline{x}}_k \tag{3.5}$$

and the information matrix

$$\mathbf{Y}_k := \mathbf{C}_k^{-1} \tag{3.6}$$

are considered, which are the quantities to be processed and updated in the prediction and filtering step, respectively. It can be shown [133] that the information filter is a log-likelihood representation of the Bayesian state estimation concept in Section 2.1.1, which is further studied in Section 4.4. The inverse covariance matrix (3.6) is equal to the Fisher information matrix (for a non-random state). Hence, minimum mean squared error estimation is related to maximizing the Fisher information about the state. Both processing steps of the Kalman filter algorithm have to be reformulated for the parameters (3.5) and (3.6).

Prediction The prediction of the information parameters $\hat{\underline{y}}_k^{\mathrm{e}}$ and $\mathbf{Y}_k^{\mathrm{e}}$ turns out to be more elaborate than the prediction step of the standard Kalman filter. We define the mapping

$$\mathbf{L}_{k+1} = \mathbf{A}_k(\mathbf{Y}_k^{\mathrm{e}})^{-1}$$

that first transforms the information vector back into state space and then applies the state transition model (3.1). The predicted information matrix thus becomes

$$
\begin{aligned}
\mathbf{Y}_{k+1}^{\mathrm{p}} &= \ (\mathbf{C}_{k+1}^{\mathrm{p}})^{-1} \\
&\overset{(2.20)}{=} \left(\mathbf{A}_k(\mathbf{Y}_k^{\mathrm{e}})^{-1}\mathbf{A}_k^{\mathrm{T}} + \mathbf{B}_k\mathbf{C}_k^u\mathbf{B}_k^{\mathrm{T}}\right)^{-1} \\
&= \left(\mathbf{L}_{k+1}\mathbf{Y}_k^{\mathrm{e}}\mathbf{L}_{k+1}^{\mathrm{T}} + \mathbf{B}_k\mathbf{C}_k^u\mathbf{B}_k^{\mathrm{T}}\right)^{-1} ,
\end{aligned}
\tag{3.7}
$$

and the predicted information vector is given by

$$\hat{\underline{y}}_{k+1}^{\mathrm{p}} = \mathbf{Y}_{k+1}^{\mathrm{p}} \underbrace{\left(\mathbf{L}_{k+1}\,\hat{\underline{y}}_k^{\mathrm{e}} + \mathbf{B}_k\,\hat{\underline{u}}_k\right)}_{\overset{(2.19)}{=}\hat{\underline{x}}_{k+1}^{\mathrm{p}}} . \tag{3.8}$$

In contrast to the Kalman filter, the computation of both parameters hence requires matrix inversions.

Filtering During the derivation of the Kalman gain (2.24), a rearrangement of the considered equations leads to the equivalent representations

$$\mathbf{K}_k = \left((\mathbf{C}_k^{\mathrm{p}})^{-1} + \mathbf{H}_k^{\mathrm{T}}(\mathbf{C}_k^z)^{-1}\mathbf{H}_k\right)^{-1}\mathbf{H}_k^{\mathrm{T}}(\mathbf{C}_k^z)^{-1} \tag{3.9}$$

and

$$\mathbf{I} - \mathbf{K}_k\mathbf{H}_k = \left((\mathbf{C}_k^{\mathrm{p}})^{-1} + \mathbf{H}_k^{\mathrm{T}}(\mathbf{C}_k^z)^{-1}\mathbf{H}_k\right)^{-1}(\mathbf{C}_k^{\mathrm{p}})^{-1} . \tag{3.10}$$

By applying these gains to (2.26), the updated covariance matrix can immediately be simplified to

$$\mathbf{C}_k^{\mathrm{e}} = \left((\mathbf{C}_k^{\mathrm{p}})^{-1} + \mathbf{H}_k^{\mathrm{T}}(\mathbf{C}_k^z)^{-1}\mathbf{H}_k\right)^{-1} ,$$

which can alternatively be derived by means of the Woodbury matrix identity. Consequently, the corresponding information matrix yields

$$\mathbf{Y}_k^{\mathrm{e}} = (\mathbf{C}_k^{\mathrm{p}})^{-1} + \mathbf{H}_k^{\mathrm{T}}(\mathbf{C}_k^z)^{-1}\mathbf{H}_k$$
$$= \mathbf{Y}_k^{\mathrm{p}} + \mathbf{H}_k^{\mathrm{T}}(\mathbf{C}_k^z)^{-1}\mathbf{H}_k \ ,$$

and the information vector is updated according to

$$\hat{\underline{y}}_k^{\mathrm{e}} = \mathbf{Y}_k^{\mathrm{e}}\,\hat{\underline{x}}_k^{\mathrm{e}}$$
$$= \mathbf{Y}_k^{\mathrm{e}}\left((\mathbf{I} - \mathbf{K}_k\mathbf{H}_k)\hat{\underline{x}}_k^{\mathrm{p}} + \mathbf{K}_k\,\hat{\underline{z}}_k\right)$$
$$= \mathbf{Y}_k^{\mathrm{e}}\left((\mathbf{C}_k^{\mathrm{p}})^{-1} + \mathbf{H}_k^{\mathrm{T}}(\mathbf{C}_k^z)^{-1}\mathbf{H}_k\right)^{-1}\left((\mathbf{C}_k^{\mathrm{p}})^{-1}\hat{\underline{x}}_k^{\mathrm{p}} + \mathbf{H}_k^{\mathrm{T}}(\mathbf{C}_k^z)^{-1}\,\hat{\underline{z}}_k\right)$$
$$= \hat{\underline{y}}_k^{\mathrm{p}} + \mathbf{H}_k^{\mathrm{T}}(\mathbf{C}_k^z)^{-1}\hat{\underline{z}}_k \ .$$

Apparently, the filtering step becomes the simple sum of information vectors and matrices.

Unlike (3.4), the processing of multiple measurements related to the sensor models (3.2) can now efficiently be carried out by means of the sums

$$\hat{\underline{y}}_k^{\mathrm{e}} = \hat{\underline{y}}_k^{\mathrm{p}} + \sum_{i=1}^{N} \underline{i}_k^i$$
$$= \hat{\underline{y}}_k^{\mathrm{p}} + \sum_{i=1}^{N} (\mathbf{H}_k^i)^{\mathrm{T}}(\mathbf{C}_k^{z,i})^{-1}\hat{\underline{z}}_k^i \tag{3.11}$$

and

$$\mathbf{Y}_k^{\mathrm{e}} = \mathbf{Y}_k^{\mathrm{p}} + \sum_{i=1}^{N} \mathbf{I}_k^i$$
$$= \mathbf{Y}_k^{\mathrm{p}} + \sum_{i=1}^{N} (\mathbf{H}_k^i)^{\mathrm{T}}(\mathbf{C}_k^{z,i})^{-1}\mathbf{H}_k^i \tag{3.12}$$

with the sensor nodes contributing the innovation vectors

$$\underline{i}_k^i = (\mathbf{H}_k^i)^{\mathrm{T}}(\mathbf{C}_k^{z,i})^{-1}\hat{\underline{z}}_k^i \tag{3.13}$$

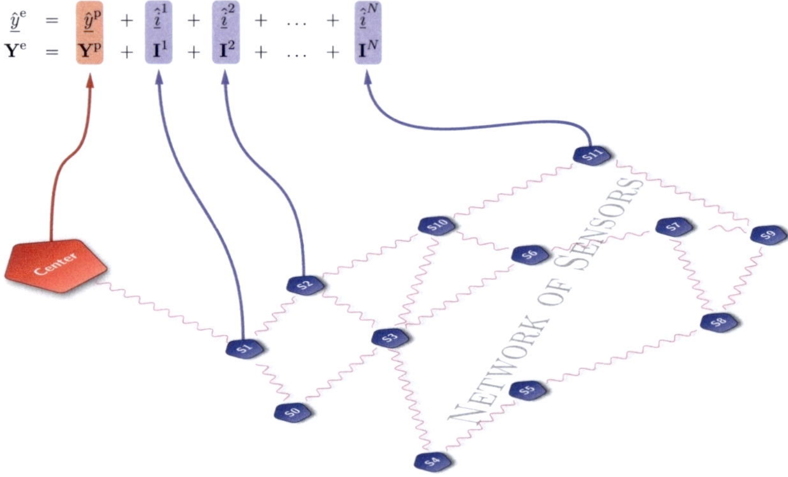

Figure 3.6: Multisensor information filtering. The innovation parameters can be preprocessed along any communication path.

and matrices

$$\mathbf{I}_k^i = (\mathbf{H}_k^i)^{\mathrm{T}}(\mathbf{C}_k^{z,i})^{-1}\mathbf{H}_k^i \ . \tag{3.14}$$

Parts of these calculations can now easily be distributed. More precisely, the measurement data along any communication path can be collected and already be condensed into a single information vector, which has the dimension of the system state, so that, in large networks, the overall communication load can significantly be reduced, as illustrated by Figure 3.6.

A further useful property of the information form is that common information can also easily be removed, simply by subtracting the corresponding information vector and matrix. The removal of common information—even if it is unknown to the fusion node—is discussed in Section 3.5.2. Although the data can be efficiently processed on each communication path, still the received sensor data of each time step need to be transmitted. A local prediction step, so that data can be communicated at arbitrary instants of time, cannot be established without losing optimality or even consistency.

For this, further rearrangements of the formulas are necessary, as explained in the following subsection.

3.3.3   An Optimally Distributed Kalman Filter

A fully distributed formulation of the Kalman filter has been presented in [68, 69, 108, 109], where communication to the center node may take place at arbitrary instants of time. This concept rests upon the insight that specific sensor observations do not enter into the equation (2.26) for the posterior covariance matrix. The formulas for initialization, prediction, and filtering have been derived by decomposing Gaussian densities into different products of Gaussian densities. The fusion of a number of N independent estimates $(\hat{\underline{x}}_k^i, \mathbf{C}_k^i)$ can be viewed as a product of N Gaussian densities that has the resulting parameters

$$\hat{\underline{x}}_k = \mathbf{C}_k \sum_{i=1}^{N} (\mathbf{C}_k^i)^{-1} \hat{\underline{x}}_k^i \tag{3.15}$$

and

$$(\mathbf{C}_k)^{-1} = \sum_{i=1}^{N} (\mathbf{C}_k^i)^{-1} \tag{3.16}$$

for mean and covariance matrix, respectively. It can be noticed that this product is related to a standard Kalman filtering step, where one Gaussian density serves as a prior. In [108], a special product decomposition is employed that yields N *globalized* estimates, i.e., each factor Gaussian density has the parameters $(\bar{\underline{x}}_k^i, \bar{\mathbf{C}}_k)$ with

$$\bar{\underline{x}}_k^i := \bar{\mathbf{C}}_k (\mathbf{C}_k^i)^{-1} \hat{\underline{x}}_k \tag{3.17}$$

and

$$(\bar{\mathbf{C}}_k)^{-1} := \frac{1}{N} \sum_{i=1}^{N} (\mathbf{C}_k^i)^{-1} . \tag{3.18}$$

The term globalization refers to the fact that each Gaussian density $\mathcal{N}(\ \cdot\ ; \bar{\underline{x}}_k^i, \bar{\mathbf{C}}_k)$ has the same shape due to identical covariance parameters $\bar{\mathbf{C}}_k$, and their product—or fusion—again yields (3.15) and (3.16).

These parameters are distributed among the sensor nodes in order to initialize their local prediction and filtering steps. For a single global estimate $(\hat{\underline{x}}_k, \mathbf{C}_k)$, an initialization complying with (3.17) and (3.18) can be achieved by setting the local parameters to

$$(\bar{\underline{x}}_k^{\mathrm{e},i}, \bar{\mathbf{C}}_k^{\mathrm{e}}) := (\hat{\underline{x}}_k, N \cdot \mathbf{C}_k) \,. \tag{3.19}$$

It is easy to recognize but important to note that the local parameters $(\bar{\underline{x}}_k^{\mathrm{e},i}, \bar{\mathbf{C}}_k^{\mathrm{e}})$ do not necessarily represent valid estimates. Especially after filtering, this is generally the case and these parameters must therefore be regarded rather as synthetic variables carrying no information. A valid estimate is obtained by fusing the local data according to (3.15) and (3.16).

Prediction The local prediction steps for the globalized estimates must warrant that the process noise is not overrated, as discussed in Section 3.2.2. The basic idea is to inflate the noise covariance matrix such that fusion does not result into an underestimated error. In place of (3.1), we consider the simplified system model[4]

$$\underline{x}_{k+1} = \mathbf{A}_k \, \underline{x}_k + \underline{w}_k \tag{3.20}$$

with $\underline{w}_k \sim \mathcal{N}(\underline{0}, \mathbf{C}_k^w)$. Each local parameter set $(\bar{\underline{x}}_k^{\mathrm{e},i}, \bar{\mathbf{C}}_k^{\mathrm{e}})$ is predicted according to

$$\bar{\underline{x}}_{k+1}^{\mathrm{p},i} = \mathbf{A}_k \, \bar{\underline{x}}_k^{\mathrm{e},i}$$

and

$$\bar{\mathbf{C}}_{k+1}^{\mathrm{p}} = \mathbf{A}_k \bar{\mathbf{C}}_k^{\mathrm{e}} \mathbf{A}_k + N \mathbf{C}_k^w \,, \tag{3.21}$$

where the N-fold of the process noise is employed. Hence, instead of (3.20), the predicted parameters correspond to the model

$$\underline{x}_{k+1} = \mathbf{A}_k \, \underline{x}_k + \tilde{\underline{w}}_k \tag{3.22}$$

with $\tilde{\underline{w}} \sim \mathcal{N}(\underline{0}, N\mathbf{C}_k^w)$, which is called the *relaxed* system model. The local prediction steps can be performed without requiring any communication

---

[4]The treatment of different local inputs would again demand for a continual communication.

and automatically guarantee that the local parameters remain globalized, i.e., the predicted covariance parameter (3.21) is the same on each sensor site. The fusion result of the local parameters $(\underline{\bar{x}}_{k+1}^{\mathrm{p},i}, \bar{\mathbf{C}}_{k+1}^{\mathrm{p}})$ equals the results of a centralized prediction step and is thus optimal.

Filtering The filtering step for the globalized parameters appears, at first sight, to be similar to the standard Kalman filter. The local parameter $\underline{\bar{x}}_k^{\mathrm{p},i}$ is updated according to

$$\underline{\bar{x}}_k^{\mathrm{e},i} = \bar{\mathbf{C}}_k^{\mathrm{e}} \left((\bar{\mathbf{C}}_k^{\mathrm{p}})^{-1} \, \underline{\bar{x}}_k^{\mathrm{p},i} + (\mathbf{H}_k^i)^{\mathrm{T}} (\mathbf{C}_k^{z,i})^{-1} \, \underline{\hat{z}}_k^i \right) , \tag{3.23}$$

where the inverse-covariance Kalman gains (3.9) and (3.10) are employed. The difference consists in the updated globalized matrix

$$\bar{\mathbf{C}}_k^{\mathrm{e}} = \left((\bar{\mathbf{C}}_k^{\mathrm{p}})^{-1} + (\bar{\mathbf{C}}_k^{z})^{-1} \right)^{-1} \tag{3.24}$$

where the globalized measurement covariance matrix

$$(\bar{\mathbf{C}}_k^{z})^{-1} = \frac{1}{N} \sum_{j=1}^{N} (\mathbf{H}_k^j)^{\mathrm{T}} (\mathbf{C}_k^{z,j})^{-1} \mathbf{H}_k^j \tag{3.25}$$

instead of only $(\mathbf{H}_k^i)^{\mathrm{T}} (\mathbf{C}_k^{z,i})^{-1} \mathbf{H}_k^i$ enters into the update equation. The Gaussian function that corresponds to $(\mathbf{H}_k^i)^{\mathrm{T}} (\mathbf{C}_k^{z,i})^{-1} \, \underline{\hat{z}}_k^i$ and $\bar{\mathbf{C}}_k^z$ is called *globalized likelihood*. Evidently, all local parameter sets $(\underline{\bar{x}}_k^{\mathrm{e},i}, \bar{\mathbf{C}}_k^{\mathrm{e}})$ feature the same globalized covariance matrix $\bar{\mathbf{C}}_k^{\mathrm{e}}$. However, the globalized parameters $(\underline{\bar{x}}_k^{\mathrm{e},i}, \bar{\mathbf{C}}_k^{\mathrm{e}})$ generally do not represent valid state estimates, but fusing them according to (3.15) and (3.16) yields the optimal Kalman-type estimation results. For the covariance matrix (3.16), this is

$$\begin{aligned}
(\mathbf{C}_k^{\mathrm{e}})^{-1} \overset{(3.16)}{=} & \sum_{i=1}^{N} (\bar{\mathbf{C}}_k^{\mathrm{e}})^{-1} \\
= & \; N (\bar{\mathbf{C}}_k^{\mathrm{p}})^{-1} + N (\bar{\mathbf{C}}_k^{z})^{-1} \\
= & \; (\mathbf{C}_k^{\mathrm{p}})^{-1} + \sum_{j=1}^{N} (\mathbf{H}_k^j)^{\mathrm{T}} (\mathbf{C}_k^{z,j})^{-1} \mathbf{H}_k^j ,
\end{aligned}$$

where the latter sum exactly corresponds to the updated covariance matrix (3.12) in its information form. This relation clarifies why the globalized error covariance matrix (3.25) is required. The update

$$
(\mathbf{C}_k^{\mathrm{e}})^{-1}\,\hat{\underline{x}}_k^{\mathrm{e}} \overset{(3.15)}{=} \sum_{i=1}^{N}(\bar{\mathbf{C}}_k^{\mathrm{e}})^{-1}\bar{\underline{x}}_k^{\mathrm{e},i}
$$

$$
= \sum_{i=1}^{N}(\bar{\mathbf{C}}_k^{\mathrm{P}})^{-1}\,\bar{\underline{x}}_k^{\mathrm{P},i} + \sum_{i=1}^{N}(\mathbf{H}_k^i)^{\mathrm{T}}(\mathbf{C}_k^{z,i})^{-1}\,\hat{\underline{z}}_k^i
$$

$$
= (\mathbf{C}_k^{\mathrm{P}})^{-1}\hat{\underline{x}}_k^{\mathrm{P}} + \sum_{i=1}^{N}(\mathbf{H}_k^i)^{\mathrm{T}}(\mathbf{C}_k^{z,i})^{-1}\,\hat{\underline{z}}_k^i
$$

again is analogous to the information state update (3.11). A different way to perform the update step is to apply the standard Kalman filtering step to the parameters $(\bar{\underline{x}}_{k+1}^{\mathrm{P},i}, \bar{\mathbf{C}}_{k+1}^{\mathrm{P}})$ and to globalize the obtained results by means of (3.17) and (3.18) just before the next prediction step takes place. In this case, the optimal distributed Kalman filter consists of three steps (besides initialization)—prediction, filtering, and globalization.

Discussion: Distributed But Not Decentralized

The considered concept overcomes the problem of common process noise among local state predictions, which has been discussed in Section 3.2.2, by employing the relaxed system model (3.22). The fact that this relaxation alone does not guarantee minimum MSE estimates states the reason for the globalization during or after the filtering step and will also later be discussed in Section 3.6.3. Yet the need for globalization shows clearly that this approach may suffer from strict requirements regarding the local availability of knowledge about utilized models. Each sensor node must be in the position to compute the globalized likelihood with covariance matrix (3.25) and therefore must be aware of the other nodes' sensor models, i.e., of the measurement matrices $\mathbf{H}_k^i$ and noise matrices $\mathbf{C}_k^{z,i}$. This problem is further compounded by the fact that the local parameters do not represent valid estimates and must be available in their entirety at the fusion center so as to obtain a valid state estimate. Hence, the failure of a single component brings down the entire state estimation system. The effects of node failures and incomplete knowledge about the participating nodes have been studied in [209], where slightly differing assumptions effect drastic deviations of the estimates. This concept depending on global knowledge about sensor models and providing noninformative local estimates is an excellent example of a distributed estimation system, which only provides in its entirety meaningful estimates.

The distributed Kalman filter has been further developed in [67] and [209, 210] in order to relax the strict requirements on global knowledge. The general idea in [209, 210] is to simultaneously keep track of a correction matrix that allows computing an unbiased estimate out of the local parameters. Also, parts of the local estimates can be fused and the result can then be transformed into an unbiased estimate. Hence, the correction matrix can be employed if the assumptions on the network are not met and an incorrect globalized matrix (3.25) is computed. With this correcting matrix, the distributed Kalman filter is improved to degrade gracefully depending on how well the assumptions on the global network are met, i.e., to which extent nodes are aware of each other. A second research direction is to employ the distributed Kalman filter for suboptimal decentralized control [208].

3.4 Stochastic and Set-membership Information Filtering

The previous section has unveiled that an efficient processing of multisensor data is significantly facilitated by focusing on inverse covariance matrices. The filtering steps of both the information filter from Section 3.3.2 and the distributed Kalman filter from Section 3.3.3 employ information state vectors and matrices, i.e., equations (3.11) and (3.12) for the information filter and equations (3.23) and (3.24) for the distributed Kalman filter. In this section, we aspire to extent the information filter in order to simultaneously treat stochastic and set-membership uncertainties. The derived stochastic and set-membership information filter is based on [184] and [204]. In particular, the results of [204] are presented in an extended fashion and are studied in more detail.

In the presence of additional set-membership uncertainties, a state estimate $\hat{\underline{x}}_k$, which is now characterized by both an error covariance matrix $\mathbf{C}_k$ and an ellipsoidal error shape matrix $\mathbf{X}_k$, must be represented in its information form. At first sight, deriving an information form of the set-membership error bound that preserves the distributable formulation of the filtering step appears to be complicated. However, Chapter 2 has proposed different interpretations of set-membership uncertainties. The Bayesian viewpoint discussed in Section 2.4.1 allows regarding the estimate $\hat{\underline{x}}_k$ and the shape matrix $\mathbf{X}_k$ as an ellipsoidal set $\mathcal{E}(\hat{\underline{x}}_k, \mathbf{X}_k)$ of means, i.e., the state estimate is characterized by a set of Gaussian densities, which share the same covariance matrix $\mathbf{C}_k$. The transformation $\mathbf{C}_k^{-1}$ into the information form is thus the same for every possible mean. Hence, the transformed set of means yields

$$
\begin{aligned}
\mathcal{Y}_k &= \mathbf{C}_k^{-1}\mathcal{E}(\hat{\underline{x}}_k, \mathbf{X}_k) \\
&\overset{(2.35)}{=} \mathcal{E}(\mathbf{C}_k^{-1}\hat{\underline{x}}_k, \mathbf{C}_k^{-1}\mathbf{X}_k\mathbf{C}_k^{-\mathrm{T}})
\end{aligned}
\tag{3.26}
$$

and is an ellipsoid of information vectors that are all related to the same information matrix $\mathbf{Y}_k = \mathbf{C}_k^{-1}$. For shorter notation, we set

$$
\mathbf{Q}_k := \mathbf{C}_k^{-1}\mathbf{X}_k\mathbf{C}_k^{-\mathrm{T}}
$$

in the following. This representation still entails an easily distributable formulation of the filtering step, as illustrated in the following subsections.

3.4.1 Processing Steps of the Stochastic and Set-membership Information Filter

As for the standard information filter in Section 3.3.2, the advantages in the filtering step again come at the expense of more elaborate formulas for the prediction step.

Prediction in Information Form   As for the predicted covariance matrix of the standard Kalman filter, the predicted information matrix is again obtained independently from the set of information vectors by means of

$$\mathbf{Y}^{\mathrm{p}}_{k+1} \overset{(3.7)}{=} \left(\mathbf{A}_k(\mathbf{Y}^{\mathrm{e}}_k)^{-1}\mathbf{A}_k^{\mathrm{T}} + \mathbf{B}_k\mathbf{C}_k^u\mathbf{B}_k^{\mathrm{T}}\right)^{-1} ,$$

where system mapping and input error covariance matrix are related to the model (3.1). Each element of the information ellipsoid $\mathcal{E}(\hat{\underline{y}}^{\mathrm{e}}_k, \mathbf{Y}^{\mathrm{e}}_k)$ has to be processed through (3.8). In particular, the midpoints $\hat{\underline{y}}^{\mathrm{p}}_{k+1}$ and $\hat{\underline{u}}_k$ are predicted according to

$$\hat{\underline{y}}^{\mathrm{p}}_{k+1} \overset{(3.8)}{=} \mathbf{Y}^{\mathrm{p}}_{k+1}\left(\mathbf{L}_{k+1}\,\hat{\underline{y}}^{\mathrm{e}}_k + \mathbf{B}_k\,\hat{\underline{u}}_k\right)$$

with

$$\mathbf{L}_{k+1} = \mathbf{A}_k(\mathbf{Y}^{\mathrm{e}}_k)^{-1} .$$

It remains to compute the shape matrix $\mathbf{Q}^{\mathrm{p}}_{k+1}$ for the outer approximation of the Minkowski sum

$$\mathcal{Y}^{\mathrm{p}}_{k+1} = \mathbf{Y}^{\mathrm{p}}_{k+1}\mathbf{L}_{k+1}\mathcal{E}(\hat{\underline{y}}^{\mathrm{e}}_k, \mathbf{Q}^{\mathrm{e}}_k) \oplus \mathbf{Y}^{\mathrm{p}}_{k+1}\mathbf{B}_k\mathcal{E}(\hat{\underline{u}}_k, \mathbf{X}^u_k)$$
$$\subseteq \mathcal{E}(\hat{\underline{y}}^{\mathrm{p}}_{k+1}, \mathbf{Q}^{\mathrm{p}}_{k+1}) ,$$

where $\mathbf{X}^u_k$ defines the set-membership uncertainty associated to the input. From the family of possible predicted shape matrices

$$\mathbf{Q}^{\mathrm{p}}_{k+1} \overset{(2.101)}{=} (1 + p^{-1})(\mathbf{Y}^{\mathrm{p}}_{k+1}\mathbf{L}_{k+1})\mathbf{Q}^{\mathrm{e}}_k(\mathbf{Y}^{\mathrm{p}}_{k+1}\mathbf{L}_{k+1})^{\mathrm{T}}$$
$$+ (1 + p)(\mathbf{Y}^{\mathrm{p}}_{k+1}\mathbf{B}_k)\mathbf{X}^u_k(\mathbf{Y}^{\mathrm{p}}_{k+1}\mathbf{B}_k)^{\mathrm{T}} ,$$

we propose to choose the element that is optimal in the sense of the trace, i.e., the sum of the squared semiaxes lengths, i.e., setting p to (2.102).

Apparently, the prediction step only requires the outer approximation of a Minkowski sum of two ellipsoids. The principal difficulty in simultaneous stochastic and set-membership information filtering consists of an efficient and distributable approximation of the filtering step for multiple measurements that are affected by both types of uncertainty, which lies in the focus of the following paragraph.

Information Filtering of Multiple Measurements   In contrast to (3.2), a measurement $\hat{\underline{z}}_k^i$ received on a sensor node i is affected by both a random error $\underline{v}_k^i \sim \mathcal{N}(\underline{0}, \mathbf{C}_k^{z,i})$ and an unknown but bounded error $\underline{e}_k^i$ enclosed by the ellipsoid $\mathcal{E}(\underline{0}, \mathbf{X}_k^{z,i})$. It is hence an outcome of

$$\underline{z}_k^i = \mathbf{H}_k^i \, \underline{x}_k + \underline{v}_k^i + \underline{e}_k^i \ .$$

By adapting the Bayesian interpretation of Section 2.4.1 and bringing $\underline{e}_k^i$ to the left side, we consider the measurement and the error bound as set of possible measurements $\mathcal{Z}_k^i = \mathcal{E}(\hat{\underline{z}}_k^i, \mathbf{X}_k^{z,i})$, where the symmetry of ellipsoids has been exploited. In compliance with Section 3.3.2, the corresponding information form has to be computed. The innovation matrix

$$\mathbf{I}_k^i \overset{(3.14)}{=} \left(\mathbf{H}_k^i\right)^{\mathrm{T}} \left(\mathbf{C}_k^{z,i}\right)^{-1} \mathbf{H}_k^i$$

is again independent of specific measurements and the set-membership uncertainty. For the ellipsoidal set $\mathcal{Z}_k^i$ of observations, we obtain the ellipsoid

$$\begin{aligned}
\mathcal{I}_k^i &= \left(\mathbf{H}_k^i\right)^{\mathrm{T}} \left(\mathbf{C}_k^{z,i}\right)^{-1} \mathcal{Z}_k^i \\
&= \left(\mathbf{H}_k^i\right)^{\mathrm{T}} \left(\mathbf{C}_k^{z,i}\right)^{-1} \mathcal{E}\left(\hat{\underline{z}}_k^i, \mathbf{X}_k^{z,i}\right) \\
&= \mathcal{E}\left(\left(\mathbf{H}_k^i\right)^{\mathrm{T}} \left(\mathbf{C}_k^{z,i}\right)^{-1} \hat{\underline{z}}_k^i, \left(\mathbf{H}_k^i\right)^{\mathrm{T}} \left(\mathbf{C}_k^{z,i}\right)^{-1} \mathbf{X}_k^{z,i} \left(\mathbf{C}_k^{z,i}\right)^{-\mathrm{T}} \mathbf{H}_k^i\right)
\end{aligned} \tag{3.27}$$

of innovation vectors (3.13). As a result, the filtering equation (3.11) becomes the Minkowski sum

$$\mathcal{Y}_k^{\mathrm{e}} = \mathcal{Y}_k^{\mathrm{p}} \oplus \mathcal{I}_k^1 \oplus \ldots \oplus \mathcal{I}_k^N \tag{3.28}$$

with $\mathcal{Y}_k^{\mathrm{p}} = \mathcal{E}(\hat{\underline{y}}_k^{\mathrm{p}}, \mathbf{Y}_k^{\mathrm{p}})$ denoting the ellipsoid of prior information vectors. An outer ellipsoidal approximation of this Minkowski sum can be computed for every part of the sum by means of (2.36), and thus it can also be implemented in a distributed fashion, but this procedure may finally result into a very conservative approximation of the total sum. The distributed computation of an ellipsoid that encloses the total sum and is optimal with respect to the trace of the shape matrix is subject of the subsequent subsection.

3.4.2   Distributed Fusion of Information Ellipsoids

Deriving a trace-optimal approximation of the total sum (3.28) may at first sight conflict with the idea to distribute its computation over the sensor network. This section therefore presents an efficient and distributable method for an outer trace-optimal approximation of the sum of more than two ellipsoids. At first, we can achieve the optimality with regard to the ellipsoid (3.26) of information vectors. An optimal approximation in the information space does not necessarily imply that the same optimality criterion is fulfilled in the state space. Therefore, we explain how to ensure that the back-transformed ellipsoid in the state space is trace-optimal.

A   Distributed Approximation of Minkowski Sums

In order to enclose the Minkowski sum (3.28) by an ellipsoid that is optimal with respect to some criterion, we have to compute an outer approximation of the entire Minkowski sum

$$\mathcal{E}(\hat{\underline{c}}_1, \mathbf{X}_1) \oplus \mathcal{E}(\hat{\underline{c}}_2, \mathbf{X}_2) \oplus \ldots \oplus \mathcal{E}(\hat{\underline{c}}_N, \mathbf{X}_N) \subseteq \mathcal{E}(\hat{\underline{c}}^{\mathrm{sum}}, \mathbf{X}^{\mathrm{sum}})$$

of N ellipsoidal sets. The parameters of the aspired enclosing ellipsoid $\mathcal{E}(\hat{\underline{c}}^{\mathrm{sum}}, \mathbf{X}^{\mathrm{sum}})$ are given by

$$\hat{\underline{c}}^{\mathrm{sum}} = \sum_{i=1}^{N} \hat{\underline{c}}_i$$

and

$$\mathbf{X}^{\mathrm{sum}} = \left(\sum_{i=1}^{N} p_i \right) \cdot \sum_{i=1}^{N} p_i^{-1} \mathbf{X}_i \tag{3.29}$$

with arbitrary $p_i > 0$ [111], where $\mathcal{E}(\hat{\underline{c}}^{\mathrm{sum}}, \mathbf{X}^{\mathrm{sum}})$ is in general not a tight bound for the sum. The set of parameters $\{p_1, \ldots, p_N\}$ has to be determined such that the chosen optimality criterion is fulfilled. Fortunately, deploying the trace as criterion, as it is done throughout this work due to its relation to the MSE, simplifies matters significantly. In [40], it has been proven that

$$p_i = \sqrt{\mathrm{trace}(\mathbf{X}_i)}, \quad \forall i = 1, \ldots, N \tag{3.30}$$

minimizes the trace[5] of (3.29). By employing the parameters (3.30), the sum (3.29) provides the particularly useful property that it is associative in the following sense: For the sum of three ellipsoids with shape matrices $\mathbf{X}_i$, $\mathbf{X}_j$ and $\mathbf{X}_k$, first

$$\mathbf{X}_{j,k} = (p_j + p_k) \cdot \left(p_j^{-1} \mathbf{X}_j + p_k^{-1} \mathbf{X}_k \right)$$

can be computed, and the final sum then yields

$$\begin{aligned}
\mathbf{X}_{i,j,k} &= (p_i + p_{j,k}) \cdot \left(p_i^{-1} \mathbf{X}_i + p_{j,k}^{-1} \mathbf{X}_{j,k} \right) \\
&= (p_i + p_{j,k}) \cdot \left(p_i^{-1} \mathbf{X}_i + p_{j,k}^{-1} \cdot (p_j + p_k) \cdot (p_j^{-1} \mathbf{X}_j + p_k^{-1} \mathbf{X}_k) \right) \\
&= (p_i + p_j + p_k) \cdot \left(p_i^{-1} \mathbf{X}_i + p_j^{-1} \mathbf{X}_j + p_k^{-1} \mathbf{X}_k \right) ,
\end{aligned}$$

where the equality

$$\begin{aligned}
p_{j,k} &= \sqrt{\text{trace}(\mathbf{X}_{j.k})} = \sqrt{\text{trace}\left((p_j + p_k) \cdot (p_j^{-1} \mathbf{X}_j + p_k^{-1} \mathbf{X}_k) \right)} \\
&= \sqrt{(p_j + p_k) \cdot \left(p_j^{-1} \text{trace}(\mathbf{X}_j) + p_k^{-1} \text{trace}(\mathbf{X}_k) \right)} \\
&= \sqrt{(p_j + p_k) \cdot (p_j + p_k)} \\
&= p_j + p_k
\end{aligned}$$

has been exploited. Thus, the matrix (3.29) does not need to be computed at once but can be computed step-by-step, which is indispensable to preprocess data along any communication path, as illustrated in Figure 3.7.

These results immediately imply that each sensor node i can autonomously compute the parameter

$$p_i = \sqrt{\text{trace}\left(\left(\mathbf{H}_k^i\right)^{\text{T}} \left(\mathbf{C}_k^{z,i}\right)^{-1} \mathbf{X}_k^{z,i} \left(\mathbf{C}_k^{z,i}\right)^{-\text{T}} \mathbf{H}_k^i \right)}$$

that corresponds to its contribution (3.27). In order to simplify the notation of (3.27), we set $\mathcal{I}_k^i = \mathcal{E}(\underline{i}_k^i, \mathbf{Z}_k^i)$ with

$$\underline{i}_k^i := \left(\mathbf{H}_k^i\right)^{\text{T}} \left(\mathbf{C}_k^{z,i}\right)^{-1} \underline{\hat{z}}_k^i ,$$

$$\mathbf{Z}_k^i := \left(\mathbf{H}_k^i\right)^{\text{T}} \left(\mathbf{C}_k^{z,i}\right)^{-1} \mathbf{X}_k^{z,i} \left(\mathbf{C}_k^{z,i}\right)^{-\text{T}} \mathbf{H}_k^i$$

---

[5]Note that this solution is identical to the trace-minimal outer approximation of (2.36) for $N = 2$.

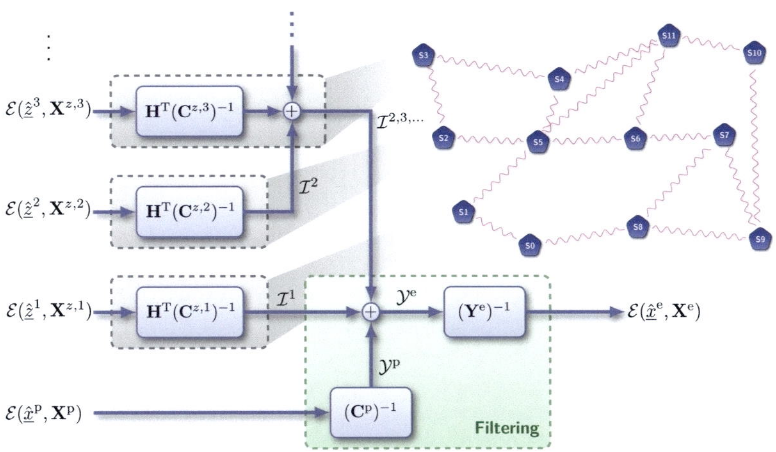

Figure 3.7: Parts of the filtering step can be performed locally on the sensor nodes, where data can be collected and already be condensed.

in the following. A node j that receives innovation parameters from node i updates its parameters according to

$$\underline{i}_k^{i,j} = \underline{i}_k^{i} + \underline{i}_k^{j} \, , \tag{3.31}$$

$$\mathbf{I}_k^{i,j} = \mathbf{I}_k^{i} + \mathbf{I}_k^{j} \, , \tag{3.32}$$

$$\mathbf{Z}_k^{i,j} = \left(p_i + p_j\right)\left(p_i^{-1}\mathbf{Z}_k^{i} + p_j^{-1}\mathbf{Z}_k^{i}\right) \tag{3.33}$$

before transmitting them to the next node on the communication path. The first two parameters correspond to (3.13) and (3.14) of the standard information filter, whereas the latter parameter is required for the ellipsoidal bound. In the data sink, these parameters are employed to update the predicted information parameters by means of

$$\hat{\underline{y}}_k^{e} = \hat{\underline{y}}_k^{p} + \underline{i}_k^{1,\dots,N} \, , \tag{3.34}$$

$$\mathbf{Y}_k^{e} = \mathbf{Y}_k^{p} + \mathbf{I}_k^{1,\dots,N} \, , \tag{3.35}$$

$$\mathbf{Q}_k^{e} = \left(p^{p} + p_{1,\dots,N}\right)\left((p^{p})^{-1}\mathbf{Q}_k^{p} + p_{1,\dots,N}^{-1}\mathbf{Z}_k^{1,\dots,N}\right) \tag{3.36}$$

with $p^{\mathrm{P}} = \sqrt{\mathrm{trace}(\mathbf{Q}_k^{\mathrm{p}})}$. The computation of the shape matrix $\mathbf{Q}_k^{\mathrm{e}}$ evidently conforms to (3.29) and is optimal with respect to the trace. This computation is the only additional effort compared to the standard information filter. However, the calculated shape matrix is only trace-optimal in the information form.

B Optimality in State Space

The procedure proposed in the preceding subsection only provides a trace-optimal ellipsoid in the information space. More specifically, due to the inequality

$$\mathrm{trace}(\mathbf{X}) \neq \mathrm{trace}(\mathbf{A}\mathbf{X}\mathbf{A}^{\mathrm{T}}) \,, \tag{3.37}$$

a trace-optimal $\mathbf{Q}_k^{\mathrm{e}}$ does not imply that the trace of

$$\mathbf{X}_k^{\mathrm{e}} = (\mathbf{Y}_k^{\mathrm{e}})^{-1}\mathbf{Q}_k^{\mathrm{e}}(\mathbf{Y}_k^{\mathrm{e}})^{-\mathrm{T}}$$

is optimal. Hence, the ellipsoid

$$\begin{aligned}
\mathcal{E}(\hat{\underline{x}}_k^{\mathrm{e}}, \mathbf{X}_k^{\mathrm{e}}) &= (\mathbf{Y}_k^{\mathrm{e}})^{-1}\mathcal{E}(\hat{\underline{y}}_k^{\mathrm{e}}, \mathbf{Q}_k^{\mathrm{e}}) \\
&= \mathcal{E}\big((\mathbf{Y}_k^{\mathrm{e}})^{-1}\hat{\underline{y}}_k^{\mathrm{e}}, (\mathbf{Y}_k^{\mathrm{e}})^{-1}\mathbf{Q}_k^{\mathrm{e}}(\mathbf{Y}_k^{\mathrm{e}})^{-\mathrm{T}}\big)
\end{aligned}$$

converted back to the state space does in general not represent an optimal outer approximation.

For the purpose of minimizing the trace of $(\mathbf{Y}_k^{\mathrm{e}})^{-1}\mathbf{Q}_k^{\mathrm{e}}(\mathbf{Y}_k^{\mathrm{e}})^{-\mathrm{T}}$ instead of $\mathbf{Q}_k^{\mathrm{e}}$, we can recognize from (3.36), and the back-transformed version

$$\begin{aligned}
(\mathbf{Y}_k^{\mathrm{e}})^{-1}\mathbf{Q}_k^{\mathrm{e}}(\mathbf{Y}_k^{\mathrm{e}})^{-\mathrm{T}} = \big(p^{\mathrm{P}} + p_{1,\dots,N}\big) \cdot \Big(&(p^{\mathrm{P}})^{-1}(\mathbf{Y}_k^{\mathrm{e}})^{-1}\mathbf{Q}_k^{\mathrm{P}}(\mathbf{Y}_k^{\mathrm{e}})^{-\mathrm{T}} \\
&+ p_{1,\dots,N}^{-1}(\mathbf{Y}_k^{\mathrm{e}})^{-1}\mathbf{Z}_k^{1,\dots,N}(\mathbf{Y}_k^{\mathrm{e}})^{-\mathrm{T}}\Big)
\end{aligned}$$

that the parameters $p^{\mathrm{P}}, p_1, \dots, p_N$ simply have to be replaced by

$$\begin{aligned}
p^{\mathrm{P}} &= \sqrt{\mathrm{trace}\big((\mathbf{Y}_k^{\mathrm{e}})^{-1}\mathbf{Q}_k^{\mathrm{P}}(\mathbf{Y}_k^{\mathrm{e}})^{-\mathrm{T}}\big)} \,, \\
p_i &= \sqrt{\mathrm{trace}\big((\mathbf{Y}_k^{\mathrm{e}})^{-1}\mathbf{Z}_k^{i}(\mathbf{Y}_k^{\mathrm{e}})^{-\mathrm{T}}\big)} \,, \quad i \in \{1, \dots, N\} \,.
\end{aligned} \tag{3.38}$$

Apparently, each node i must be aware of the estimated information matrix $\mathbf{Y}_k^{\mathrm{e}}$ before it is in the position compute its parameter p_i. This can be achieved by first transmitting the innovation matrices $\mathbf{I}_k^i$ to the data sink, where $\mathbf{Y}_k^{\mathrm{e}}$ can be determined, and thereafter broadcasting the information matrix $\mathbf{Y}_k^{\mathrm{e}}$ back to the fusion nodes so that the parameters (3.38) can be computed locally. With these parameters, the shape matrices $\mathbf{Z}_k^i$ can then be communicated to other nodes, preprocessed along any communication path, and finally be fused with $\mathbf{Q}_k^{\mathrm{p}}$ according to (3.36). In summary, the procedure is

1. first to transmit the information matrices $\mathbf{I}_k^i$ to the data sink. At each intermediate node, the matrix is updated according to (3.32), as illustrated in Figure 3.8(a).

2. At the data sink, $\mathbf{Y}_k^{\mathrm{e}}$ can then be computed by means of (3.35). The determined information matrix $\mathbf{Y}_k^{\mathrm{e}}$ is communicated back to each fusing sensor node that needs to know $\mathbf{Y}_k^{\mathrm{e}}$ in order to determine the parameters (3.38), as shown by Figure 3.8(b).

3. The parameters p_i can then locally be calculated according to (3.38), and the matrices $\mathbf{Z}_k^i$ are transmitted to the data sink according to (3.33). The data sink computes $\mathbf{Q}_k^{\mathrm{e}}$ by means of (3.36), as depicted in Figure 3.8(c).

4. In state space, the fused estimate is given by the midpoint $\hat{\underline{x}}_k^{\mathrm{e}} = (\mathbf{Y}_k^{\mathrm{e}})^{-1}\hat{\underline{y}}_k^{\mathrm{e}}$, the covariance matrix $\mathbf{C}_k^{\mathrm{e}} = (\mathbf{Y}_k^{\mathrm{e}})^{-1}$, and the shape matrix $\mathbf{X}_k^{\mathrm{e}} = (\mathbf{Y}_k^{\mathrm{e}})^{-1}\mathbf{Q}_k^{\mathrm{e}}(\mathbf{Y}_k^{\mathrm{e}})^{-\mathrm{T}}$, which is summarized in Figure 3.8(d).

For this communication strategy, an additional communication load consisting of the transmission of $\mathbf{Y}_k^{\mathrm{e}}$ is required. The communication and computation of $\hat{\underline{y}}_k^{\mathrm{e}}$ can be done either in step 1 or step 3 by means of (3.31) and (3.34).

In spite of the additional transmission of $\mathbf{Y}_k^{\mathrm{e}}$, this procedure is more favorable than to directly employ the multisensor Kalman filter formulas from Section 3.3.1. Multisensor Kalman filtering even poses a further difficulty in the presence of set-membership uncertainties. In the case of

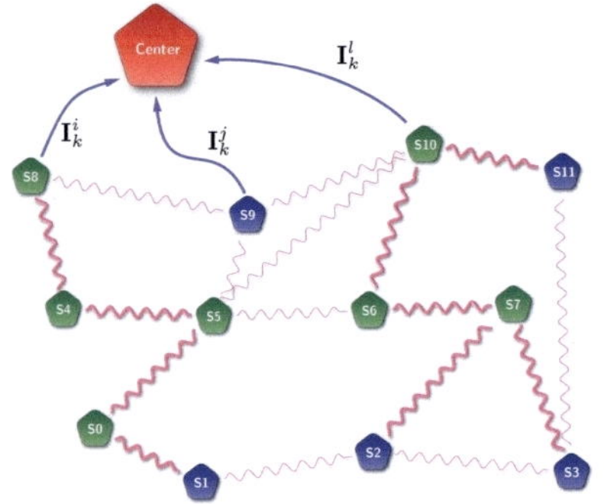

(a) Transmission of innovation matrices $\mathbf{I}_k^i$ to the data sink.

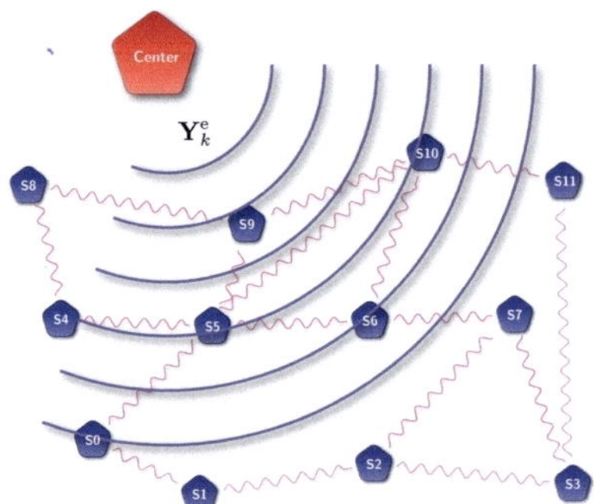

(b) Computation of final information matrix $\mathbf{Y}_k^e$ and broadcasting it back to the sensor nodes.

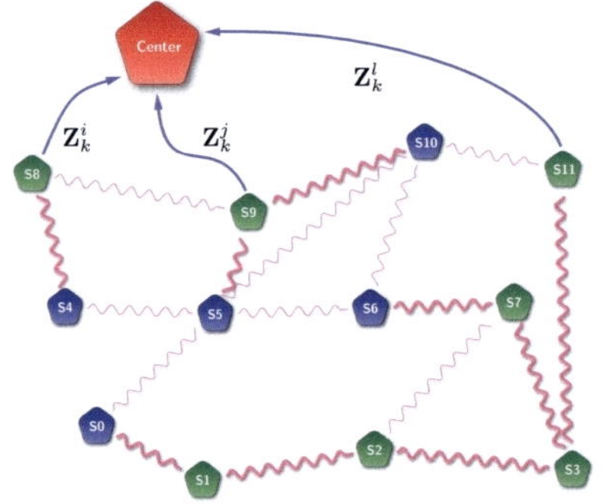

(c) Being aware of $\mathbf{Y}_k^e$, the parameter p_i can be determined for optimal processing of shape matrices $\mathbf{Z}_k^i$.

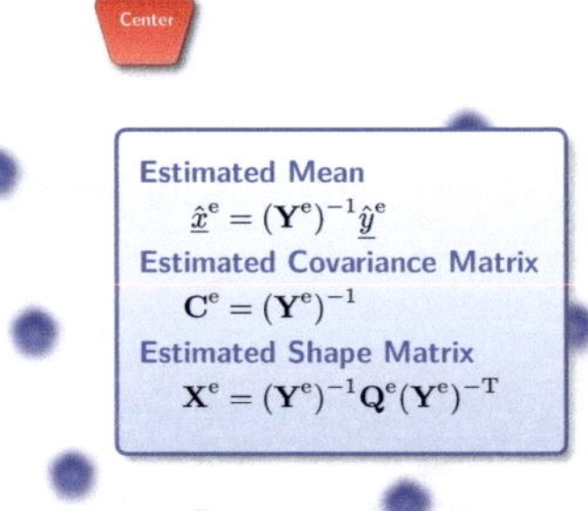

(d) Back-transformation of fused information parameters.

Figure 3.8: Data transmission strategy for stochastic and set-membership information filtering. The green nodes preprocess obtained data according to (3.31), (3.32), and (3.33).

two measurement devices and a sequential processing scheme according to (3.3), an enclosing ellipsoid for

$$(\mathbf{I} - \mathbf{K}_k^2\mathbf{H}_k^2)\Big((\mathbf{I} - \mathbf{K}_k^1\mathbf{H}_k^1)\mathcal{E}(\hat{\underline{x}}_k^\mathrm{p}, \mathbf{X}_k^\mathrm{p}) \oplus \mathbf{K}_k^1\mathcal{E}(\hat{\underline{z}}_k^1, \mathbf{X}_k^{z,1})\Big)$$
$$\oplus \mathbf{K}_k^2\mathcal{E}(\hat{\underline{z}}_k^2, \mathbf{X}_k^{z,2}) \subseteq \mathcal{E}(\hat{\underline{x}}_k^\mathrm{e}, \mathbf{X}_k^\mathrm{e})$$

has to be calculated. Approximating first the inner sum optimally and afterwards the outer sum would not provide a trace-optimal result because of inequality (3.37). It becomes apparent that each shape matrix requires a different linear transformation before a trace-optimal fusion result can be determined. In the above scenario, this is $(\mathbf{I} - \mathbf{K}_k^2\mathbf{H}_k^2)(\mathbf{I} - \mathbf{K}_k^1\mathbf{H}_k^1)$ for the first ellipsoid, $(\mathbf{I} - \mathbf{K}_k^2\mathbf{H}_k^2)\mathbf{K}_k^1$ for the second ellipsoid, and $\mathbf{K}_k^2$ for the third ellipsoid. This stands in contrast to the information filter where each fusing node only needs to receive the same $\mathbf{Y}_k^\mathrm{e}$ to compute the optimal parameter according to (3.38). These issues are further discussed in the subsequent simulated scenario.

3.4.3 Discussion and Simulation

The proposed concept is evaluated in a hierarchical network with one intermediate layer consisting of 9 nodes. The situation is similar to the scenario in [204], but in the here considered setup a total of 99 nodes and arbitrary noise matrices is employed. The nodes strive for the common goal to track a mobile object, which is related to the true system model

$$\underline{x}_{k+1} = \underline{x}_k + \underbrace{\begin{bmatrix} \cos(\gamma) & -\sin(\gamma) \\ \sin(\gamma) & \cos(\gamma) \end{bmatrix}}_{=\mathbf{B}_k} \hat{\underline{u}}_k$$

where the control inputs are determined according to $\hat{\underline{u}}_0 = [0.05, 0]^\mathrm{T}$ and $\underline{u}_k = \hat{\underline{x}}_k - \hat{\underline{x}}_{k-1}$ for $k > 0$. The angle is set to $\gamma = \frac{\pi}{15}$ for $k < 20$ and to $\gamma = \frac{3\pi}{40}$ afterwards. To the estimator, a differing model is known with $\gamma = \frac{\pi}{10}$ and the control input is determined by $\hat{\underline{u}}_k = \hat{\underline{x}}_k^\mathrm{e} - \hat{\underline{x}}_{k-1}^\mathrm{e}$. The observed area is subdivided into 9 segments, each of which is monitored by 10 nodes. One node in each segment serves as an intermediate node to the data sink.

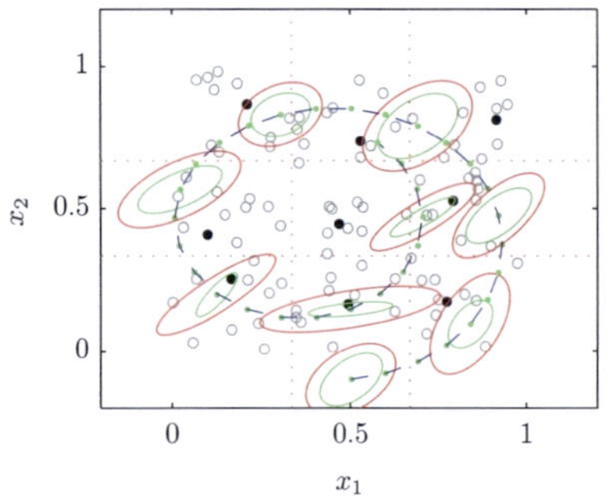

(a) The true trajectory is drawn blue. Ellipses are the set-membership error bounds of the estimates. Green: Trace-optimal result. Red: Result of naive information filtering.

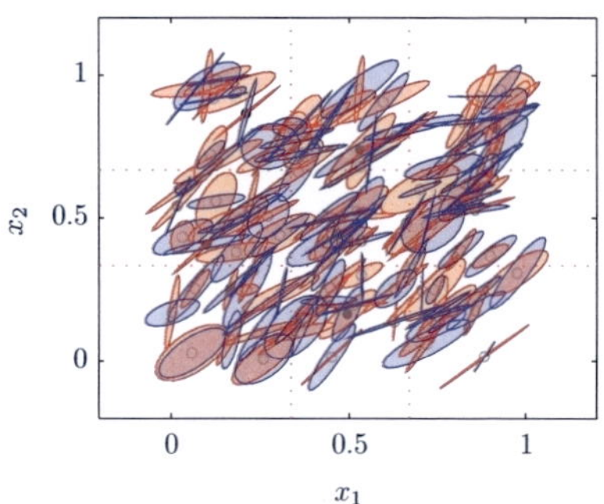

(b) Blue: Covariance ellipses of stochastic measurement noise. Red: Bounding ellipses for unknown but bounded measurement errors.

Figure 3.9: A network of 99 sensor nodes observes an object to be tracked. The 9 filled black nodes can communicate to the data sink; the other nodes have to send their data to these intermediate nodes first.

The state is directly observed by each sensor node, i.e., the matrix $\mathbf{H}_k^i$ is the identity. The sensor noise covariance and shape matrices are both randomly determined at each time step, as illustrated in Figure 3.9(b). A trace-optimal fusion result at each time step can be achieved either by sequential or blockwise Kalman filtering or by the information filter employing the aforementioned communication strategy from Section 3.4.2-B. The green ellipses in Figure 3.9(a) are the estimated trace-optimal bounds for set-membership errors. The Kalman filter formulation requires the awareness of every covariance and shape matrix, which must first be transmitted to the intermediate nodes ( ) and then passed on to the center node together with the data of each intermediate node ( ). This must be done in each segment ( ) and for each of both types of error matrices ( ). Hence, a total of

$$(10 + 11) \cdot 9 \cdot 2 = 378$$

matrix transmissions is required at each time step. In this regard, the proposed information filtering scheme reveals a definite advantage: The intermediate nodes can comprise the obtained matrices into a single one before sending them to the center node. In order to minimize the trace of the final shape matrix $\mathbf{Q}_k^e$, the center node has to send the fused information matrix $\mathbf{Y}_k^e$ back to the intermediate nodes ( ) so that the required parameters (3.38) can be determined prior to the transmission of $\mathbf{Q}_k^i$. The amount of matrices to be transmitted can significantly be reduced to

$$(10 + 1) \cdot 9 \cdot 2 + 9 = 207 \ .$$

A naive information filtering scheme that only provides optimality in the information space only saves the 9 matrix transfers back to the intermediate node and requires

$$(10 + 1) \cdot 9 \cdot 2 = 198$$

transmissions but the set-membership errors bounds become continually more conservative as shown in Figure 3.9(a) by the red ellipses. The traces

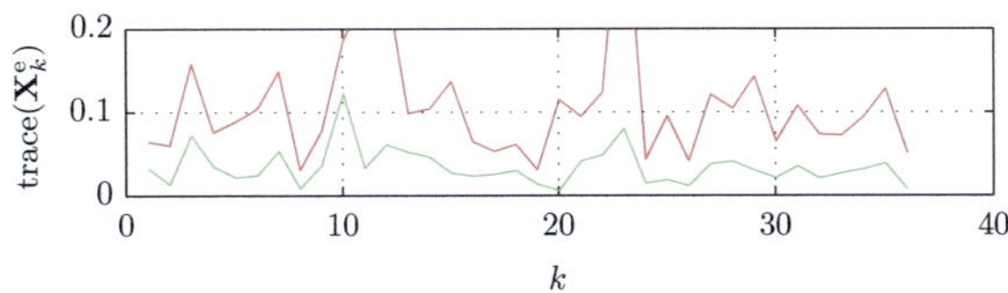

Figure 3.10: The trace of the shape matrix obtained by means of the communication strategy from Section 3.4.2-B is, at each time step, lower than the trace that corresponds to naive information filtering. They correspond to the ellipsoidal error bounds in Figure 3.9(a).

of the red and green ellipses are compared in Figure 3.10. Of course, the amount of additional data transfers for the optimal strategy depends on the network topology. As a rule of thumb, flat hierarchies, where nodes on intermediate levels can collect and condense much data, are particularly suited.

3.5   Towards Decentralized Estimation

Large-scale networks may render it difficult or even impossible to sustain a certain communication strategy and network topology. As a consequence, each node or a group of nodes should be capable of operating autonomously without being reliant on a central processing unit. In this regard, a fully decentralized processing is evidently the best solution so that each node is in the position to determine a valid state estimate on its own. By exchanging and fusing local estimates, they become more informative when possible dependencies are taken into account correctly. This section commences with an optimal fusion method of local estimates, which relies on the complete knowledge of the entire joint covariance matrix. Unfortunately, this is only applicable in rare and exceptional cases. Therefore, conservative algorithms are presented that are robust to dependency effects on the state estimate. However, in some cases, they can be very conservative.

3.5.1   Processing and Fusion of Local Estimates

Processing information in a network of sensor nodes without involving a central computer system offers clear advantages but also poses many additional challenges. Each node should be able to operate independently and to remain unaffected by unreliable communication links, varying network topologies, or failures of other nodes. As studied in Section 3.2, an independent information processing does by no means imply that the local estimation results are independent. Even in the early initialization phase, the sensor nodes may share the same prior information in order to be in the position to compute informative estimates locally. In a simple two-node situation, the prior information can be distributed to the nodes according to

$$
\begin{bmatrix} \hat{x}_A^{\mathrm{P}} \\ \hat{x}_B^{\mathrm{P}} \end{bmatrix} = \begin{bmatrix} \mathbf{I} \\ \mathbf{I} \end{bmatrix} \hat{x}_0^{\mathrm{P}} , \tag{3.39}
$$

which entails the joint error covariance matrix

$$
\mathrm{E}\left[\left(\begin{bmatrix} \hat{x}_A \\ \hat{x}_B \end{bmatrix} - \begin{bmatrix} \mathbf{I} \\ \mathbf{I} \end{bmatrix} x \right) \left(\begin{bmatrix} \hat{x}_A \\ \hat{x}_B \end{bmatrix} - \begin{bmatrix} \mathbf{I} \\ \mathbf{I} \end{bmatrix} x \right)^{\mathrm{T}} \right] = \begin{bmatrix} \mathbf{C}_0^{\mathrm{P}} & \mathbf{C}_0^{\mathrm{P}} \\ \mathbf{C}_0^{\mathrm{P}} & \mathbf{C}_0^{\mathrm{P}} \end{bmatrix} . \tag{3.40}
$$

Hence, the local estimates $\hat{\underline{x}}_A$ and $\hat{\underline{x}}_B$ are fully correlated with cross-covariance matrix $\mathbf{C}_{AB} = \mathbf{C}_0^{\mathrm{p}}$. Another possibility is to initialize each estimator with the very first measurement that is received. However, the estimates become correlated through the prediction phase at the latest. The state transition model (3.1) is to be applied on each local "copy" of the state, i.e.,

$$\begin{bmatrix} \underline{x}_{k+1} \\ \underline{x}_{k+1} \end{bmatrix} = \begin{bmatrix} \mathbf{A}_k & \mathbf{0} \\ \mathbf{0} & \mathbf{A}_k \end{bmatrix} \begin{bmatrix} \underline{x}_k \\ \underline{x}_k \end{bmatrix} + \begin{bmatrix} \mathbf{I} \\ \mathbf{I} \end{bmatrix} \underline{w}_k \ , \tag{3.41}$$

where we disregard possible control inputs in order to simplify matters. In each local prediction model, it is the same process noise $\underline{w}_k$ that affects the state. The predicted local estimates are then given by

$$\begin{bmatrix} \hat{\underline{x}}_A^{\mathrm{p}} \\ \hat{\underline{x}}_B^{\mathrm{p}} \end{bmatrix} = \begin{bmatrix} \mathbf{A}_k \, \hat{\underline{x}}_A^{\mathrm{e}} \\ \mathbf{A}_k \, \hat{\underline{x}}_B^{\mathrm{e}} \end{bmatrix} \ , \tag{3.42}$$

and the predicted joint covariance matrix yields

$$\begin{bmatrix} \mathbf{C}_A^{\mathrm{p}} & \mathbf{C}_{AB}^{\mathrm{p}} \\ \mathbf{C}_{BA}^{\mathrm{p}} & \mathbf{C}_B^{\mathrm{p}} \end{bmatrix} = \begin{bmatrix} \mathbf{A}_k & \mathbf{0} \\ \mathbf{0} & \mathbf{A}_k \end{bmatrix} \begin{bmatrix} \mathbf{C}_A^{\mathrm{e}} & \mathbf{C}_{AB}^{\mathrm{e}} \\ \mathbf{C}_{BA}^{\mathrm{e}} & \mathbf{C}_B^{\mathrm{e}} \end{bmatrix} \begin{bmatrix} \mathbf{A}_k & \mathbf{0} \\ \mathbf{0} & \mathbf{A}_k \end{bmatrix}^{\mathrm{T}} + \begin{bmatrix} \mathbf{C}_k^{w} & \mathbf{C}_k^{w} \\ \mathbf{C}_k^{w} & \mathbf{C}_k^{w} \end{bmatrix} \ . \tag{3.43}$$

The added noise covariance matrix is exactly the common process noise we discussed in Section 3.2.2. The cross-covariance matrices $\mathbf{C}_{AB}^{\mathrm{p}} = (\mathbf{C}_{BA}^{\mathrm{p}})^{\mathrm{T}}$ have to be adjusted in the prediction step, but the predicted estimates (3.42) can be computed locally, unlike the filtering step, where the incorporation of local measurements also updates the local estimates of the other nodes and hence requires a communication: A local measurement $\hat{\underline{z}}_A$ received at sensor node A is related to the model

$$\underline{z}_A = \begin{bmatrix} \mathbf{H}_A & \mathbf{0} \end{bmatrix} \begin{bmatrix} \underline{x}_k \\ \underline{x}_k \end{bmatrix} + \underline{v}_A$$

with local measurement noise $\underline{v}_A \sim \mathcal{N}(\underline{0}, \mathbf{C}_A^z)$. The Kalman gain for the joint state estimate is given by

$$\mathbf{K} = \begin{bmatrix} \mathbf{C}_A^{\mathrm{p}} & \mathbf{C}_{AB}^{\mathrm{p}} \\ \mathbf{C}_{BA}^{\mathrm{p}} & \mathbf{C}_B^{\mathrm{p}} \end{bmatrix} \begin{bmatrix} \mathbf{H}_A^{\mathrm{T}} \\ \mathbf{0} \end{bmatrix} \cdot \left(\mathbf{C}^z + \begin{bmatrix} \mathbf{H}_A & \mathbf{0} \end{bmatrix} \begin{bmatrix} \mathbf{C}_A^{\mathrm{p}} & \mathbf{C}_{AB}^{\mathrm{p}} \\ \mathbf{C}_{BA}^{\mathrm{p}} & \mathbf{C}_B^{\mathrm{p}} \end{bmatrix} \begin{bmatrix} \mathbf{H}_A^{\mathrm{T}} \\ \mathbf{0} \end{bmatrix} \right)^{-1}$$

$$= \begin{bmatrix} \mathbf{C}_A^{\mathrm{p}} \mathbf{H}_A^{\mathrm{T}} \left(\mathbf{C}^z + \mathbf{H}_A \mathbf{C}_A^{\mathrm{p}} \mathbf{H}_A^{\mathrm{T}} \right)^{-1} \\ \mathbf{C}_{BA}^{\mathrm{p}} \mathbf{H}_A^{\mathrm{T}} \left(\mathbf{C}^z + \mathbf{H}_A \mathbf{C}_A^{\mathrm{p}} \mathbf{H}_A^{\mathrm{T}} \right)^{-1} \end{bmatrix} =: \begin{bmatrix} \mathbf{K}_A \\ \mathbf{K}_B \end{bmatrix},$$
(3.44)

in compliance with (2.24). The local prior or predicted estimates $\underline{\hat{x}}_A^{\mathrm{p}}$ and $\underline{\hat{x}}_B^{\mathrm{p}}$ are then updated by means of

$$\begin{bmatrix} \underline{\hat{x}}_A^{\mathrm{e}} \\ \underline{\hat{x}}_B^{\mathrm{e}} \end{bmatrix} \overset{(2.26)}{=} \begin{bmatrix} \underline{\hat{x}}_A^{\mathrm{p}} \\ \underline{\hat{x}}_B^{\mathrm{p}} \end{bmatrix} + \begin{bmatrix} \mathbf{K}_A \\ \mathbf{K}_B \end{bmatrix} \left(\underline{\hat{z}}_A - \begin{bmatrix} \mathbf{H}_A & \mathbf{0} \end{bmatrix} \begin{bmatrix} \underline{\hat{x}}_A^{\mathrm{p}} \\ \underline{\hat{x}}_B^{\mathrm{p}} \end{bmatrix} \right)$$

$$= \begin{bmatrix} \underline{\hat{x}}_A^{\mathrm{p}} + \mathbf{K}_A (\underline{\hat{z}}_A - \mathbf{H}_A \underline{\hat{x}}_A) \\ \underline{\hat{x}}_B^{\mathrm{p}} + \mathbf{K}_B (\underline{\hat{z}}_A - \mathbf{H}_A \underline{\hat{x}}_A) \end{bmatrix}.$$
(3.45)

Due to nonzero cross-covariance matrix $\mathbf{C}_{BA}^{\mathrm{p}} = (\mathbf{C}_{AB}^{\mathrm{p}})^{\mathrm{T}}$ and hence nonzero matrix $\mathbf{K}_B$, both estimates $\underline{\hat{x}}_A^{\mathrm{p}}$ and $\underline{\hat{x}}_B^{\mathrm{p}}$ are updated, although only sensor A observes the state.

If we initialize the local estimators with (3.39) and (3.40), then $\mathbf{C}_{BA}^{\mathrm{p}} = \mathbf{C}_A^{\mathrm{p}}$ and hence $\mathbf{K}_B = \mathbf{K}_A$ holds. Eventually, each estimate is affected by the incoming measurement in the same way and each sensor node is again equipped with the same estimate, i.e., $\underline{\hat{x}}_A^{\mathrm{e}} = \underline{\hat{x}}_B^{\mathrm{e}}$, that is equal to a centralized estimation result, where all measurements are processed by a single computer system. In this particular case, the cross-correlations effectuate a redundant processing of the globally optimal state estimate, i.e., each node has a copy of the globally optimal state estimate.

Of course, maintaining a constant communication link between all nodes is least favorable as the network would provide no substantial advantages anymore compared to gathering all measurements at a data sink. In order to avoid interchanging sensor data, the Kalman gain (3.44) is only computed for the local estimate, i.e., $\mathbf{K}_A$, and only the first row of the vector transformation (3.45) is carried out. Although $\underline{\hat{x}}_B^{\mathrm{p}}$ and $\mathbf{C}_B^{\mathrm{p}}$ do not undergo the filtering step, still the cross-covariance matrices have to be updated and need to be revised to

$$
\begin{aligned}
\mathbf{C}_{AB}^{\mathrm{e}} &= \mathrm{E}\left[(\hat{\underline{x}}_A^{\mathrm{e}} - \underline{x}_k)(\hat{\underline{x}}_B^{\mathrm{P}} - \underline{x}_k)^{\mathrm{T}}\right] \\
&= \mathrm{E}\left[\left((\mathbf{I} - \mathbf{K}_A\mathbf{H}_A)(\hat{\underline{x}}_A^{\mathrm{P}} - \underline{x}_k) + \mathbf{K}_A\,\underline{v}_A\right)(\hat{\underline{x}}_B^{\mathrm{P}} - \underline{x}_k)^{\mathrm{T}}\right] \\
&= \mathrm{E}\left[(\mathbf{I} - \mathbf{K}_A\mathbf{H}_A)(\hat{\underline{x}}_A^{\mathrm{P}} - \underline{x}_k)(\hat{\underline{x}}_B^{\mathrm{P}} - \underline{x}_k)^{\mathrm{T}}\right] \\
&= (\mathbf{I} - \mathbf{K}_A\mathbf{H}_A)\mathbf{C}_{AB}^{\mathrm{P}}\ .
\end{aligned}
\tag{3.46}
$$

Against the background of target tracking applications, the treatment and combination of locally computed estimates is referred to as the *track-to-track fusion* problem [7–9]. In order to fuse the information provided by different sensor nodes, the cross-correlations need to be taken into account properly. Assuming independence can be hazardous and misleading, since full correlation, for instance, imply that the estimates have to be considered to be the same, as discussed before. The following paragraph is dedicated to the question of how to exploit known cross-covariance matrices (3.46) when estimates are to be fused.

A Bar-Shalom/Campo Formulas

For the fusion of locally processed estimates, the required Kalman gains can be derived pursuant to the filtering step in Section 2.1.1-B. So, the aim is to derive the gain that combines two estimates $\hat{\underline{x}}_A$ and $\hat{\underline{x}}_B$ according to

$$
\hat{\underline{x}}_{\mathrm{fus}} = (\mathbf{I} - \mathbf{K})\,\hat{\underline{x}}_A + \mathbf{K}\,\hat{\underline{x}}_B\ .
$$

Unlike Kalman filtering of conditionally independent measurements, the cross-correlations between the estimates to be fused must be addressed properly. In contrast to the Kalman filter covariance matrix (2.26), the fused covariance matrix yields

$$
\begin{aligned}
\mathbf{C}_{\mathrm{fus}} &= \mathrm{E}\left[(\hat{\underline{x}}_{\mathrm{fus}} - \underline{x}_k)(\hat{\underline{x}}_{\mathrm{fus}} - \underline{x}_k)^{\mathrm{T}}\right] \\
&= (\mathbf{I} - \mathbf{K})\mathbf{C}_A(\mathbf{I} - \mathbf{K})^{\mathrm{T}} + (\mathbf{I} - \mathbf{K})\mathbf{C}_{AB}\mathbf{K}^{\mathrm{T}} \\
&\quad + \mathbf{K}\mathbf{C}_{BA}(\mathbf{I} - \mathbf{K})^{\mathrm{T}} + \mathbf{K}\mathbf{C}_B\mathbf{K}^{\mathrm{T}}\ .
\end{aligned}
\tag{3.47}
$$

Minimizing its trace results into the gain

$$
\mathbf{K} = (\mathbf{C}_A - \mathbf{C}_{AB}) \cdot (\mathbf{C}_A + \mathbf{C}_B - \mathbf{C}_{AB} - \mathbf{C}_{BA})^{-1}\ ,
$$

and we obtain the *Bar-Shalom/Campo* formulas [8]

$$\hat{\underline{x}}_{\text{fus}} = (\mathbf{C}_B - \mathbf{C}_{BA})(\mathbf{C}_A + \mathbf{C}_B - \mathbf{C}_{AB} - \mathbf{C}_{BA})^{-1}\hat{\underline{x}}_A \\ + (\mathbf{C}_A - \mathbf{C}_{AB})(\mathbf{C}_A + \mathbf{C}_B - \mathbf{C}_{AB} - \mathbf{C}_{BA})^{-1}\hat{\underline{x}}_B \tag{3.48}$$

and

$$\mathbf{C}_{\text{fus}} = \mathbf{C}_A - (\mathbf{C}_A - \mathbf{C}_{AB})(\mathbf{C}_A + \mathbf{C}_B - \mathbf{C}_{AB} - \mathbf{C}_{BA})^{-1}(\mathbf{C}_A - \mathbf{C}_{AB})^{\mathrm{T}} \tag{3.49}$$

for the fusion of correlated tracks. This fusion of two estimates can be interpreted as a Kalman filtering step with correlated sensor noise, where $\hat{\underline{z}} = \hat{\underline{x}}_B$ is the observation, the measurement matrix $\mathbf{H}$ is the identity, and the error covariance matrix $\mathbf{C}_B$ characterizes the sensor noise. The estimate $(\hat{\underline{x}}_A, \mathbf{C}_A)$ serves as prior knowledge. However, this interpretation as Bayes' rule can be delusive, and it is a common misconception to believe that storing and updating of cross-correlations with a subsequent Bar-Shalom/Campo fusion renders an optimal strategy, as the following example illustrates.

Example 3.3: Bar-Shalom/Campo formulas

In order to demonstrate the performance degradation with respect to a centralized processing, we compute the fused covariance matrix of two one-dimensional estimates each of which incorporates an observation locally. The prior joint covariance is set to

$$\begin{bmatrix} C_A^{\mathrm{p}} & C_{AB}^{\mathrm{p}} \\ C_{BA}^{\mathrm{p}} & C_B^{\mathrm{p}} \end{bmatrix} = \begin{bmatrix} 6 & 6 \\ 6 & 6 \end{bmatrix} ,$$

i.e., the corresponding global estimate $\hat{x}^{\mathrm{p}}$ has variance 6 (cf. equations (3.39) and (3.40)). Sensor A receives a measurement with error variance $C_A^z = 3$, and, at sensor node B, the state is observed with error variance $C_B^z = 6$. For both sensors, the measurement mapping is the identity. The local Kalman gain for the former node yields

$$K_A = C_A^{\mathrm{p}}(C_A^{\mathrm{p}} + C_A^z)^{-1} = 6(6 + 3)^{-1} = \frac{2}{3}$$

such that the updated variance is

$$C_A^{\mathrm{e}} = C_A^{\mathrm{p}} - K_A C_A^{\mathrm{p}} = 6 - \frac{2}{3}6 = 2 .$$

For sensor node B, the variance is updated with the gain

$$K_B = C_B^{\mathrm{p}}(C_B^{\mathrm{p}} + C_B^z)^{-1} = 6(6+6)^{-1} = \frac{1}{2}$$

and results into

$$C_B^{\mathrm{e}} = C_B^{\mathrm{p}} - K_B C_B^{\mathrm{p}} = 6 - \frac{1}{2}6 = 3 \;.$$

In order to be capable of applying the Bar-Shalom/Campo formulas, the cross-correlations need to be adapted according to (3.46), which yields

$$C_{AB}^{\mathrm{e}} = \mathrm{E}\left[\left((1-K_A)(\hat{x}_A^{\mathrm{p}}-\boldsymbol{x})\right)\left((1-K_B)(\hat{x}_B^{\mathrm{p}}-\boldsymbol{x})\right)^{\mathrm{T}}\right] = \frac{1}{3}\cdot C_{AB}^{\mathrm{p}} \cdot \frac{1}{2} = 1 \;.$$

With this cross-covariance, the Bar-Shalom/Campo formula (3.49) can be applied to obtain the fusion result

$$C_{\mathrm{fus}} = C_A^{\mathrm{e}} - \frac{(C_A^{\mathrm{e}} - C_{AB}^{\mathrm{e}})^2}{C_A^{\mathrm{e}} + C_B^{\mathrm{e}} - C_{AB}^{\mathrm{e}} - C_{AB}^{\mathrm{e}}} = 2 - \frac{(2-1)^2}{2+3-1-1} = 1\frac{2}{3} \;.$$

We compare the fused variance with a centralized processing of the estimate and measurements. The order in which the measurements are processed is, of course, irrelevant. So, first we update the estimate with the measurement of sensor A and obtain the error variance

$$C_1^{\mathrm{e}} = C^{\mathrm{p}} - K_A C^{\mathrm{p}} = 6 - \frac{2}{3}6 = 2 \;.$$

For the update with the second measurement, the gain $K = C_1^{\mathrm{e}}(C_1^{\mathrm{e}} + C_B^z)^{-1} = \frac{2}{2+6}$ has to be used and the final fused error variance is

$$C_2^{\mathrm{e}} = C_1^{\mathrm{e}} - K C_1^{\mathrm{e}} = 2 - \frac{2}{8}2 = 1\frac{1}{2} \;.$$

Evidently, a global processing of the measurements reports an estimate with lower error variance than a distributed processing with correctly updated cross-correlations and subsequent Bar-Shalom/Campo fusion.

This discrepancy has first been studied in [150] and can be traced back to the fact that the Bar-Shalom/Campo formulas rather characterize an estimate that is optimal in the maximum likelihood sense [37]. They should not be considered to be optimal in the Bayesian sense that the fused estimate optimally incorporates the measurement history. The globally optimal strategy is hence to persevere in a centralized fusion of measurements, as described in Section 3.3. In spite of that, the Bar-Shalom/Campo fusion algorithm still embodies the best strategy for combining two tracks, i.e., estimates, with known cross-correlations.

By means of Millman's formulas [2,157], the Bar-Shalom/Campo fusion rule can be generalized to the instantaneous combination of multiple estimates, which of course requires to employ the entire joint cross-covariance matrix. In general, attaining and sustaining valid cross-covariance matrices is a central issue, which makes this approach unattractive. For a vast number of sensor nodes, large matrices must be stored and processed, which clearly overshadows any benefits a decentralized processing may provide. In many networks such as mobile ad-hoc networks, it will not even be possible to obtain the actual cross-correlations, in particular, if a node is not aware of its neighborhood.

B Extension to Set-membership Uncertainties

An incorporation of set-membership error characteristics into the Bar-Shalom/Campo formulas can be achieving by following the derivations in Section 2.4.2. In order to minimize the bound for the total MSE associated with the fused estimate, the trace of

$$
\begin{aligned}
\overline{\mathrm{E}}\left[(\hat{\underline{x}}_{\mathrm{fus}} - \underline{x}_k)(\hat{\underline{x}}_{\mathrm{fus}} - \underline{x}_k)^{\mathrm{T}}\right] = {} & (\mathbf{I} - \mathbf{K})\mathbf{C}_A(\mathbf{I} - \mathbf{K})^{\mathrm{T}} + (\mathbf{I} - \mathbf{K})\mathbf{C}_{AB}\mathbf{K}^{\mathrm{T}} \\
& + \mathbf{K}\mathbf{C}_{BA}(\mathbf{I} - \mathbf{K})^{\mathrm{T}} + \mathbf{K}\mathbf{C}_B\mathbf{K}^{\mathrm{T}} \\
& + (1 + p^{-1})(\mathbf{I} - \mathbf{K})\mathbf{X}_A(\mathbf{I} - \mathbf{K})^{\mathrm{T}} \\
& + (1 + p)\mathbf{K}\mathbf{X}_B\mathbf{K}^{\mathrm{T}}
\end{aligned}
$$

needs to be considered instead of (3.47), where $\mathbf{X}_A$ and $\mathbf{X}_B$ are the corresponding shape matrices. Analogously to (2.91), the gain can be derived to the expression

$$\mathbf{K}(p) = \left((1 + p^{-1})\mathbf{X}_A + \mathbf{C}_A - \mathbf{C}_{AB} \right) \cdot$$
$$\left((1 + p^{-1})\mathbf{X}_A + (1 + p)\mathbf{X}_B + \mathbf{C}_A + \mathbf{C}_B - \mathbf{C}_{AB} - \mathbf{C}_{BA} \right)^{-1} .$$
$$(3.50)$$

A further extension to an adjustable gain according to Section 2.5 is straight-forward. As mentioned earlier, dependencies between set-membership error characteristics do not require any attention.

C — Feasibility in Large-scale Sensor Networks

The aforementioned difficulty to compute a cross-covariance matrix for the entirety of participating sensor nodes constitutes the reason why the Bar-Shalom/Campo formulas can scarcely be applied to large-scale networks. In its original formulation, a communication between nodes, local fusion of state vectors, and a corresponding update of the cross-covariance matrix have not been considered, and the question arises where to store and to keep track of the correlations. In particular, often changing neighborhoods and network topology render it almost impossible to keep the joint cross-covariance matrix up-to-date. For these reasons, a state vector fusion architecture that maintains cross-correlations does not comprise any clear advantages with respect to centralized architectures, where data is transmitted to and managed by a single processing system.

3.5.2 — Removal of Common Information and Ellipsoidal Intersection

As stated above, in a network that is subject to often changes in sensor configuration and topology, it can be hardly possible to maintain the knowledge that is required to fuse estimates optimally, i.e., the cross-covariance matrices. Dependencies between locally computed estimates cannot simply be ignored without underestimating the actual MSE. In the presence of possible unknown correlations, only suboptimal information processing strategies are realizable. This subsection is devoted to the treatment of unknown data that is shared by two local estimates to be fused.

Cycles in communication paths of a network can cause double-counting of sensor data, as discussed in Section 3.2.1. For the information form from Section 3.3.2, this implies that the locally computed information parameters

$$\hat{\underline{y}}_A^{\mathrm{e}} = \hat{\underline{y}}_A^{\mathrm{p}} + \sum_{i=1}^{N_A} \underline{i}_A^i \quad \text{and} \quad \mathbf{Y}_A^{\mathrm{e}} = \mathbf{Y}_A^{\mathrm{p}} + \sum_{i=1}^{N_A} \mathbf{I}_A^i$$

for sensor A and

$$\hat{\underline{y}}_B^{\mathrm{e}} = \hat{\underline{y}}_B^{\mathrm{p}} + \sum_{i=1}^{N_B} \underline{i}_B^i \quad \text{and} \quad \mathbf{Y}_B^{\mathrm{e}} = \mathbf{Y}_B^{\mathrm{p}} + \sum_{i=1}^{N_B} \mathbf{I}_B^i$$

for sensor B have, w.l.o.g., the information vector and matrix

$$\hat{\underline{y}}_{A \cap B} = \sum_{i=1}^{M} \underline{i}^i \quad \text{and} \quad \mathbf{Y}_{A \cap B} = \sum_{i=1}^{M} \mathbf{I}^i \tag{3.51}$$

in common. If the latter parameters are known to the fusion node and the prior estimates $\hat{\underline{y}}_A^{\mathrm{p}}$ and $\hat{\underline{y}}_B^{\mathrm{p}}$ are independent, they can simply be subtracted from the fusion result, i.e.,

$$\hat{\underline{y}}_{\mathrm{fus}} = \hat{\underline{y}}_A^{\mathrm{e}} + \hat{\underline{y}}_B^{\mathrm{e}} - \hat{\underline{y}}_{A \cap B} \tag{3.52}$$

and

$$\mathbf{Y}_{\mathrm{fus}} = \mathbf{Y}_A^{\mathrm{e}} + \mathbf{Y}_B^{\mathrm{e}} - \mathbf{Y}_{A \cap B} , \tag{3.53}$$

so that they are only incorporated once. In networks that are acyclic, for instance, tree-connected without loops, the *channel filter* [27, 71, 123] stores the information parameters (3.51) that are shared by neighboring nodes and removes them from the fusion results. If the shared information parameters are unknown, only suboptimal fusion results are attainable, as discussed in the next paragraph.

A   Unknown Common Information

In the situation that the common parameters (3.51) are unknown and there is no other source of dependent information existent, a conservative fusion result can be achieved by means of the *ellipsoidal intersection* algorithm that has been derived in [158, 160] and evaluated in [159]. The idea is to remove the "maximum" possible common information from the fusion results, given by the information vector $\bar{\underline{y}}$ and matrix $\bar{\mathbf{Y}}$. Hence, each local estimate

$$\hat{\underline{y}}_A^{\mathrm{e}} = \hat{\underline{y}}_A^{\mathrm{I}} + \bar{\underline{y}} \quad \text{and} \quad \mathbf{Y}_A^{\mathrm{e}} = \mathbf{Y}_A^{\mathrm{I}} + \bar{\mathbf{Y}} \tag{3.54}$$

and

$$\hat{\underline{y}}_B^{\mathrm{e}} = \hat{\underline{y}}_B^{\mathrm{I}} + \bar{\underline{y}} \quad \text{and} \quad \mathbf{Y}_B^{\mathrm{e}} = \mathbf{Y}_B^{\mathrm{I}} + \bar{\mathbf{Y}} \tag{3.55}$$

is composed of an independent information vector denoted by the superscript I and the common information vector $\bar{\underline{y}}$. In order to prevent an overconfident estimate, the removal of the information vector $\bar{\underline{y}}$ to be determined cannot provide a better fusion result than for any true but unknown $\hat{\underline{y}}_{A\cap B}$, which implies that, for the error covariance matrix and the true but unknown $\mathbf{Y}_{A\cap B}$,

$$\mathbf{C}_{\mathrm{fus}} = \left(\mathbf{Y}_A^{\mathrm{e}} + \mathbf{Y}_B^{\mathrm{e}} - \mathbf{Y}_{A\cap B}\right)^{-1} \leq \left(\mathbf{Y}_A^{\mathrm{e}} + \mathbf{Y}_B^{\mathrm{e}} - \bar{\mathbf{Y}}\right)^{-1} \tag{3.56}$$

must hold, where the inequality denotes the Löwner partial order [13]

$$\mathbf{C}^1 \leq \mathbf{C}^2 \iff \underline{z}^{\mathrm{T}} \mathbf{C}^1 \underline{z} \leq \underline{z}^{\mathrm{T}} \mathbf{C}^2 \underline{z}$$

for every nonzero $\underline{z}$ such that $\underline{z}^{\mathrm{T}}\underline{z} = 1$, i.e., $\mathbf{C}^2 - \mathbf{C}^1$ is positive semidefinite. This means we aspire to compute a *conservative* estimate $\hat{\underline{x}}_{\mathrm{fus}}$ with covariance matrix $\mathbf{C}_{\mathrm{fus}}$ that preserves covariance consistency

$$\mathbf{C}_{\mathrm{fus}} \geq \widetilde{\mathbf{C}}_{\mathrm{fus}} , \tag{3.57}$$

where $\widetilde{\mathbf{C}}_{\mathrm{fus}}$ is the actual MSE matrix $\mathrm{E}[(\hat{\underline{x}}_{\mathrm{fus}} - \underline{x})(\hat{\underline{x}}_{\mathrm{fus}} - \underline{x})^{\mathrm{T}}]$. From (3.56), we can deduce the inequality

$$\mathbf{Y}_A^{\mathrm{e}} + \mathbf{Y}_B^{\mathrm{e}} - \mathbf{Y}_{A\cap B} \geq \mathbf{Y}_A^{\mathrm{e}} + \mathbf{Y}_B^{\mathrm{e}} - \bar{\mathbf{Y}}$$

$$\iff \qquad -\mathbf{Y}_{A\cap B} \geq -\bar{\mathbf{Y}}$$

$$\iff \qquad \mathbf{Y}_{A\cap B} \leq \bar{\mathbf{Y}} . \tag{3.58}$$

Since $\mathbf{Y}_A^{\mathrm{e}}$ and $\mathbf{Y}_B^{\mathrm{e}}$ also comprise independent information, we additionally have

$$\mathbf{Y}_A^{\mathrm{e}} \geq \bar{\mathbf{Y}} \quad \text{and} \quad \mathbf{Y}_B^{\mathrm{e}} \geq \bar{\mathbf{Y}} \,. \tag{3.59}$$

In the state space, the common information corresponds to a state estimate $\bar{x} = \bar{\mathbf{C}}\,\bar{y}$ with covariance matrix $\bar{\mathbf{C}} = \bar{\mathbf{Y}}^{-1}$. Relations (3.58) and (3.59) then imply

$$\mathbf{C}_A^{\mathrm{e}} \leq \bar{\mathbf{C}}, \quad \mathbf{C}_B^{\mathrm{e}} \leq \bar{\mathbf{C}}, \quad \text{and} \quad \mathbf{C}_{A \cap B} \geq \bar{\mathbf{C}}$$

for the covariance matrices. By considering the corresponding covariance ellipsoids, $\bar{\mathbf{C}}$ can now be determined as the shape matrix of the Löwner-John ellipsoid $\mathcal{E}(\underline{0}, \bar{\mathbf{C}})$ that is the smallest ellipsoid containing $\mathcal{E}(\underline{0}, \mathbf{C}_A^{\mathrm{e}})$ and $\mathcal{E}(\underline{0}, \mathbf{C}_B^{\mathrm{e}})$. The required formulas for the Löwner-John ellipsoid are explained in [158, Sec. 4.4].

With the derived "maximum" common information, the Bar-Shalom/Campo formulas can be employed: In state space, the local estimates are given by

$$\hat{\underline{x}}_A^{\mathrm{e}} = \mathbf{C}_A\big((\mathbf{C}_A^{\mathrm{I}})^{-1}\hat{\underline{x}}_A^{\mathrm{I}} + \bar{\mathbf{C}}^{-1}\bar{x}\big) \quad \text{with} \quad \mathbf{C}_A^{\mathrm{e}} = \big((\mathbf{C}_A^{\mathrm{I}})^{-1} + \bar{\mathbf{C}}^{-1}\big)^{-1}$$

and

$$\hat{\underline{x}}_B^{\mathrm{e}} = \mathbf{C}_B\big((\mathbf{C}_B^{\mathrm{I}})^{-1}\hat{\underline{x}}_B^{\mathrm{I}} + \bar{\mathbf{C}}^{-1}\bar{x}\big) \quad \text{with} \quad \mathbf{C}_B^{\mathrm{e}} = \big((\mathbf{C}_B^{\mathrm{I}})^{-1} + \bar{\mathbf{C}}^{-1}\big)^{-1} \,,$$

for which the cross-covariance matrix yields

$$\begin{aligned}
\mathbf{C}_{AB} &= \mathrm{E}\left[(\hat{\underline{x}}_A - \underline{x})(\hat{\underline{x}}_B - \underline{x})^{\mathrm{T}}\right] \\
&= \mathrm{E}\left[\mathbf{C}_A(\bar{\mathbf{C}}^{-1}\bar{x} - \bar{\mathbf{C}}^{-1}\underline{x})(\bar{\mathbf{C}}^{-1}\bar{x} - \bar{\mathbf{C}}^{-1}\underline{x})^{\mathrm{T}}\mathbf{C}_B^{\mathrm{T}}\right] \\
&= \mathbf{C}_A\bar{\mathbf{C}}^{-1}\mathbf{C}_B \,.
\end{aligned} \tag{3.60}$$

With this matrix, the Bar-Shalom/Campo formulas (3.48) and (3.49) can be utilized to fuse the local estimates $\hat{\underline{x}}_A^{\mathrm{e}}$ and $\hat{\underline{x}}_B^{\mathrm{e}}$. In particular, (3.49) gives the right-hand side of inequality (3.56). The advantage of this state space fusion is that $\bar{x}$, $\hat{\underline{x}}_A^{\mathrm{I}}$, and $\hat{\underline{x}}_B^{\mathrm{I}}$ do not need to be explicitly computed. It is important to note that the ellipsoidal intersection algorithm is prone to numerical instabilities [158], which have to be taken care of by slightly modified formulas.

The previous considerations have shown that the ellipsoidal intersection algorithm can account for common information, but it remains an open question how to fuse estimates when the sources of possible dependencies are unknown and also common process noise must be taken into consideration.

B Incorporation of Additional Unknown But Bounded Errors

In case of the channel filter, the fusion formulas (3.52) and (3.53) have to be generalized for the purpose of incorporating additional set-membership errors. More specifically, (3.52) becomes a Minkowski sum of ellipsoidal sets of information vectors according to

$$\mathcal{E}(\hat{\underline{y}}_{\text{fus}}, \mathbf{Q}_{\text{fus}}) = \mathcal{E}(\hat{\underline{y}}_A, \mathbf{Q}_A) \oplus \mathcal{E}(\hat{\underline{y}}_B, \mathbf{Q}_B) \oplus \mathcal{E}(-\hat{\underline{y}}_{A \cap B}, \mathbf{Q}_{A \cap B}) \, ,$$

where the formulas for outer approximations of multiple ellipsoids in Section 3.4 can be employed.

In case of unknown common information (and no other dependent information), the fusion can be conducted by means of the gain (3.50), which has been derived for the Bar-Shalom/Campo rule in the presence of set-membership uncertainties. The required cross-covariance matrix $\mathbf{C}_{AB}$ is given by (3.60). In doing so, we implicitly assume that no set-membership uncertainties are associated to the common information because it has no effect on the fusion result whether the unknown but bounded errors belong to the independent or dependent parts of the local estimates (3.54) and (3.55).

3.5.3 Covariance Intersection and Bounds

The most popular strategy guaranteeing consistent and conservative fusion results is discussed in this subsection. Irrespective of the sources of dependencies, reliable estimation results are provided, and hence no specific knowledge about the underlying network topology is required. However, these advantages are payed with less informative estimates.

Let $\hat{\underline{x}}_A$ and $\hat{\underline{x}}_B$ denote two local estimates on a state $\underline{x}_k$ with error covariance matrices $\mathbf{C}_A$ and $\mathbf{C}_B$, and we assume that no further set-membership uncertainties are present. In the situation that the cross-covariance matrix $\mathbf{C}_{AB} = \mathrm{E}[(\hat{\underline{x}}_A - \underline{x}_k)(\hat{\underline{x}}_B - \underline{x}_k)^\mathrm{T}]$ is unknown and the Bar-Shalom/Campo formulas (3.48) and (3.49) cannot be deployed, the *covariance intersection* (CI) algorithm [96, 99, 167] enables us to compute a conservative estimate $\hat{\underline{x}}_{\mathrm{CI}}$ with covariance matrix $\mathbf{C}_{\mathrm{CI}}$ that preserves the inequality (3.57) with

$$\mathbf{C}_{\mathrm{CI}} \geq \tilde{\mathbf{C}}_{\mathrm{CI}}$$

and $\tilde{\mathbf{C}}_{\mathrm{CI}}$ being the true MSE matrix $\mathrm{E}[(\hat{\underline{x}}_{\mathrm{CI}} - \underline{x})(\hat{\underline{x}}_{\mathrm{CI}} - \underline{x})^\mathrm{T}]$. Interestingly, we are already familiar with the combination rule exploited by the CI algorithm, which yields the estimate

$$\hat{\underline{x}}_{\mathrm{CI}} = \mathbf{C}_{\mathrm{CI}}\big(\omega \mathbf{C}_A^{-1}\hat{\underline{x}}_A + (1-\omega)\mathbf{C}_B^{-1}\hat{\underline{x}}_B\big) \tag{3.61}$$

and error covariance matrix

$$\mathbf{C}_{\mathrm{CI}} = \big(\omega \mathbf{C}_A^{-1} + (1-\omega)\mathbf{C}_B^{-1}\big)^{-1} \tag{3.62}$$

with weighting factor $\omega \in [0, 1]$. More precisely, this formula matches the centered intersection (2.92) for sets. The CI algorithm yields covariance consistent estimates, i.e.,

$$\mathbf{C}_{\mathrm{CI}} \geq \mathrm{E}\left[(\hat{\underline{x}}_{\mathrm{CI}} - \underline{x}_k)(\hat{\underline{x}}_{\mathrm{CI}} - \underline{x}_k)^\mathrm{T}\right], \tag{3.63}$$

irrespective of the actual cross-covariance matrix $\mathbf{C}_{AB}$ and choice of ω, provided that $(\hat{\underline{x}}_A, \mathbf{C}_A)$ and $(\hat{\underline{x}}_B, \mathbf{C}_B)$ are consistent estimates. Figure 3.11 illustrates the CI fusion result of two estimates depicted by their corresponding covariance ellipsoids, with ω being determined to minimize the determinant of (3.62) and thus to minimize the volume of the CI covariance ellipsoid. Also, some results of the Bar-Shalom/Campo formulas are depicted, where the cross-covariance matrices $\mathbf{C}_{AB}$ have been chosen arbitrarily.

In the considered situation of unknown correlations, it is also possible to directly derive a Kalman gain that provides a conservative fusion result and concurrently minimizes the bound (3.63) on the MSE matrix [38]. Similarly to (2.91), determining such a gain $\mathbf{K} \in \mathbb{R}^{n_x \times n_x}$ is an $(n_x)^2$-dimensional optimization problem that fortunately can be reduced to a one-dimensional one. More precisely, it is equivalent to minimizing the trace of (3.62). Hence,

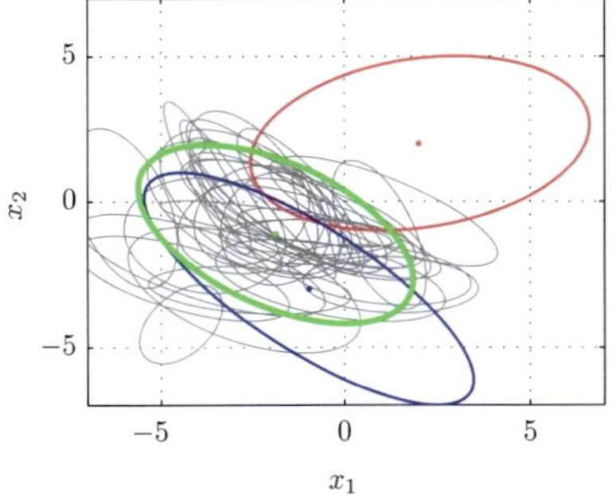

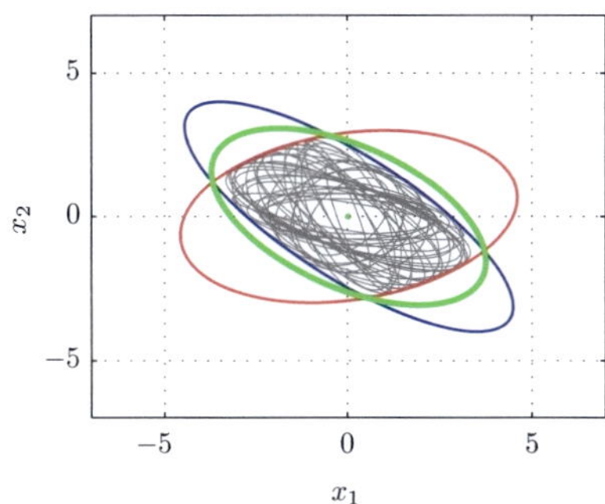

(a) Covariance intersection is applied to the red and blue estimate. The green ellipsoid represents a conservative estimate.

(b) Every ellipsoid has been shifted to the origin. The gray fusion results for arbitrarily chosen cross-covariance matrices lie in the intersection that is bounded by the CI result.

Figure 3.11: In Fig. (a), covariance intersection applied to the red and blue estimate yields the green ellipsoid, where the determinant has been minimized. The gray ellipsoids represent Bar-Shalom/Campo fusion results for some arbitrarily chosen cross-covariance matrices $\mathbf{C}_{AB}$. In Fig. (b), all covariance ellipsoids are centered around origin.

the trace-optimal covariance matrix in the set of all possible conservative fusion results lies in the set of CI fusion results.

A Covariance Bounds

The CI algorithm can alternatively be formulated in terms of the joint error covariance matrix: Before fusing two estimates $\hat{\underline{x}}_A$ and $\hat{\underline{x}}_B$, the *covariance bounds* (CB) algorithm [74, 75]—also called *covariance inflation* [148]—computes an upper bound for the joint covariance matrix, i.e.,

$$\begin{bmatrix} \frac{1}{\omega}\mathbf{C}_A & \mathbf{0} \\ \mathbf{0} & \frac{1}{1-\omega}\mathbf{C}_B \end{bmatrix} \geq \begin{bmatrix} \mathbf{C}_A & \mathbf{C}_{AB} \\ \mathbf{C}_{BA} & \mathbf{C}_B \end{bmatrix} \tag{3.64}$$

with $\omega \in (0,1)$. This bound holds for every possible cross-covariance matrix $\mathbf{C}_{AB}$. Apparently, by employing the bound as the current joint MSE matrix, the estimates can be considered to be uncorrelated, and the fusion rule for uncorrelated estimates, which corresponds to the Kalman filter formula (2.26), can be utilized to calculate the estimated covariance matrix

$$\mathbf{C}_{\mathrm{CB}} = \tfrac{1}{\omega}\mathbf{C}_A - \left(\tfrac{1}{\omega}\mathbf{C}_A\right)\left(\tfrac{1}{\omega}\mathbf{C}_A + \tfrac{1}{1-\omega}\mathbf{C}_B\right)^{-1}\left(\tfrac{1}{\omega}\mathbf{C}_B\right)$$
$$= \left(\omega\mathbf{C}_A^{-1} + (1-\omega)\mathbf{C}_B^{-1}\right)^{-1},$$

where the last equation is a result of the Woodbury matrix identity [177]. Accordingly, the estimate $\hat{\underline{x}}_{\mathrm{CB}}$ is identical to (3.61). The joint state space of estimates encompasses a very useful representation to identify and conservatively bound unknown correlations.

B   Comments and Analogies

The CI algorithm has extensively been studied and further developed by analyzing the effect of the weighting parameter, formulating it in terms of the joint state space, or extending it to the instantaneous fusion of multiple estimates. As already implied by its name, a particularly interesting aspect is also its relation to set-membership state estimation.

Relation to Set-membership Estimation We have already become acquainted with equations (3.61) and (3.62) in Section 2.4.2-C, which occur in case of vanishing stochastic uncertainties and represent the centralized intersection of the set-membership uncertainty descriptions. Evidently, it is the set-membership fusion methodology that is employed by the CI algorithm and gives its name to it. The intuition behind this connection is that cross-correlations introduce a systematic behavior to the estimation error. In case of independence, the MSE is reduced into every direction, but, in case of strong correlations, high errors of one estimate may be interconnected with high errors of the other estimate, and the MSE cannot be reduced in some directions. In Figure 3.11(b), some gray ellipsoids therefore touch the boundaries of the intersection. This becomes most conspicuous when the errors even have the same covariance matrices and are fully correlated, i.e., $\mathbf{C}_A = \mathbf{C}_B = \mathbf{C}_{AB}$. Then, no update takes place, which corresponds to the same result as a set-membership estimator would provide for identical sets. Due to full correlation, one estimate conditioned on the other is purely deterministic, and no random zero-mean behavior can be exploited to reduce the estimation error. So, a lack of independence implies an unknown systematic behavior that is to be bounded, preferably by means of set-membership methods.

Fusion of Multiple Estimates The CI algorithm was designed to fuse two estimates but can easily be extended to the combination of multiple estimates at once. For N local estimates $(\underline{\hat{x}}_1, \mathbf{C}_1), \ldots, (\underline{\hat{x}}_N, \mathbf{C}_N)$ to be fused, the convex combinations

$$\underline{\hat{x}}_{\mathrm{CI}} = \mathbf{C}_{\mathrm{CI}} \left(\omega_1 \mathbf{C}_1^{-1} \underline{\hat{x}}_1 + \ldots + \omega_N \mathbf{C}_N^{-1} \underline{\hat{x}}_N \right)$$

for the mean and

$$\mathbf{C}_{\mathrm{CI}}^{-1} = \omega_1 \mathbf{C}_1^{-1} + \ldots + \omega_N \mathbf{C}_N^{-1}$$

for the conservative error covariance matrix can be employed with $\omega_1 + \ldots + \omega_N = 1$ and $\omega_i \geq 0$. Interestingly, CI is a convex combination of information vectors and information matrices. Minimizing the trace or determinant of $\mathbf{C}_{\mathrm{CI}}$ requires to solve an $(N - 1)$-dimensional optimization problem but can yield a less conservative fusion result than fusing the estimates sequentially by means of (3.61) and (3.62).

Optimal Weighting Parameter Criteria for determining ω have gained considerable attention. In general, ω is determined in such way that the determinant or trace of $\mathbf{C}_{\mathrm{CI}}$ becomes minimal. As stated earlier, the trace is a reasonable choice due to its relation to the MSE [38], but it requires a numerical optimization. Therefore, approximate closed-form solutions have been proposed that also consider the traces or determinants of the matrices [59, 136] or employ an information-theoretic measure like the Kullback-Leibler distance [170].

Especially when using the trace or determinant, one of the estimates to be fused will be rejected if $\mathbf{C}_A - \mathbf{C}_B$ or vice versa is positive definite, i.e., if the covariance ellipsoid $\mathcal{E}(\underline{0}, \mathbf{C}_A)$ contains $\mathcal{E}(\underline{0}, \mathbf{C}_B)$. This means, more precisely, that the fusion of $(\hat{\underline{x}}_A, \mathbf{C}_A)$ and $(\hat{\underline{x}}_B, \mathbf{C}_B)$ yields the estimate with the smaller covariance matrix. In one-dimensional setups, this happens at every fusion step, which is, of course, undesirable, because a node would place greater trust in a single possible outlier with small variance than in many sources that report estimates with high variances. In this regard, information-theoretic [84, 170] and set-theoretic [41, 171] optimization criteria for ω have been proposed, where the choice of ω not only depends on the error covariance matrices but also on the means $\hat{\underline{x}}_A$ and $\hat{\underline{x}}_B$. The concept proposed in [206] follows the pattern of Section 2.3, and the unknown correlation coefficient is regarded to parameterize a set of Gaussian densities. Each of these densities can be considered to be conditioned on the correlation parameter. By modeling the correlation coefficient to be uniformly distributed, it can be marginalized out, and a single density remains. The result is then employed as an optimization criterion for the CI algorithm, providing a closed-form solution for ω. Also, an extension to multidimensional state spaces has been proposed in [206].

C Covariance Intersection in the Presence of Unknown But Bounded Perturbations

In Chapter 2, we have pointed out several possibilities to incorporate set-membership error descriptions. Two estimates $\hat{\underline{x}}_A$ and $\hat{\underline{x}}_B$ with unknown cross-covariance matrix $\mathbf{C}_{AB}$ can be fused by means of the CI formulas (3.61) and (3.62), and, in the additional presence of unknown but bounded errors being characterized by the shape matrices $\mathbf{X}_A$ and $\mathbf{X}_B$, the simplest way is to use the gain that is related to the CI algorithm, namely

$$\mathbf{K}_{\mathrm{CI}} = \frac{1}{\omega} \mathbf{C}_A \mathbf{C}_{\mathrm{CI}}^{-1} \quad \text{and} \quad \mathbf{I} - \mathbf{K}_{\mathrm{CI}} = \frac{1}{1-\omega} \mathbf{C}_B \mathbf{C}_{\mathrm{CI}}^{-1},$$

in order to compute the updated shape matrix (2.74). These gains can easily be derived[6] from the bounding covariance matrix (3.64). With them, it can be ensured that the actual MSE is not underestimated in order to

---

[6]Often, the inverse covariance forms $\mathbf{K}_{\mathrm{CI}} = \omega \mathbf{C}_{\mathrm{CI}} \mathbf{C}_A^{-1}$ and $\mathbf{I} - \mathbf{K}_{\mathrm{CI}} = (1-\omega) \mathbf{C}_B \mathbf{C}_{\mathrm{CI}}^{-1}$ are considered.

take possible cross-correlations between the local (stochastic) errors into account, and an additional set-membership error is properly bounded.

However, as it is done in Section 2.4.1, the gain $\mathbf{K}_{\mathrm{CI}}$ plainly ignores the impact of the set-membership errors when minimizing the MSE. So, in order to compute a gain that also incorporates the shape matrices similarly to Section 2.4.2, the overall MSE needs to be considered. The investigations in [38] have revealed that for any pair of gains $\mathbf{K}_A$ and $\mathbf{K}_B$, not necessarily fulfilling the unbiased constraint, a conservative bound on the covariance matrix can be achieved by setting

$$\mathbf{C}_{\mathrm{fus}} = (1 + \gamma^{-1})\mathbf{K}_A \mathbf{C}_A \mathbf{K}_A^{\mathrm{T}} + (1 + \gamma)\mathbf{K}_B \mathbf{C}_B \mathbf{K}_B^{\mathrm{T}} \tag{3.65}$$

with $\gamma > 0$. According to inequality (2.90) and with (3.65), the total MSE can be bounded from above by virtue of

$$\begin{aligned}
\overline{\mathrm{E}}\left[(\hat{\underline{x}}_k^{\mathrm{e}} - \underline{x})(\hat{\underline{x}}_k^{\mathrm{e}} - \underline{x})\right] &= (1 + \gamma^{-1})\,\mathrm{trace}\left(\mathbf{K}_A \mathbf{C}_A \mathbf{K}_A^{\mathrm{T}}\right) \\
&\quad + (1 + \gamma)\,\mathrm{trace}\left(\mathbf{K}_B \mathbf{C}_B \mathbf{K}_B^{\mathrm{T}}\right) \\
&\quad + (1 + p^{-1})\,\mathrm{trace}\left(\mathbf{K}_A \mathbf{X}_A \mathbf{K}_A^{\mathrm{T}}\right) \\
&\quad + (1 + p)\,\mathrm{trace}\left(\mathbf{K}_B \mathbf{X}_B \mathbf{K}_B^{\mathrm{T}}\right).
\end{aligned}$$

By requiring unbiasedness and setting $\gamma = p$, the gain

$$\mathbf{K} = \left((1 + p^{-1})(\mathbf{X}_A + \mathbf{C}_A)\right)\left((1 + p^{-1})(\mathbf{X}_A + \mathbf{C}_A) + (1 + p)(\mathbf{X}_B + \mathbf{C}_B)\right)^{-1}$$

can be derived, which is a straightforward generalization of the Kalman gain (2.91). The simplification $\gamma = p$ traces back to Theorem 2 in [38], where the calculation of a trace-optimal Kalman gain is reduced to a one-dimensional optimization problem. As expected, this result clearly confirms that CI treats the covariance matrix like a set-membership error matrix.

3.6   Decentralized Estimation with Additional Knowledge

Fully decentralized state estimation techniques are of particular interest in the field of sensor networks, simultaneous localization and mapping (SLAM), and cooperative multi-robot systems. If not taken into account properly, complex dependency structures between local estimates can cause inconsistent estimation results. Covariance intersection and bounds always maintain consistency by providing conservative fusion results, with the consequence of possibly much less informative estimates. On this account, this section names techniques that allow us to explicitly identify and exploit independencies in order to achieve estimates with tighter bounds on the errors. According to previous considerations, set-membership uncertainties do not need to be considered. Their incorporation into the presented methods can be accomplished by means of the results of Chapter 2, Section 3.4, and the previous section.

3.6.1 Automatic Exploitation of Independencies

The concept proposed in [196] is designed to exploit the conditional independence of measurements, especially against the background of SLAM applications, where the total state vector

$$\underline{x}_k = \left[(\underline{x}_k^1)^T, \ldots, (\underline{x}_k^N)^T\right]^T$$

consists of a large number of substates $\underline{x}_k^i$ that represent the position and attributes of landmarks or mobile robots. Of course, each substate can also be a representative of a single state that is observed by N different nodes. The techniques from [196] are predestined for estimation tasks, where two substates $\underline{x}_k^i$ and $\underline{x}_k^j$ are related to each other by a measurement $\hat{\underline{z}}_k$ that is a realization of

$$\underline{z}_k = \begin{bmatrix} \mathbf{H}_k^i & \mathbf{H}_k^j \end{bmatrix} \begin{bmatrix} \underline{x}_k^i \\ \underline{x}_k^j \end{bmatrix} + \underline{v}_k \ .$$

This is, for instance, the case whenever distances between nodes have to be measured or when the state can only be observed in a cooperative fashion. Since $\underline{v}_k$ is independent from $\underline{x}_k$, only the correlations of the substates need to be bounded. For this purpose, we employ the covariance bounds algorithm from Section 3.5.3-A and replace $\mathbf{C}_k^{\mathrm{p}}$ by the upper bound (3.64). The filtering step is then carried out by means of

$$
\begin{bmatrix} \hat{\underline{x}}_k^{\mathrm{e},i} \\ \hat{\underline{x}}_k^{\mathrm{e},j} \end{bmatrix} = \begin{bmatrix} \hat{\underline{x}}_k^{\mathrm{p},i} \\ \hat{\underline{x}}_k^{\mathrm{p},j} \end{bmatrix} + \begin{bmatrix} \frac{1}{\omega}\mathbf{C}_k^{\mathrm{p},i} & \mathbf{0} \\ \mathbf{0} & \frac{1}{1-\omega}\mathbf{C}_k^{\mathrm{p},j} \end{bmatrix} \begin{bmatrix} \left(\mathbf{H}_k^i\right)^T \\ \left(\mathbf{H}_k^j\right)^T \end{bmatrix} \cdot
$$

$$
\left(\mathbf{C}_k^z + \begin{bmatrix} \mathbf{H}_k^i & \mathbf{H}_k^j \end{bmatrix} \begin{bmatrix} \frac{1}{\omega}\mathbf{C}_k^{\mathrm{p},i} & \mathbf{0} \\ \mathbf{0} & \frac{1}{1-\omega}\mathbf{C}_k^{\mathrm{p},j} \end{bmatrix} \begin{bmatrix} \left(\mathbf{H}_k^i\right)^T \\ \left(\mathbf{H}_k^j\right)^T \end{bmatrix} \right)^{-1}
$$

$$
\cdot \left(\hat{\underline{z}}_k - \begin{bmatrix} \mathbf{H}_k^i & \mathbf{H}_k^j \end{bmatrix} \begin{bmatrix} \hat{\underline{x}}_k^{\mathrm{p},i} \\ \hat{\underline{x}}_k^{\mathrm{p},j} \end{bmatrix} \right)
$$

for the estimate and

$$
\begin{bmatrix} \mathbf{C}_k^{\mathrm{e},i} & \mathbf{C}_k^{\mathrm{e},i,j} \\ \mathbf{C}_k^{\mathrm{e},j,i} & \mathbf{C}_k^{\mathrm{e},j} \end{bmatrix} = \begin{bmatrix} \frac{1}{\omega}\mathbf{C}_k^{\mathrm{p},i} & \mathbf{0} \\ \mathbf{0} & \frac{1}{1-\omega}\mathbf{C}_k^{\mathrm{p},j} \end{bmatrix} - \begin{bmatrix} \frac{1}{\omega}\mathbf{C}_k^{\mathrm{p},i} & \mathbf{0} \\ \mathbf{0} & \frac{1}{1-\omega}\mathbf{C}_k^{\mathrm{p},j} \end{bmatrix} \begin{bmatrix} \left(\mathbf{H}_k^i\right)^T \\ \left(\mathbf{H}_k^j\right)^T \end{bmatrix}
$$

$$
\left(\mathbf{C}_k^z + \begin{bmatrix} \mathbf{H}_k^i & \mathbf{H}_k^j \end{bmatrix} \begin{bmatrix} \frac{1}{\omega}\mathbf{C}_k^{\mathrm{p},i} & \mathbf{0} \\ \mathbf{0} & \frac{1}{1-\omega}\mathbf{C}_k^{\mathrm{p},j} \end{bmatrix} \begin{bmatrix} \left(\mathbf{H}_k^i\right)^T \\ \left(\mathbf{H}_k^j\right)^T \end{bmatrix} \right)^{-1}
$$

$$
\cdot \begin{bmatrix} \mathbf{H}_k^i & \mathbf{H}_k^j \end{bmatrix} \begin{bmatrix} \frac{1}{\omega}\mathbf{C}_k^{\mathrm{p},i} & \mathbf{0} \\ \mathbf{0} & \frac{1}{1-\omega}\mathbf{C}_k^{\mathrm{p},j} \end{bmatrix}
$$

for the joint error covariance matrix, where the obtained cross-covariance matrices $\mathbf{C}_k^{\mathrm{e},i,j} = (\mathbf{C}_k^{\mathrm{e},j,i})^{\mathrm{T}}$ are usually discarded. The local estimate of substate $\underline{x}^i$ is then given by

$$
\hat{\underline{x}}_k^{\mathrm{e},i} = \hat{\underline{x}}_k^{\mathrm{p},i} + \frac{1}{\omega}\mathbf{C}_k^{\mathrm{p},i}\left(\mathbf{H}_k^i\right)^T \cdot \left(\mathbf{C}_k^v + \mathbf{H}_k^i\left(\frac{1}{\omega}\mathbf{C}_k^{\mathrm{p},i}\right)\left(\mathbf{H}_k^i\right)^T + \right.
$$

$$
\left. \mathbf{H}_k^j\left(\frac{1}{1-\omega}\mathbf{C}_k^{\mathrm{p},j}\right)\left(\mathbf{H}_k^j\right)^T \right)^{-1} \left(\hat{\underline{z}}_k - \mathbf{H}_k^i\hat{\underline{x}}_k^{\mathrm{p},i} - \mathbf{H}_k^j\hat{\underline{x}}_k^{\mathrm{p},j} \right)
$$

with error covariance matrix

$$
\mathbf{C}_k^{\mathrm{e},i} = \frac{1}{\omega}\mathbf{C}_k^{\mathrm{p},i} - \frac{1}{\omega}\mathbf{C}_k^{\mathrm{p},i}\left(\mathbf{H}_k^i\right)^T \cdot
$$

$$
\left(\mathbf{C}_k^v + \mathbf{H}_k^i\left(\frac{1}{\omega}\mathbf{C}_k^{\mathrm{p},i}\right)\left(\mathbf{H}_k^i\right)^T + \mathbf{H}_k^j\left(\frac{1}{1-\omega}\mathbf{C}_k^{\mathrm{p},j}\right)\left(\mathbf{H}_k^j\right)^T\right)^{-1}\mathbf{H}_k^i\frac{1}{\omega}\mathbf{C}_k^{\mathrm{p},i},
$$

and $\underline{\hat{x}}_k^{\mathrm{e},j}$ and $\mathbf{C}_k^{\mathrm{e},j}$ are computed analogously. These formulas can be also employed to directly fuse two estimates and to exploit additional independent measurement knowledge. The following concluding example illustrates the benefits from exploiting independent information.

Example 3.4: Localization by exploiting independent measurement data
We consider the example from [196], where a robot computes an estimate on its own position and the positions of four landmarks, which are the corners of a wall. We consider a simplified measurement model that is a linearization of the Euclidean distances around the true values, which are, of course, in practical applications unknown. With the idealized model, we intend to avoid additional effects from linearization errors. The robot performs several measurements and can achieve the optimal, centralized result depicted in Figure 3.12(a), when cross-correlation are known. Ignoring dependencies, as shown in Figure 3.12(b), results into biased estimates, which can especially be seen by the green and the right magenta ellipse. Covariance intersection astonishingly does not provide any improvements in Figure 3.12(c) because only measurements between substates take place, and CI provides also a bound for the worst case, i.e., full correlation, which does not permit new insights to be gained. By exploiting the knowledge that the measurement errors are independent, the result in Figure 3.12(d) is obtained, which is impressively close to the centralized solution.

3.6.2 Split Covariance Intersection

While in the previous subsection covariance bounds provide the means to exploit independent information, the covariance intersection algorithm can

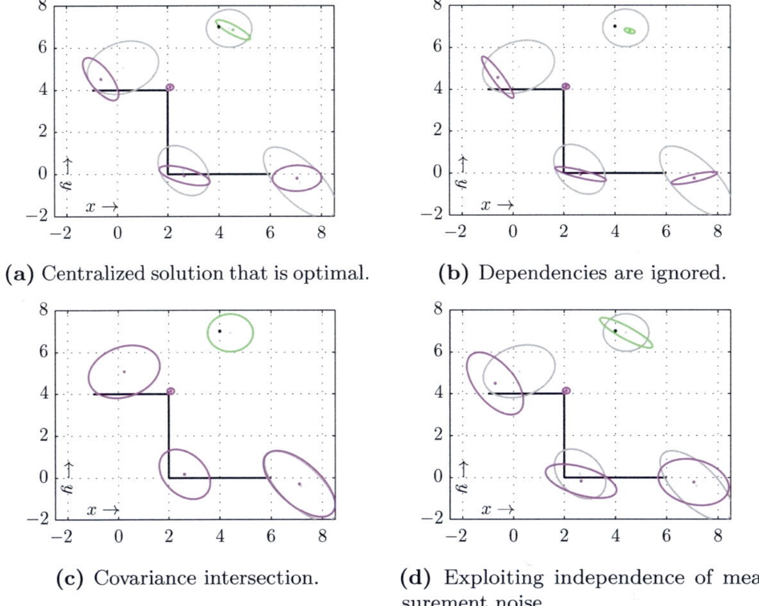

(a) Centralized solution that is optimal.

(b) Dependencies are ignored.

(c) Covariance intersection.

(d) Exploiting independence of measurement noise.

Figure 3.12: The stationary robot in the north measures distances to the corners of a wall [196]. The state vectors is composed of the robot's position and the corner positions as landmarks. The initial uncertainties are drawn gray.

similarly be generalized to *split covariance intersection* (split CI) [92, 94, 98]. In the case that the errors are known to be partially independent, a fusion result with tighter error covariance matrix can be accomplished. We assume that the error covariance matrices of sensors A and B can be written as

$$\mathbf{C}_A = \mathbf{C}_A^{\mathrm{I}} + \mathbf{C}_A^{\mathrm{D}}$$

and

$$\mathbf{C}_B = \mathbf{C}_B^{\mathrm{I}} + \mathbf{C}_B^{\mathrm{D}} \,,$$

where the superscript I, in each case, labels the independent part. In order to apply split CI, it is not necessary to also split the according estimates

$\hat{\underline{x}}_A$ and $\hat{\underline{x}}_B$ into interdependent and independent parts. To simplify the presentation, we consider the information form of split CI, which yields the mean

$$\mathbf{C}_{\mathrm{sCI}}^{-1}\,\hat{\underline{x}}_{\mathrm{sCI}} = \omega\left(\omega\mathbf{C}_A^{\mathrm{I}} + \mathbf{C}_A^{\mathrm{D}}\right)^{-1}\hat{\underline{x}}_A + (1-\omega)\left((1-\omega)\mathbf{C}_B^{\mathrm{I}} + \mathbf{C}_B^{\mathrm{D}}\right)^{-1}\hat{\underline{x}}_B$$

and error covariance matrix

$$\mathbf{C}_{\mathrm{sCI}}^{-1} = \omega\left(\omega\mathbf{C}_A^{\mathrm{I}} + \mathbf{C}_A^{\mathrm{D}}\right)^{-1} + (1-\omega)\left((1-\omega)\mathbf{C}_B^{\mathrm{I}} + \mathbf{C}_B^{\mathrm{D}}\right)^{-1}$$

$$= \underbrace{\left(\mathbf{C}_A^{\mathrm{I}} + \frac{1}{\omega}\mathbf{C}_A^{\mathrm{D}}\right)^{-1}}_{=:(\mathbf{C}_A^*)^{-1}} + \underbrace{\left(\mathbf{C}_B^{\mathrm{I}} + \frac{1}{1-\omega}\mathbf{C}_B^{\mathrm{D}}\right)}_{=:(\mathbf{C}_B^*)^{-1}},$$

where $\omega \in [0,1]$ is chosen to fulfill a predefined optimality criterion. With these formulas, we can also implement the approach from the previous Subsection 3.6.1, although we here have to explicitly distinguish between dependent and independent data.

At this point, one may be inclined to presume that again a strong relationship between split CI and the concepts from Chapter 2 can be spotted. Indeed, with $\mathbf{C}_A^*$ and $\mathbf{C}_B^*$ defined above and by setting $1+p^{-1} = \frac{1}{\omega}$, the Kalman gain

$$\mathbf{K}_k(p) = \mathbf{C}_A^* \cdot \left(\mathbf{C}_A^* + \mathbf{C}_B^*\right)^{-1}$$

$$= \left(\mathbf{C}_A^{\mathrm{I}} + (1+p^{-1})\mathbf{C}_A^{\mathrm{D}}\right)\left(\mathbf{C}_A^{\mathrm{I}} + (1+p^{-1})\mathbf{C}_A^{\mathrm{D}} + \mathbf{C}_B^{\mathrm{I}} + (1+p)\mathbf{C}_B^{\mathrm{D}}\right)^{-1}$$

can be derived, which equals the gain (2.91) for $\mathbf{H}_k = \mathbf{I}$, where the independent parts still represent stochastic error characteristics, but the dependent parts are now viewed as shape matrices for set-membership error bounds. In conclusion, the conservative treatment of dependent and independent information bears a strong resemblance to combined stochastic and set-membership state estimation in Chapter 2.

3.6.3 Federated Kalman Filtering

The federated Kalman filter [33–35] can be viewed as the predecessor of the optimal distributed Kalman filter of Section 3.3.3 and, as such, it requires less assumptions to be met. It moreover allows for a decentralized processing of estimates as the locally computed parameters represent valid state estimates. Each node must only be aware of the number of participating nodes in order to be able to determine the inflation parameter for the relaxed system model (3.22). Initializing each sensor node with the same prior, as in (3.39), results into a fully occupied joint covariance matrix (3.40). The federated Kalman filter instead employs the inflated joint covariance matrix

$$
\begin{bmatrix}
\frac{1}{\omega_1}\mathbf{C}_0^{\mathrm{P}} & \mathbf{0} & \cdots & \mathbf{0} \\
\mathbf{0} & \frac{1}{\omega_2}\mathbf{C}_0^{\mathrm{P}} & \ddots & \vdots \\
\vdots & \ddots & \ddots & \mathbf{0} \\
\mathbf{0} & \cdots & \mathbf{0} & \frac{1}{\omega_N}\mathbf{C}_0^{\mathrm{P}}
\end{bmatrix}
\geq
\begin{bmatrix}
\mathbf{C}_0^{\mathrm{P}} & \mathbf{C}_0^{\mathrm{P}} & \cdots & \mathbf{C}_0^{\mathrm{P}} \\
\mathbf{C}_0^{\mathrm{P}} & \mathbf{C}_0^{\mathrm{P}} & \ddots & \vdots \\
\vdots & \ddots & \ddots & \mathbf{C}_0^{\mathrm{P}} \\
\mathbf{C}_0^{\mathrm{P}} & \cdots & \mathbf{C}_0^{\mathrm{P}} & \mathbf{C}_0^{\mathrm{P}}
\end{bmatrix}
$$

with $\sum_{i=1}^{N}\omega_i = 1$ and $\omega_i \geq 0$, which implies that each sensor node is initialized with $(\hat{\underline{x}}_0^{\mathrm{P},i}, \mathbf{C}_0^{\mathrm{P},i}) := (\hat{\underline{x}}_0^{\mathrm{P}}, \frac{1}{\omega_i}\mathbf{C}_0^{\mathrm{P}})$, which is equal to the initialization (3.19) for $\omega_i = \frac{1}{N}$ and guarantees that the fusion of the local estimates still yields $(\hat{\underline{x}}_0^{\mathrm{P}}, \mathbf{C}_0^{\mathrm{P}})$. The prime aim of federated Kalman filtering is to eliminate the effects of common process noise. Employing a standard system model (3.1) would implicate a fully correlated process noise according to (3.43). For this reason, again an inflated version

$$
\begin{bmatrix}
\frac{1}{\omega_1}\mathbf{C}_k^{w} & \mathbf{0} & \cdots & \mathbf{0} \\
\mathbf{0} & \frac{1}{\omega_2}\mathbf{C}_k^{w} & \ddots & \vdots \\
\vdots & \ddots & \ddots & \mathbf{0} \\
\mathbf{0} & \cdots & \mathbf{0} & \frac{1}{\omega_N}\mathbf{C}_k^{w}
\end{bmatrix}
\geq
\begin{bmatrix}
\mathbf{C}_k^{w} & \mathbf{C}_k^{w} & \cdots & \mathbf{C}_k^{w} \\
\mathbf{C}_k^{w} & \mathbf{C}_k^{w} & \ddots & \vdots \\
\vdots & \ddots & \ddots & \mathbf{C}_k^{w} \\
\mathbf{C}_k^{w} & \cdots & \mathbf{C}_k^{w} & \mathbf{C}_k^{w}
\end{bmatrix}
\tag{3.66}
$$

of the common process noise matrix is used, which is directly related to the relaxed system model (3.22). After a prediction, the fusion still yields the optimal centralized result due to

$$
\left(\mathbf{A}_k\,\hat{\underline{x}}_0^{\mathrm{P},i}, \mathbf{A}_k\mathbf{C}_0^{\mathrm{P},i}\mathbf{A}_k^{\mathrm{T}} + \frac{1}{\omega_i}\mathbf{C}_k^{w}\right) = \left(\mathbf{A}_k\,\hat{\underline{x}}_0^{\mathrm{P}}, \frac{1}{\omega_i}(\mathbf{A}_k\mathbf{C}_0^{\mathrm{P}}\mathbf{A}_k^{\mathrm{T}} + \mathbf{C}_k^{w})\right),
$$

i.e., we can first predict the estimate and can then perform the initialization step. Unfortunately, local filtering steps prevent the final fusion result from being optimal in the MSE sense since they are carried out by means of the standard Kalman update step and no globalization, as in Section 3.3.3, takes place. However, the advantage is that local estimates are conservative estimates so that nodes may also fail and still a valid estimate is attainable. The fusion result provides a conservative estimate unless no more than N local estimates are combined and no common sensor information is present. Thus, for a fully decentralized processing, cycles in the network topology have to be avoided.

3.6.4 Delayed State Filtering

Delayed state information filters are widely used in simultaneous localization and mapping (SLAM) applications and cooperative tracking problems [5, 31, 32, 81]. The basic property that is exploited is the conditional independence of all measurements given the entire trajectory, i.e.,

$$f(\mathcal{Z}_{0:k} \,|\, \underline{x}_{0:k}) = f(\underline{\hat{z}}_k^1 \,|\, \underline{x}_{0:k}) \cdot \ldots \cdot f(\underline{\hat{z}}_k^N \,|\, \underline{x}_{0:k}) \cdot f(\underline{\hat{z}}_{k-1}^1 \,|\, \underline{x}_{0:k}) \cdot$$
$$\ldots \cdot f(\underline{\hat{z}}_0^1 \,|\, \underline{x}_{0:k}) \cdot \ldots \cdot f(\underline{\hat{z}}_0^N \,|\, \underline{x}_{0:k}) \,, \tag{3.67}$$

which requires to process and store the entire state trajectory $\underline{\hat{x}}^{\mathrm{e}} = [(\underline{\hat{x}}_0^{\mathrm{e}})^{\mathrm{T}} \ldots (\underline{\hat{x}}_k^{\mathrm{e}})^{\mathrm{T}}]^{\mathrm{T}}$ and measurement sequence. Of course, this is infeasible, not only because the error covariance matrix of the state vector is fully occupied; its size grows without bound. However, in some applications, parts of the history are stored, for instance, in order to deal with delayed measurements, and fortunately the information form provides a sparse representation of the trajectory. We consider a linear system model like (3.20) without control inputs in order to simplify matters. A part of the trajectory from time step l to time step k, i.e., $\underline{\hat{x}}_{l:k}^{\mathrm{P}} = [(\underline{\hat{x}}_l^{\mathrm{P}})^{\mathrm{T}} \ldots (\underline{\hat{x}}_k^{\mathrm{P}})^{\mathrm{T}}]^{\mathrm{T}}$, has a very simple representation[7] in its information form, which is given by

$$\underline{\hat{y}}_{l:k}^{\mathrm{P}} = \begin{bmatrix} (\mathbf{C}_l^{\mathrm{P}})^{-1}\, \underline{\hat{x}}_l^{\mathrm{P}} \\ \underline{0} \\ \vdots \\ \underline{0} \end{bmatrix} = \begin{bmatrix} \underline{\hat{y}}_l^{\mathrm{P}} \\ \underline{0} \\ \vdots \\ \underline{0} \end{bmatrix},$$

---

[7]The lower entries generally also become nonzero for affine models or linearized models.

where no measurements have been incorporated yet. The measurement sequence can be stored separately in its information form

$$
\underline{i}_{l:k} = \begin{bmatrix} \underline{i}_l \\ \vdots \\ \underline{i}_k \end{bmatrix} = \begin{bmatrix} (\mathbf{H}_l)^{\mathrm{T}}(\mathbf{C}_l^z)^{-1}\underline{\hat{z}}_l \\ \vdots \\ (\mathbf{H}_k)^{\mathrm{T}}(\mathbf{C}_k^z)^{-1}\underline{\hat{z}}_k \end{bmatrix},
$$

which can be fused with the trajectory by means of a simple addition. The corresponding matrix $\mathbf{I}_{l:k}$ is simply block diagonal due to (3.67). The information matrix $\mathbf{Y}_{l:k}^{\mathrm{P}}$ related to the trajectory $\hat{\underline{y}}_{l:k}^{\mathrm{P}}$ remarkably offers also the sparse representation

$$
\begin{bmatrix}
\mathbf{Y}_l^{\mathrm{P}} + \mathbf{A}_l^{\mathrm{T}}(\mathbf{C}_l^w)^{-1}\mathbf{A}_l & -(\mathbf{C}_l^w)^{-1}\mathbf{A}_l & \mathbf{0} \\
-\mathbf{A}_l^{\mathrm{T}}(\mathbf{C}_l^w)^{-1} & (\mathbf{C}_l^w) + \mathbf{A}_{l+1}^{\mathrm{T}}(\mathbf{C}_{l+1}^w)^{-1}\mathbf{A}_{l+1} & \ddots \\
\mathbf{0} & -\mathbf{A}_{l+1}^{\mathrm{T}}(\mathbf{C}_{l+1}^w)^{-1} & \ddots & \cdots \\
\vdots & & & \\
\mathbf{0} & \cdots & \mathbf{0} & \\
\end{bmatrix}
$$

$$
\begin{bmatrix}
\cdots & & \mathbf{0} \\
& & \vdots \\
\cdots & -(\mathbf{C}_{k-1}^w)^{-1}\mathbf{A}_{k-1} & \mathbf{0} \\
& (\mathbf{C}_{k-1}^w)^{-1} + \mathbf{A}_k^{\mathrm{T}}(\mathbf{C}_k^w)^{-1}\mathbf{A}_k & -(\mathbf{C}_k^w)^{-1}\mathbf{A}_k \\
& -\mathbf{A}_k^{\mathrm{T}}(\mathbf{C}_k^w)^{-1} & (\mathbf{C}_k^w)^{-1}
\end{bmatrix}.
$$

It is important to note, that prior states must be marginalized out before the information vector for a certain instant of time can be transformed back to state space. The incorporation of additional set-membership uncertainties can be accomplished by employing the approach from Section 3.4.

The information form of the delayed state vector provides several advantages in view of a decentralized network. The innovation vectors $\underline{i}_{l:k}^i$ and matrices $\mathbf{I}_{l:k}^i$ from different sensor nodes $i \in \{1, \ldots, N\}$ characterize the independent measurement information and can be fused by simple addition.

In particular, this means that, in comparison with Subsection 3.6.1, a longer measurement history can be exploited. The dependent information is associated with the trajectory $\hat{\underline{y}}_{l:k}^{\mathrm{P},i}$. Furthermore, the information matrices $\mathbf{Y}_{l:k}^{\mathrm{P},i}$ and $\mathbf{Y}_{l:k}^{\mathrm{P},j}$ are equal except of the upper left block, i.e., $\mathbf{Y}_{l}^{\mathrm{P},i}$ and $\mathbf{Y}_{l}^{\mathrm{P},j}$, which comprise the entire locally computed trajectories prior to time step l and are dependent due to common process noise and possibly common sensor data. Hence, a convex combination (3.62) for a conservative fusion, as it is done by the CI algorithm in Section 3.5.3, only manipulates the upper left block with $\mathbf{Y}_{l}^{\mathrm{CI}} = \omega \mathbf{Y}_{l}^{\mathrm{P},i} + (1-\omega)\mathbf{Y}_{l}^{\mathrm{P},j}$. The sensor nodes therefore only need to interchange this part for fusing the trajectories. By means of the explained splitting into dependent and independent information, the split CI algorithm from Subsection 3.6.2 can be employed for fusion. Of course, delayed state estimation comes at the expense of higher data transfer volume and storage requirements. However, both aspects only dependent linearly on the length of the considered horizon $l : k$ if the information form is utilized.

3.7   Conclusions from Chapter 3

This chapter has elucidated several additional difficulties that networked systems impose to the estimation problem. In particular, we singled out dependencies among locally processed data as the major issue to be addressed. The first named possibility is to maintain a centralized network architecture and to focus on an efficient preprocessing of measurement data, in order to reduce the amount of data to be communicated and to relieve workload of the center station. In this regard, the information form of the Kalman filter proves to be particularly valuable as the filtering step becomes the simple addition of information vectors and matrices. The information form is also employed to formulate an optimal distributed Kalman filter, where each node must be aware of all employed sensor models in order to compute a globalized likelihood. However, the local parameters have to be altered to synthetic variables and cannot serve as estimates anymore. If one node fails, even no valid estimate can be reconstructed. As one of the main contributions, this chapter focuses on the incorporation of additional set-membership uncertainty descriptions into the information filter. Fortunately, approximating a Minkowski sum of multiple error bounds in a distributed fashion turned out to be an easy task. In order to obtain a trace-optimal error bound, a certain communication strategy has to be employed that consists of first transmitting the information matrices, sending back the result, and computing the shape matrices with the correct parameters.

In many applications, the presence of a center node is prohibitive or even impossible. The necessity to continually send data to the center station can restrict the overall lifetime of battery-powered nodes significantly and can prevent nodes from operating and enacting decisions independently. However, in order to enable the nodes to be capable of solving a higher-level estimation problem, a fusion of local estimates must still provide additional insights to the state to be estimated. Possible correlations between local estimates introduce a systematic error behavior to the estimation problem but, unfortunately, cannot be tracked over time or traced back to their origin in a fully decentralized network. With the covariance intersection/covariance bounds algorithm, a solution has been proposed that is robust to any effects of correlated errors. This method is closely linked to the combined concepts of Chapter 2 as the possibles correlations are treated as unknown systematic errors that are to be bounded by means of the set-membership methodology. The drawback is that estimates are often highly conservative and therefore less informative. For this reason, we have shown several possibilities to exploit additional knowledge in order to tighten the conservative bounds. Employing the independent sensor noise of currently received measurements appears to be most evident, which can automatically be achieved by the covariance bounds algorithm. Split covariance intersection can be applied to estimates that are explicitly divided into dependent and independent parts. When the total amount of participating nodes is known and cycles of communication paths can be avoided, the federated Kalman filter can be employed that decorrelates the process noise by inflating the joint noise covariance matrix. The last considered concept is delayed state filtering, where sparse representations of trajectories in the information form can be exploited, so that the conditional independence of measurements over a predefined time horizon is guaranteed.

The variety of named approaches in this chapter is, of course, not all-encompassing. Internal covariance approximations [16] and consensus Kalman filtering [138] comprise also well-known concepts. Unfortunately, many concepts lack a guarantee that they provide covariance-consistent estimates and are therefore not considered here.

CHAPTER
4

Nonlinear State Estimation under Nonlinear Dependencies

Overview of Chapter 4

Nonlinear state estimation is generally a challenging task. The techniques discussed in the preceding chapters are then only applicable to a limited extent. Of course, uncertain quantities can be characterized by their first and second order error statistics and standard Kalman filtering techniques can be employed. However, mean and covariance matrix often do not suffice to take the underlying uncertainty fully into account and can even be deceptive, in particular, when multi-modal probability densities have to be dealt with. Since closed-form computations of the actual densities are, in general, not possible, a lot of effort has been expended on approximate solutions to nonlinear state estimation. Extended [163] and linear regression Kalman filtering, of which the unscented Kalman filter [100] is a well-known candidate, explicitly or implicitly linearize system and sensor models and therefore suffer from the aforementioned problem: Due to the underlying Gaussian assumption, they only provide very limited capabilities for capturing multi-modalities. Approximations of predicted and posterior densities are far better suited to account for multi-modalities. In the past years, the advances in state estimation theory provide us with a wide variety of density approximation techniques, such as, inter alia, particle representations [49], orthonormal bases representations like truncated Fourier series [12,30], Gaussian mixture densities [3], and exponential densities [63,147]. Apparently, nonlinear estimation is in itself challenging but becomes even more complicated in the context of distributed and decentralized sensor networks.

In nonlinear contexts, dependencies can neither easily be parameterized nor associated with a certain family of possible dependency structures. Accordingly, they do not convey an intuitive understanding of what a conservative fusion result has to be. For linear estimation problems, conservativeness can be achieved by means of the covariance intersection (CI) algorithm. Like the Kalman filter, CI only relies on the assumption of linearity and the first two moments of the estimates to be fused. It can hence also be applied in nonlinear setups, but we have to reckon with the same aforementioned difficulties that, for instance, multi-modalities are neglected. This chapter is dedicated to distributed and decentralized nonlinear information fusion. Section 4.1 provides a deeper analysis of nonlinear dependencies among local estimates and describes how nonlinear fusion is conducted when dependencies are entirely known. In contrast, Section 4.2 considers the situation that no knowledge about dependencies can be exploited. Conservative fusion results have then to be determined, which requires that a proper notion of conservativeness is definable at all. In some estimation problems, dependencies become tractable and parameterizable when considered in a transformed state space. Such transformations are employed in Section 4.3. A generalization of the information filter is presented in Section 4.4 in order to ease efficient processing of multiple sensor data. This formulation is rather useful in distributed networks than in decentralized ones. If the computation of local tracks is inevitable, but a central fusion node is present, the nonlinear federated filter in Section 4.5

can be employed that only computes a conservative bound on the process noise and thereby reduces conservatism.

Set-membership uncertainties are not considered in this chapter, although this can be achieved by means of the concepts from Chapter 2. The incorporation of set-membership uncertainties can be solved independently from the question of how to cope with nonlinear dependencies. Again, interdependencies between unknown but bounded quantities do not have any affect on the computed bounds and are hence not in the focus of this chapter.

4.1   Nonlinear Dependencies

In nonlinear estimation problems, estimates are characterized by conditional probability densities, as explained in Section 2.1.1. Dependencies then emerge as joint probability density functions, and local estimates can be regarded as marginal densities. Unfortunately, as in the linear case, this implies that a far higher-dimensional state space needs to be considered in order to model dependencies. This section elucidates the arbitrariness of joint densities in nonlinear estimation problems and analyzes how dependencies can be exploited to fuse locally computed densities.

4.1.1   Lack of Sole Linearity

The preceding Chapter 3 has revealed a variety of approaches to cope with unknown dependencies in linear estimation problems. Each involved uncertain quantity—the state estimate, process noise, and sensor noise— is characterized by means of an error covariance matrix, and dependencies between local estimates are entirely characterized by cross-correlations between the estimation errors. The corresponding cross-covariance matrices are only linearly manipulated by linear system and sensor models. Hence, linear state estimation ensures that only linear dependencies, i.e., cross-correlations, between Gaussian estimates are present. Unfortunately, this valuable property cannot be exploited in nonlinear estimation problems anymore. Figure 4.1(a) shows a Gaussian density that characterizes the

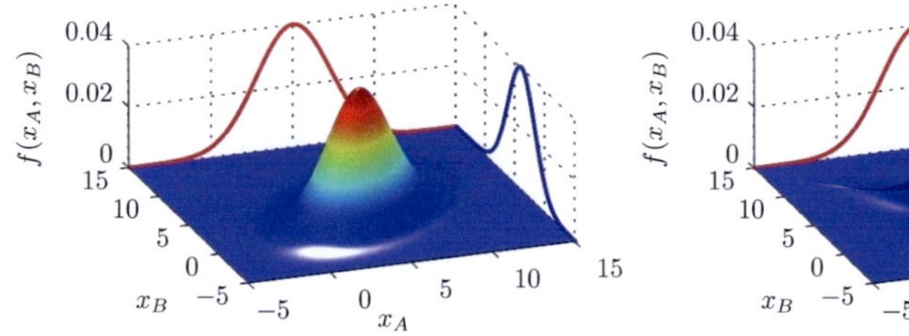

(a) Joint Gaussian density with Gaussian marginals.

(b) Non-Gaussian joint density with Gaussian marginals.

Figure 4.1: Dependencies between the red and blue local Gaussian estimates are not necessarily represented by a Gaussian density.

dependencies between two Gaussian local densities, but—in nonlinear estimation problems—also the density depicted in Figure 4.1(b) is a possible candidate to characterize the dependency structure of the same Gaussian marginals. While the density in Figure 4.1(a) can easily be parameterized by means of the corresponding joint covariance matrix, approximations might be inevitable to model dependency structures like the one in Figure 4.1(b). Studies that are concerned with parameterizing nonlinear dependencies are scarce. A concept that gained unfortunate notoriety with regard to the 2008 global financial crisis [152] is to employ *copulas* [134, 145]. Although copulas allow to characterize dependencies over a hypercube $[0, 1]^{2n_x}$, still approximations are required for complex dependency structures and it has not been studied yet how to embed copulas into the Bayesian estimation framework from Section 2.1.1. A second possibility is to ascribe nonlinear dependencies to correlations in a higher-dimensional state space, which will be considered in Section 4.3.

If dependencies between estimates are anyway unknown, it is more important to define a proper notion of consistency and conservativeness than to determine parameterizations. The issue to be addressed is then how to conservatively bound densities like the one plotted in Figure 4.1(b). However, the question how to treat unknown dependencies has not been sufficiently solved yet since a joint density may take arbitrary and unexpected forms, as discussed in the following subsection. A general possibility to obtain conservative estimates will be discussed in Section 4.2. In some estimation problems, it is possible to limit the arbitrariness of joint densities for given marginals, which is exploited in Section 4.3.

4.1.2 Fusion of Dependent Information

The problem of fusing dependent estimates is discussed in this subsection by means of two sensor nodes A and B that communicate their estimates of an uncertain system state $\underline{x}_k$ at a given time instant k. Parts of the following considerations have been published in [197]. To simplify matters, we omit the time index in the following considerations. The sets $\mathcal{Z}_A$ and $\mathcal{Z}_B$ denote the data collected at each local node. Each set consists of the measurement sequences obtained through own sensor observations and through communication with other nodes. The local state estimates are represented by the conditional densities $f(\underline{x}\,|\,\mathcal{Z}_A)$ and $f(\underline{x}\,|\,\mathcal{Z}_B)$. If $\mathcal{Z}_A$ and $\mathcal{Z}_B$ are conditionally independent, i.e.,

$$f(\mathcal{Z}_A \cup \mathcal{Z}_B\,|\,\underline{x}) = f(\mathcal{Z}_A\,|\,\underline{x}) \cdot f(\mathcal{Z}_B\,|\,\underline{x}) \,,$$

and prior information $f(\underline{x})$ is available, the local estimates can be fused [42, 118] according to

$$
\begin{aligned}
f(\underline{x}\,|\,\mathcal{Z}_A \cup \mathcal{Z}_B) &= \frac{f(\mathcal{Z}_A \cup \mathcal{Z}_B|\underline{x}) \cdot f(\underline{x})}{f(\mathcal{Z}_A \cup \mathcal{Z}_B)} \\
&= \frac{f(\mathcal{Z}_A\,|\,\underline{x}) \cdot f(\mathcal{Z}_B\,|\,\underline{x}) \cdot f(\underline{x})}{f(\mathcal{Z}_A \cup \mathcal{Z}_B)} \\
&= \frac{f(\mathcal{Z}_A) \cdot f(\mathcal{Z}_B)}{f(\mathcal{Z}_A \cup \mathcal{Z}_B)} \cdot \frac{f(\underline{x}\,|\,\mathcal{Z}_A) \cdot f(\underline{x}\,|\,\mathcal{Z}_B)}{f(\underline{x})} \\
&= c \cdot \frac{f(\underline{x}\,|\,\mathcal{Z}_A) \cdot f(\underline{x}\,|\,\mathcal{Z}_B)}{f(\underline{x})} \,,
\end{aligned}
\tag{4.1}
$$

where the constant factors are collected in the term c. When no prior information $f(\underline{x})$ on $\underline{x}$ is available, the fusion formula reduces to

$$f(\underline{x}\,|\,\mathcal{Z}_A \cup \mathcal{Z}_B) = c \cdot f(\underline{x}\,|\,\mathcal{Z}_A) \cdot f(\underline{x}\,|\,\mathcal{Z}_B) \,,
\tag{4.2}$$

which implicitly indicates that a non-informative prior is used. The conditional densities $f(\underline{x}\,|\,\mathcal{Z}_A)$ and $f(\underline{x}\,|\,\mathcal{Z}_B)$ can further be considered as the marginals of the joint density

$$f(\underline{x}_A, \underline{x}_B\,|\,\mathcal{Z}_A \cup \mathcal{Z}_B) = f(\underline{x}_A\,|\,\mathcal{Z}_A) \cdot f(\underline{x}_B\,|\,\mathcal{Z}_B) \,. \tag{4.3}$$

It is worth to note that this joint density can also be derived similarly to (4.1) and (4.2) when no prior densities $f(\underline{x}_A)$, $f(\underline{x}_B)$, and $f(\underline{x}_A, \underline{x}_B)$ are given. The joint density (4.3) is related to the fusion result (4.2) by

$$\begin{aligned} f(\underline{x}\,|\,\mathcal{Z}^A \cup \mathcal{Z}^B) &= c \cdot f(\underline{x}\,|\,\mathcal{Z}^A) \cdot f(\underline{x}\,|\,\mathcal{Z}^B) \\ &= c \cdot f(\underline{x}, \underline{x}\,|\,\mathcal{Z}^A \cup \mathcal{Z}^B) \,. \end{aligned} \tag{4.4}$$

Thus, the fused density (4.2) is obtained from the joint density conditioned on the event $\mathcal{E} = \{[\underline{x}_A, \underline{x}_B] \mid \underline{0} = \underline{x}_A - \underline{x}_B\}$ that $\underline{x}_A$ and $\underline{x}_B$ are equal. This relation can also be seen from the fact that the estimates are, in general, computed for the joint state space model

$$\begin{bmatrix} \boldsymbol{x}_A \\ \boldsymbol{x}_B \end{bmatrix} = \begin{bmatrix} \mathbf{I} \\ \mathbf{I} \end{bmatrix} \underline{x} \,,$$

as, for instance, in (3.39) and (3.41). The density of $\underline{x}$ is then, by construction, related to the joint state space by

$$f(\underline{x}) = c \cdot f(\underline{x}, \underline{x}) \,,$$

which corresponds to the same equality constraint. Hence, $f(\underline{x}_A, \underline{x}_B)$ is evaluated on its "diagonal" in order to obtain the fusion result.

In general, the sets $\mathcal{Z}_A$ and $\mathcal{Z}_B$ do not represent conditionally independent information. If no state transition takes place, dependencies can generally be expressed in terms of the non-empty intersection $\mathcal{Z}_A \cap \mathcal{Z}_B$ [91, 139, 139]. More precisely, this common information is related to the estimated density $f(\underline{x} \mid \mathcal{Z}_A \cap \mathcal{Z}_B)$, which must not be double-counted when $f(\underline{x} \mid \mathcal{Z}_A)$ and $f(\underline{x} \mid \mathcal{Z}_B)$ are fused. With

$$
\begin{aligned}
f(\underline{x} \mid \mathcal{Z}^A \cup \mathcal{Z}^B) &= \frac{f(\mathcal{Z}_A \cup \mathcal{Z}_B \mid \underline{x}) \cdot f(\underline{x})}{f(\mathcal{Z}_A \cup \mathcal{Z}_B)} \\
&= \frac{f(\mathcal{Z}_A \backslash \mathcal{Z}_B \mid \underline{x}) \cdot f(\mathcal{Z}_B \backslash \mathcal{Z}_A \mid \underline{x}) \cdot f(\mathcal{Z}_A \cap \mathcal{Z}_B \mid \underline{x}) \cdot f(\underline{x})}{f(\mathcal{Z}_A \cup \mathcal{Z}_B)} \\
&= \frac{f(\mathcal{Z}_A \backslash \mathcal{Z}_B \mid \underline{x}) \cdot f(\mathcal{Z}_A \cap \mathcal{Z}_B \mid \underline{x}) \cdot f(\mathcal{Z}_B \backslash \mathcal{Z}_A \mid \underline{x}) \cdot f(\mathcal{Z}_A \cap \mathcal{Z}_B \mid \underline{x}) \cdot f(\underline{x})}{f(\mathcal{Z}_A \cup \mathcal{Z}_B) \cdot f(\mathcal{Z}_A \cap \mathcal{Z}_B \mid \underline{x})} \\
&= \frac{f(\mathcal{Z}_A \mid \underline{x}) \cdot f(\mathcal{Z}_B \mid \underline{x}) \cdot f(\underline{x})}{f(\mathcal{Z}_A \cup \mathcal{Z}_B) \cdot f(\mathcal{Z}_A \cap \mathcal{Z}_B \mid \underline{x})} \\
&= \frac{f(\mathcal{Z}_A) \cdot f(\mathcal{Z}_B)}{f(\mathcal{Z}_A \cup \mathcal{Z}_B)} \cdot \frac{f(\underline{x} \mid \mathcal{Z}_A) \cdot f(\underline{x} \mid \mathcal{Z}_B)}{f(\underline{x} \mid \mathcal{Z}_A \cap \mathcal{Z}_B)} \\
&= c \cdot \frac{f(\underline{x} \mid \mathcal{Z}_A) \cdot f(\underline{x} \mid \mathcal{Z}_B)}{f(\underline{x} \mid \mathcal{Z}_A \cap \mathcal{Z}_B)} ,
\end{aligned}
$$

$$(4.5)$$

the common information is only incorporated once. Consequently, by dividing by $f(\underline{x} \mid \mathcal{Z}_A \cap \mathcal{Z}_B)$, common information is removed from the fusion result (4.4) so as to avoid double-counting. It is important to note that this fusion methodology is employed by the channel filter [71, 140], which corresponds to the subtraction of common information vectors and matrices, i.e., equations (3.52) and (3.53), in the linear case.

However, especially when local tracks are computed within a decentralized network, local estimates may become fully dependent. In line with (4.1), the fused estimated probability density is then given by

$$
\begin{aligned}
f(\underline{x} \mid \mathcal{Z}_A \cup \mathcal{Z}_B) &= \frac{f(\mathcal{Z}_A \cup \mathcal{Z}_B \backslash \mathcal{Z}_A \mid \underline{x}) \cdot f(\underline{x})}{f(\mathcal{Z}_A \cup \mathcal{Z}_B)} \\
&= \frac{f(\mathcal{Z}_A) \cdot f(\mathcal{Z}_B \backslash \mathcal{Z}_A)}{f(\mathcal{Z}_A \cup \mathcal{Z}_B)} \cdot \frac{f(\underline{x} \mid \mathcal{Z}_A) \cdot f(\underline{x} \mid \mathcal{Z}_B \backslash \mathcal{Z}_A)}{f(\underline{x})} ,
\end{aligned}
$$

and, according to (4.3), the joint density is

$$
f(\underline{x}_A, \underline{x}_B \mid \mathcal{Z}_A \cup \mathcal{Z}_B) := f(\underline{x}_A \mid \mathcal{Z}_A) \cdot f(\underline{x}_B \mid \mathcal{Z}_B \backslash \mathcal{Z}_A) . \tag{4.6}
$$

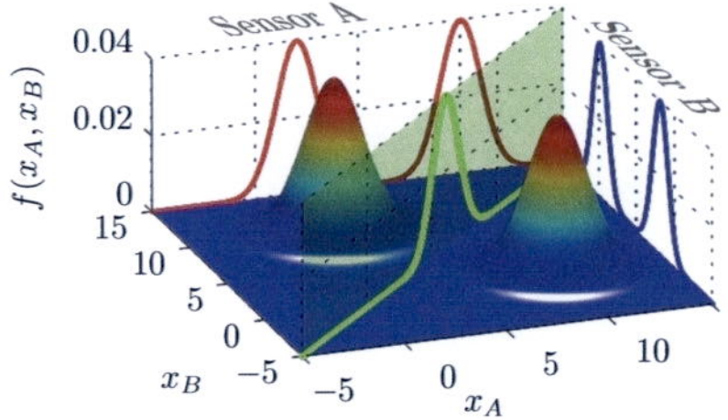

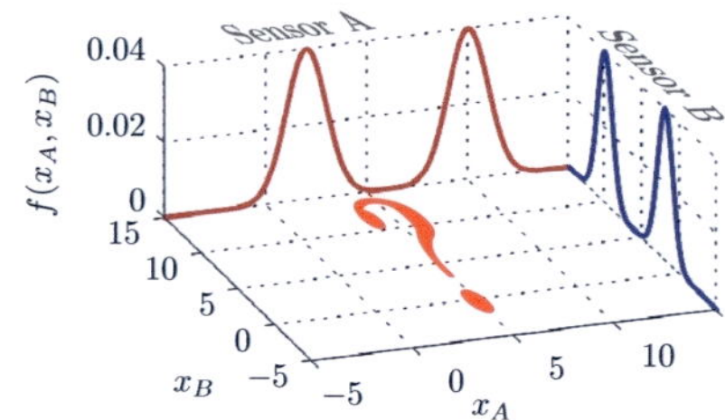

(a) If joint density is known to sensor nodes, it can be exploited for fusion.

(b) However, in many situations, the joint density remains hidden to the sensor nodes.

Figure 4.2: The red and blue bimodal densities that are identical can be fused when their dependency structure is known. The fusion result (green) lies on the "diagonal". (Example based on [197].)

Essentially, the conditional independence of $\mathcal{Z}_A$ and $\mathcal{Z}_B \backslash \mathcal{Z}_A$ only holds for static or deterministic systems. For dynamic stochastic systems, we can draw the same conclusions by replacing $\mathcal{Z}_B \backslash \mathcal{Z}_A$ with an artificially introduced set $\tilde{\mathcal{Z}}_B$ that encompasses the conditionally independent information. Again the fusion result lies on the "diagonal"

$$f(\underline{x} \,|\, \mathcal{Z}^A \cup \mathcal{Z}^B) = c \cdot f(\underline{x} \,|\, \mathcal{Z}^A) \cdot f(\underline{x} \,|\, \mathcal{Z}^B \backslash \mathcal{Z}^A)$$
$$= c \cdot f(\underline{x}, \underline{x} \,|\, \mathcal{Z}^A \cup \mathcal{Z}^B)$$

of the joint density (4.6). An example is depicted in Figure 4.2(a), where the fusion of two bimodal and even equal densities $f(\underline{x} \,|\, \mathcal{Z}^A)$ and $f(\underline{x} \,|\, \mathcal{Z}^B)$ unexpectedly yields a unimodal density. The corresponding fusion result is plotted in Figure 4.3(a). Hence, the local information may differ severely from the global estimate. Especially in a fully decentralized network, only the marginals $f(\underline{x} \,|\, \mathcal{Z}^A)$ and $f(\underline{x} \,|\, \mathcal{Z}^B)$ are known, i.e., the dependency structure between local estimates remains hidden, as indicated in Figure 4.2(b). The joint density or respectively the conditional density $f(\underline{x}|\mathcal{Z}^B \backslash \mathcal{Z}^A)$ therefore cannot be reconstructed uniquely, and assuming independence, as it is done for Figure 4.3(b), can be hazardous and misleading. In other words, finding the joint density for given marginals is an ill-posed inverse problem.

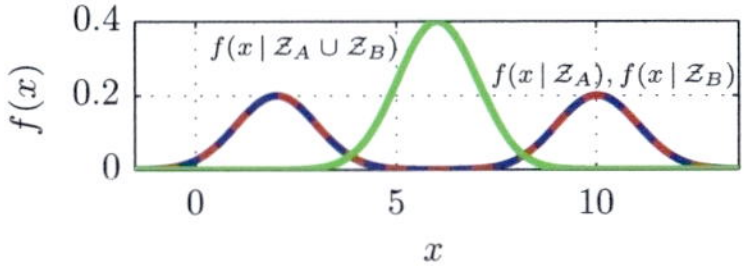

(a) Optimal fusion result.

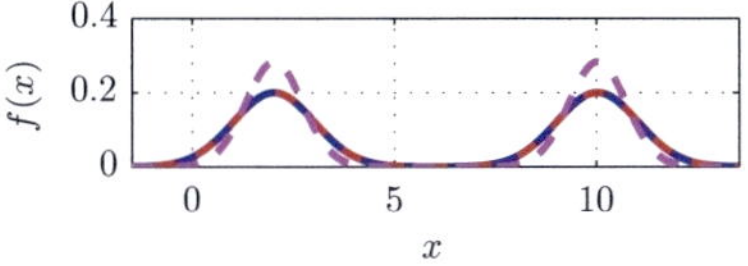

(b) Naive fusion result.

Figure 4.3: Fusion results for the dependency structure depicted in Figure 4.2. In Fig. (a), the unexpected green density is obtained. In Fig. (b), the red and blue density are fused (magenta, dashed) by assuming independence, and no information is preserved between the modes. (Left example based on [197].)

Even for local Gaussian estimates, the fusion result can be far from being Gaussian anymore [153]. In the following section, a general conservative fusion technique for nonlinear estimation problems is discussed.

4.2   Exponential Mixture Densities

The results of the previous section indicate that the development of conservative fusion strategies for nonlinear estimation problems is an arduous task. Even simple Gaussian densities may possess cumbersome dependency structures [153] that appear to be too arbitrary to be utilizable. This issue stands in stark contrast to linear estimation problems, where the family of possible dependencies can easily be parameterized. Even so, it is fortunately possible to derive conservative fusion results in the presence of unknown nonlinear dependencies. This section states a fusion rule that is a generalization of the covariance intersection (CI) algorithm and enables us to take into account dependencies that arise from common sensor information as well as common process noise.

4.2.1   Generalization of Covariance Intersection

In linear estimation problems, unknown dependencies can conservatively be bounded with the aid of the CI algorithm described in Section 3.5.3.

Especially from the analogous covariance bound formulation (3.64), we can see that the CI fusion result is related to the joint Gaussian density

$$
\mathcal{N}(\underline{x}; \hat{\underline{x}}_{\mathrm{CI}}, \mathbf{C}_{\mathrm{CI}}) \stackrel{(4.4)}{=} c \cdot f(\underline{x}, \underline{x} \mid \mathcal{Z}_A \cup \mathcal{Z}_B)
$$

$$
= c \cdot \mathcal{N}\left(\begin{bmatrix} \underline{x} \\ \underline{x} \end{bmatrix}; \begin{bmatrix} \hat{\underline{x}}_A \\ \hat{\underline{x}}_B \end{bmatrix}, \begin{bmatrix} \frac{1}{\omega}\mathbf{C}_A & \mathbf{0} \\ \mathbf{0} & \frac{1}{1-\omega}\mathbf{C}_B \end{bmatrix}\right)
$$

$$
= c \cdot \mathcal{N}(\underline{x}; \hat{\underline{x}}_A, \tfrac{1}{\omega}\mathbf{C}_A) \cdot \mathcal{N}(\underline{x}; \hat{\underline{x}}_B, \tfrac{1}{1-\omega}\mathbf{C}_B)
$$

$$
= c \cdot \mathcal{N}(\underline{x}; \hat{\underline{x}}_A, \mathbf{C}_A)^{\omega} \cdot \mathcal{N}(\underline{x}; \hat{\underline{x}}_B, \mathbf{C}_B)^{1-\omega} \ ,
$$

which has been simplified to an exponential mixture of Gaussian densities with $\omega \in [0, 1]$. The latter product gives exactly the CI fusion result (3.61) and (3.62) for mean and covariance matrix, respectively. Based on this realization, [84] and [120] have independently proposed to generalize the CI fusion rule to the *exponential mixture density* (EMD)

$$
f_{\mathrm{EMD}}(\underline{x} \mid \mathcal{Z}_A \cup \mathcal{Z}_B) = \frac{f^{\omega}(\underline{x} \mid \mathcal{Z}_A) \cdot f^{(1-\omega)}(\underline{x} \mid \mathcal{Z}_B)}{\int_{\mathbb{R}^N} f^{\omega}(\underline{x} \mid \mathcal{Z}_A) \cdot f^{(1-\omega)}(\underline{x} \mid \mathcal{Z}_B) \, \mathrm{d}\underline{x}} \tag{4.7}
$$

for arbitrary, locally estimated densities $f(\underline{x} \mid \mathcal{Z}_A)$ and $f(\underline{x} \mid \mathcal{Z}_B)$. In [6], this fusion rule is also named the *normalized weighted geometric mean*. Promising results of the EMD fusion rule have been presented with regard to Gaussian mixtures [90], exponentials of polynomials [165], and multi-object densities [43]. Nevertheless, it still remains an open question what is to be considered as a conservative estimate in nonlinear state estimation, which is discussed in the following subsection.

4.2.2 Conservativeness for Non-Gaussian Densities

For given local estimates, an optimal Bayesian fusion result is only attainable by accessing the underlying joint dependency structure. However, bookkeeping of the joint density is in general not an option since this task requires tremendous effort to compute, store, and communicate the dependencies. A fully decentralized network generally offers no other alternative than to discard at least some information about underlying dependencies. Only suboptimal fusion strategies can then be employed, and the primary aim is to still gain insight without becoming overconfident. In this section, we study in which way the EMD (4.7) can be considered to be a conservative fusion rule. The use of EMDs in decentralized state estimation has first been considered in [93] and, in [6], the following notion of conservativeness for arbitrary probability densities has been introduced that requires two properties to be satisfied: f_c is a conservative approximation of a probability density f if

1. for the differential entropy, the inequality $H(f(\underline{x})) \leq H(f_c(\underline{x}))$ holds,

2. and the ordering

$$f_c(\underline{x}_i) \leq f_c(\underline{x}_j) \iff f(\underline{x}_i) \leq f(\underline{x}_j)$$

   holds for all $\underline{x}_i \in \mathbb{R}^{n_x}$ and $\underline{x}_j \in \mathbb{R}^{n_x}$.

For given local estimates, it is desirable that a conservative fusion rule provides results that fulfills these two properties with respect to any possible true Bayesian fusion result. [6] presents the following two conditions on a conservative fusion rule: A fusion rule is conservative if and only if it

1. does not double-count common information and

2. replaces each component of independent information with a conservative approximation.

The EMD fusion rule (4.7) proves to be a conservative fusion rule, and the basic idea behind these conditions is clarified by considering common sensor information and common process noise.

Conservativeness for Common Sensor Information Common information $\mathcal{Z}_A \cap \mathcal{Z}_B$ represented by the conditional density $f(\underline{x} \mid \mathcal{Z}_A \cap \mathcal{Z}_B)$ can be prevented from being double-counted by means of (4.5). If the common information is unknown, the EMD fusion rule can be employed [91], which yields

$$f_{\text{EMD}}(\underline{x} \mid \mathcal{Z}^A \cup \mathcal{Z}^B) = c \cdot f^\omega(\underline{x} \mid \mathcal{Z}_A) \cdot f^{(1-\omega)}(\underline{x} \mid \mathcal{Z}_B)$$
$$= \tilde{c} \cdot f^\omega(\mathcal{Z}_A \mid \underline{x}) \cdot f^\omega(\underline{x}) \cdot f^{(1-\omega)}(\mathcal{Z}_B \mid \underline{x}) \cdot f^{(1-\omega)}(\underline{x})$$

$$= \tilde{c} \cdot \left(f(\mathcal{Z}_A \backslash \mathcal{Z}_B \,|\, \underline{x}) f(\mathcal{Z}_A \cap \mathcal{Z}_B \,|\, \underline{x}) \right)^{\omega}$$

$$\cdot \left(f(\mathcal{Z}_B \backslash \mathcal{Z}_A \,|\, \underline{x}) f(\mathcal{Z}_A \cap \mathcal{Z}_B \,|\, \underline{x}) \right)^{(1-\omega)} \cdot f(\underline{x})$$

$$= \tilde{c} \cdot f^{\omega}(\mathcal{Z}_A \backslash \mathcal{Z}_B \,|\, \underline{x}) \cdot f^{(1-\omega)}(\mathcal{Z}_B \backslash \mathcal{Z}_a \,|\, \underline{x})$$

$$\cdot f(\mathcal{Z}_A \cap \mathcal{Z}_B \,|\, \underline{x}) \cdot f(\underline{x}) \,,$$

where $\mathcal{Z}_A$ has been split into the conditionally independent subsets $\mathcal{Z}_A \backslash \mathcal{Z}_B$ and $\mathcal{Z}_A \cap \mathcal{Z}_B$ and the same has been done with $\mathcal{Z}_B$. As can be seen in the last row of the equation, the common part $f(\underline{x} \,|\, \mathcal{Z}_A \cap \mathcal{Z}_B)$ is only incorporated once such that condition 1 is satisfied. The independent parts $f(\mathcal{Z}_A \backslash \mathcal{Z}_B \,|\, \underline{x})$ and $f(\mathcal{Z}_B \backslash \mathcal{Z}_A \,|\, \underline{x})$ are raised to the powers ω and $1 - \omega$, respectively. According to [6], raising a probability density to the power $\omega \in (0, 1]$ yields a conservative approximation in terms of the above definition. Condition 2 is hence also fulfilled.

Conservativeness for Common Process Noise The prediction of local estimates and a subsequent fusion also conceals the risk of being overconfident due to common process noise. In nonlinear estimation problems, this means that transition densities are erroneously assumed to be independent and are incorporated multiple times into the fusion result. By employing the EMD fusion rule (4.7), the following relation

$$f_{\mathrm{EMD}}^{\mathrm{P}}(\underline{x}_{k+1} \,|\, \mathcal{Z}^A \cup \mathcal{Z}^B) = c \cdot \left(f^{\mathrm{P}}(\underline{x}_{k+1} \,|\, \mathcal{Z}_A) \right)^{\omega} \cdot \left(f^{\mathrm{P}}(\underline{x}_{k+1} \,|\, \mathcal{Z}_B) \right)^{1-\omega}$$

$$= c \cdot \left(\int_{\mathbb{R}^{n_x}} f(\underline{x}_{k+1} \,|\, \underline{x}_k) f^{\mathrm{e}}(\underline{x}_k \,|\, \mathcal{Z}^A) \, \mathrm{d}\underline{x}_k \right)^{\omega}$$

$$\cdot \left(\int_{\mathbb{R}^{n_x}} f(\underline{x}_{k+1} \,|\, \underline{x}_k) f^{\mathrm{e}}(\underline{x}_k \,|\, \mathcal{Z}^B) \, \mathrm{d}\underline{x}_k \right)^{1-\omega}$$

$$\geq c \cdot \int_{\mathbb{R}^{n_x}} \left(f(\underline{x}_{k+1} \,|\, \underline{x}_k) f^{\mathrm{e}}(\underline{x}_k \,|\, \mathcal{Z}^A) \right)^{\omega}$$

$$\left(f(\underline{x}_{k+1} \,|\, \underline{x}_k) f^{\mathrm{e}}(\underline{x}_k \,|\, \mathcal{Z}^B) \right)^{1-\omega} \, \mathrm{d}\underline{x}_k$$

$$= c \cdot \int_{\mathbb{R}^{n_x}} f(\underline{x}_{k+1} \,|\, \underline{x}_k) \left(f^{\mathrm{e}}(\underline{x}_k \,|\, \mathcal{Z}^A) \right)^{\omega} \left(f^{\mathrm{e}}(\underline{x}_k \,|\, \mathcal{Z}^B) \right)^{1-\omega} \, \mathrm{d}\underline{x}_k$$

holds, which is derived with the aid of Hölder's inequality by setting $p = \frac{1}{\omega}$ and $q = \frac{1}{1-\omega}$. Hence, the EMD fusion of predicted densities preserves probability mass where the global prediction of the product $\left(f^{\mathrm{e}}(\underline{x}_k \mid \mathcal{Z}^A)\right)^{\omega} \left(f^{\mathrm{e}}(\underline{x}_k \mid \mathcal{Z}^B)\right)^{1-\omega}$ is also nonzero. The transition density enters the lower bound only once, and $f^{\mathrm{e}}(\underline{x}_k \mid \mathcal{Z}^A)$ and $f^{\mathrm{e}}(\underline{x}_k \mid \mathcal{Z}^B)$ are replaced by conservative approximations. Furthermore, the normalization constant has always the property $c \geq 0$ and is a convex function of ω [6]. From this, it can also be concluded that the EMD result is bounded from below according to

$$
\begin{aligned}
f_{\mathrm{EMD}}^{\mathrm{P}}(\underline{x}_{k+1} \mid \mathcal{Z}^A \cup \mathcal{Z}^B) \geq \min \Bigg\{ & \int_{\mathbb{R}^{n_x}} f(\underline{x}_{k+1} \mid \underline{x}_k) f^{\mathrm{e}}(\underline{x}_k \mid \mathcal{Z}^A)\, \mathrm{d}\underline{x}_k, \\
& \int_{\mathbb{R}^{n_x}} f(\underline{x}_{k+1} \mid \underline{x}_k) f^{\mathrm{e}}(\underline{x}_k \mid \mathcal{Z}^B)\, \mathrm{d}\underline{x}_k \Bigg\} \\
= \min \big\{ & f^{\mathrm{P}}(\underline{x}_{k+1} \mid \mathcal{Z}_A), f^{\mathrm{P}}(\underline{x}_{k+1} \mid \mathcal{Z}_B) \big\} ,
\end{aligned}
$$

i.e., the lower bound is related to the locally predicted densities. This important property demonstrates that EMDs do not report overconfident results in a point-wise sense, which might be the case if dependencies between the transition densities were not taken into account.

The EMD formula definitely shows the potential to encompass a general conservative fusion rule. However, for the situation depicted in figures 4.2 and 4.3, it remains unanswered how a conservative fusion result has to look like. Maybe, this discrepancy can be regarded as an indication that dependencies are not as arbitrary as they might appear at first sight, and that EMDs can tackle all kind of dependencies that may arise in state estimation problems. We will further focus on this question in Section 4.3.

4.2.3   On the Optimal Selection of the Weighting Parameter

The parameter ω has been determined in Section 3.5.3 by considering the covariance matrices in order to reduce the error that is associated to the point estimate. In case of arbitrary densities, the second moment does not appear to be an appropriate criterion anymore. For instance, [120] proposed to maximize the "peakiness", i.e., supremum, of the EMD fusion result.

In [84, 90], the Chernoff information is considered instead such that the entire domain is taken into account. Against the background of multitarget tracking, a comparison of different information measures has been carried out [168]. The results have suggested that also the simple choice $\omega = 0.5$ guarantees a good performance. If, for later decision-making purposes, a specific information measure is employed, it is sensible to utilize the same measure as an optimization criterion for ω. In general, the decision for a certain criterion depends on the intended further processing of the estimated densities.

4.3 Covariance Intersection with Pseudo Gaussian Densities

Although the EMD fusion rule provides promising results, it remains an open question whether EMDs can tackle the arbitrariness of possible dependencies. For example, applying the EMD formula (4.7) to fuse the blue and red density in Figure 4.3 yields again the same bimodal density, which hardly captures the mode of the green true fusion result. As primarily discussed in Chapter 3, there are manifold reasons for a lack of independence. However, as already stated in the previous section, the good news is that dependencies are not always as arbitrary as they might appear at first sight: Our knowledge about the system, sensors, and fusion models can be utilized to significantly limit the arbitrariness, and we know from Section 3.5.3 how to systematically deal with unknown correlations in linear estimation problems. Hence, the idea behind this section is to employ the insights from linear estimation theory by transforming the state space into a different space, where the models become linear and only linear dependencies between estimates, i.e., correlations, can arise. The problem of consistently fusing nonlinearly dependent estimates boils down to fusion under unknown correlations. Accordingly, the probability densities being considered in the transformed state space become normally distributed. These *pseudo Gaussian densities* then allow the notion of covariance consistency to be used in decentralized nonlinear state estimation.

The concept derived here has been published in [197] and is explicated in more detail in the following.

4.3.1 State-space Transformations

The general idea behind this section is to regard the true density to be estimated as a (pseudo) Gaussian density in a different state space [73]. Against the background of set-membership state estimation, this approach has been employed to represent complicated sets as so-called pseudo ellipsoids [72]. In particular, this concept is strongly related to Carleman (bi-)linearizations [110], which are also the basis for the similar polynomial extended Kalman filter [62–64]. In [115], this technique is referred to as non-minimal state Kalman filtering since, in general, a higher-dimensional state space is required to represent the state estimation problem by a linear one.

The underlying mechanism rests upon finding a proper transformation $\underline{T} : \mathcal{S} \to \mathcal{S}^*$ with $\mathcal{S} \subseteq \mathbb{R}^N$, $\mathcal{S}^* \subseteq \mathbb{R}^M$, and

$$\underline{x}^* = \underline{T}(\underline{x}) = [T_1(\underline{x}), \ldots, T_M(\underline{x})]^{\mathrm{T}} \tag{4.8}$$

so that the estimated probability density becomes normally distributed in $\mathcal{S}^*$ with mean $\hat{\underline{x}}^*$ and covariance matrix $\mathbf{C}^*$. The fundamental prerequisite is that the fusion of two estimates as well as the measurement update turn into linear operations. The former requirement is ensured if dependencies between the estimated pseudo Gaussian densities are solely linear. This in turn is fulfilled if the models are linear and hence, also the latter condition is met, i.e., it must be possible to rewrite a nonlinear sensor model

$$\underline{z}_k = \underline{h}_k(\underline{x}_k) + \underline{v}_k \tag{4.9}$$

into a linear update equation

$$\underline{z}_k^* = \mathbf{H}^* \, \underline{x}_k^* + \underline{v}_k^* \tag{4.10}$$

in the transformed state space $\mathcal{S}^*$, where $\underline{v}_k$ is an additive noise term that becomes a normally distributed perturbation $\underline{v}_k^* \sim \mathcal{N}(\hat{\underline{v}}_k^*, \mathbf{C}_k^{*,z})$. Due to the transformation, the noise is not assumed to be necessarily zero-mean.

A   Static Systems

For static systems, where observations are only perturbed by additive Gaussian noise, an appropriate transformation (4.8) can easily be determined. Let $\underline{h}^i$ with $i \in \{1, \ldots, L\}$ denote the L different observation models for the sensors that are used in a given network. Then, the estimation process becomes linear by considering the transformed state

$$\underline{x}^* = \underline{T}(\underline{x}) := [\underline{h}^1(\underline{x}), \ldots, \underline{h}^L(\underline{x})]^{\mathrm{T}} , \tag{4.11}$$

where a single node with sensor model $\underline{h}^i(\underline{x})$ directly observes the ith component of $\underline{x}^*$. More precisely, the local sensor model becomes

$$\underline{z}^i = \underline{h}^i(\underline{x}) + \underline{v}^i = \mathbf{H}^* \, \underline{x}^* + \underline{v}^i ,$$

where $\mathbf{H}^* = [\mathbf{0}, \ldots, \mathbf{I}, \ldots, \mathbf{0}]$ picks out the ith component of $\underline{x}^*$. Measurement and noise are not altered by the transformation (4.11). Such a transformation ensures that an exchange of estimates can only cause linear dependencies between fusion results. If the sensor models are similar, a transformation of dimension less than L can be derived as shown in the following simple example.

Example 4.1: Simple sensor network—Transformed state space
A one-dimensional setup with four closely positioned sensor nodes at $P^1 = -1$, $P^2 = -1.2$, $P^3 = -0.5$, and $P^3 = 0.5$ is considered. At each time step k, they measure the distances

$$z_k^i = (x - P^i)^2 + v^i , \; i \in \{1, 2, 3, 4\} ,$$

where v^i are zero-mean Gaussian noise terms with high variances $C^{z,1} = 10$, $C^{z,2} = 9$, $C^{z,3} = 7$, and $C^{z,4} = 20$. The true state is located at 1. The measurement equation can be transformed to the affine sensor model

$$z_k^{*,i} = \mathbf{H}^{*,i} \, \underline{x}_k^* + (P^i)^2 + v^i$$

with the linear mapping

$$\mathbf{H}^{*,i} = [-2 \, P^i, 1] \tag{4.12}$$

and the transformed state

$$\underline{x}_k^* = \begin{bmatrix} \underline{x}_1^* \\ \underline{x}_2^* \end{bmatrix} = \begin{bmatrix} \underline{x}_k \\ \underline{x}_k^2 \end{bmatrix} . \tag{4.13}$$

The measurement noise $\boldsymbol{v}^{*,i} = (P^i)^2 + \boldsymbol{v}^i$ is still normally distributed according to $\mathcal{N}((P^i)^2, C^{z,i})$. In this simple scenario, the measurements z_k^* in $\mathcal{S}^*$ are identical to z_k in the original space $\mathcal{S}$.

In general, the dimension of the transformed state space is higher than the dimension of the original space and, of course, the choice of $\underline{T}$ is not unique since $\underline{x}^*$ can, for instance, be expanded by any linear combination of partial states $\underline{x}_i^*$ and $\underline{x}_j^*$. In a situation where combinations of the state $\underline{x}$ remain unobserved, the transformed state space can even be of lower dimension. For example, when a two-dimensional state is observed by a single distance sensor, a transformation to a one-dimensional state is sufficient.

B Dynamic Systems

When dealing with dynamic systems, we aspire to find a mapping $\underline{T}$ that also transforms a nonlinear system model

$$\underline{x}_{k+1} = a(\underline{x}_k, \hat{\underline{u}}_k, \underline{w}_k) \tag{4.14}$$

to a linear state transition model

$$\underline{x}_{k+1}^* = \mathbf{A}^* \, \underline{x}_k^* + \mathbf{B}_k^* \left(\hat{\underline{u}}_k^* + \underline{w}_k^* \right) ,$$

where $\hat{\underline{u}}_k$ and $\hat{\underline{u}}_k^*$ comprise possible input quantities in the original and the transformed state space, respectively. The system noise $\underline{w}_k$ becomes a normally distributed disturbance term $\underline{w}_k^* \sim \mathcal{N}(\hat{\underline{w}}_k, \mathbf{C}_k^{*,u})$ after the transformation. On the assumption that a mapping $\underline{T}$ exists that can simultaneously linearize the system and sensor equations, exclusively linear dependencies between the local estimates can arise in the transformed state space. Hence, the CI algorithm from Section 3.5.3 can be employed for suboptimal decentralized fusion of pseudo Gaussian densities, which is considered more closely in Paragraph 4.3.2-B of the following subsection. Unfortunately, such a transformation $\underline{T}$ to a finite-dimensional space, where both the system dynamics and the sensor models are linear, cannot be constructed in general. Therefore, linear approximations of systems through state transformations are discussed in Paragraph 4.3.2-C. All told, even if no optimal transformation to a linear form of the system model is obtained, the measurement update and fusion are now linear.

C   Transformation Back to Original State Space

The original nonlinearities come into play whenever an estimate in the original state space $\mathcal{S}$ is to be computed. Of course, the reverse transformation may involve complicated nonlinear estimation techniques, but the required calculations can be performed independently from the state estimation process, which solely relies on the pseudo Gaussian densities and the transformed models. Hence, the back transformation is only required at time steps when a certain estimate in the original state space is requested, for instance, for decision making. With regard to sensor networks, this further implies that every node can process the pseudo Gaussian estimates locally and linearly. In the data sink, where in general higher computational power is available, the true density and the desired parameters can be computed from the pseudo Gaussian estimates.

The original density $f(\underline{x}_k)$ is related to the estimate $(\hat{\underline{x}}_k^*, \mathbf{C}_k^*)$ by

$$
\begin{aligned}
f(\underline{x}_k) &= c \cdot \mathcal{N}(\underline{T}(\underline{x}_k), \hat{\underline{x}}^*, \mathbf{C}^*) \\
&= \tilde{c} \cdot \exp\left(-\frac{1}{2}\big(\underline{T}(\underline{x}_k) - \hat{\underline{x}}_k^*\big)^{\mathrm{T}} (\mathbf{C}_k^*)^{-1} \big(\underline{T}(\underline{x}_k) - \hat{\underline{x}}_k^*\big) \right)
\end{aligned}
\tag{4.15}
$$

with c and $\tilde{c}$ denoting the corresponding normalization factors. This calculation can be regarded as a measurement update with $\hat{\underline{x}}_k^* = \underline{T}(\underline{x}_k)$ where no prior is available, and therefore standard nonlinear filtering methods can be applied to obtain the original density.

4.3.2 Nonlinear Decentralized State Estimation with Pseudo-Gaussian Densities

By employing a linearizing transformation (4.8), distributed information processing can be eased significantly. The advantages of linear models come at the expense of generally higher-dimensional state spaces. In particular, linear dependencies are an unsurpassable advantage when estimates have to be fused and only possible correlations need to be taken into account. The preceding considerations have shown that, for filtering and fusion, according state-space transformations can easily identified. The prediction step, in general, does not allow for a linear, higher-dimensional representation and unfortunately requires approximations. Similarly, the representation of a prior Gaussian estimate is commonly not Gaussian anymore after applying the state-space transformation and has to be reapproximated, for instance, by computing the first two moments. The subsequent paragraphs focus on the measurement update and fusion of pseudo Gaussian estimates, but also provides an an approximate solution to the prediction of pseudo Gaussian estimates

A   Filtering with Pseudo Gaussian Likelihoods

Since the nonlinear sensor model (4.9) is represented by the linear one (4.10), the corresponding likelihood for the transformed parameters has the Gaussian representation

$$
\begin{aligned}
f\big(\hat{\underline{z}}_k^* \,|\, \underline{T}(\underline{x}_k)\big) =\ & f(\hat{\underline{z}}_k^* \,|\, \underline{x}_k^*) \\
\overset{(2.11)}{=}\ & f^{v^*}\big(\hat{\underline{z}}_k^* - \mathbf{H}_k^*\,\underline{x}_k^*\big) \\
=\ & c \cdot \mathcal{N}\big(\hat{\underline{z}}_k^* - \mathbf{H}_k^*\,\underline{x}_k^*;\ \hat{\underline{v}}_k^*,\ \mathbf{C}_k^{*,z}\big)\ .
\end{aligned}
$$

If the prior estimate on $\underline{x}_k$ has the conditional probability density function $f^{\mathrm{P}}(\underline{x}_k) = \mathcal{N}\big(\underline{T}(\underline{x}_k);\ \hat{\underline{x}}_k^{\mathrm{P},*},\ \mathbf{C}_k^{\mathrm{P},*}\big)$, the filtering step becomes the product of Gaussian densities, as in Section 2.1.1-B, and can be carried out by means of Kalman filter formulas, i.e.,

$$
\hat{\underline{x}}_k^{\mathrm{e},*} \overset{(2.25)}{=} (\mathbf{I} - \mathbf{K}_k^*\mathbf{H}_k^*)\hat{\underline{x}}_k^{\mathrm{P},*} + \mathbf{K}_k^*\big(\hat{\underline{z}}_k^* - \hat{\underline{v}}_k^*\big)
$$

for the mean and

$$\mathbf{C}_k^{\mathrm{e},*} \overset{(2.26)}{=} \mathbf{C}_k^{\mathrm{p},*} - \mathbf{K}_k^* \mathbf{H}_k^* \mathbf{C}_k^{\mathrm{p},*}$$

for the covariance matrix with the Kalman gain

$$\mathbf{K}_k^* = \mathbf{C}_k^{\mathrm{p},*}(\mathbf{H}_k^*)^{\mathrm{T}}\big(\mathbf{C}_k^{*,z} + \mathbf{H}_k^*\mathbf{C}_k^{\mathrm{p},*}(\mathbf{H}_k^*)^{\mathrm{T}}\big)^{-1} .$$

Hence, the filtering step can completely be formulated in terms of the transformed parameters and the estimated density in the original state space exactly yields $f^{\mathrm{e}}(\underline{x}_k) = c \cdot \mathcal{N}\big(\underline{T}(\underline{x}_k); \hat{\underline{x}}_k^{\mathrm{e},*}, \mathbf{C}_k^{\mathrm{e},*}\big)$.

B Covariance Intersection for Pseudo Gaussian Densities

At a fixed time step k, two local estimates $f_A^{\mathrm{e}}(\underline{x}) = f(\underline{x}\,|\,\mathcal{Z}_A)$ and $f_B^{\mathrm{e}}(\underline{x}) = f(\underline{x}\,|\,\mathcal{Z}_B)$ conditioned on different measurement histories $\mathcal{Z}_A$ and $\mathcal{Z}_B$ can be fused by considering the corresponding pseudo Gaussian densities $\mathcal{N}(T(\underline{x}); \hat{\underline{x}}_A^*, \mathbf{C}_A^*)$ and $\mathcal{N}(T(\underline{x}); \hat{\underline{x}}_B^*, \mathbf{C}_B^*)$. In the rather unlikely case that the correlations are known, the Bar-Shalom/Campo formulas (3.48) and (3.49) can be deployed. If the dependencies are unknown, we can apply the CI equations (3.61) and (3.62) to the parameters of the pseudo Gaussian densities. By virtue of the preceding considerations, the dependencies are linear and CI provides a covariance consistent suboptimal fusion result $(\hat{\underline{x}}_{\mathrm{CI}}^*, \mathbf{C}_{\mathrm{CI}}^*)$ in the transformed state space. The following example gives an impression of the presented concept.

Example 4.2: Simple sensor network—Fusion of estimates
We continue the considerations in Example 4.1, which is similar to the example in [197]. Each sensor platform now performs 5 measurement steps by employing (4.12) and exchanges its results with the other nodes at every time step. The prior estimate at every node is set to $\hat{x}_0^i = -2$ with error variance $\mathbf{C}_0^i = 5$, and the fusion of estimates is carried out by means of the CI algorithm. Figure 4.4(a) shows the final bivariate pseudo Gaussian estimate $(\hat{\underline{x}}_5^{*,1}, \mathbf{C}_5^{*,1})$ of sensor node 1. The density of the estimate in the original state space lies on the manifold, which is depicted explicitly in Figure 4.4(b) and corresponds to the green density in Figure 4.5. There, different fusion results are compared. The blue density represents the optimal Bayesian fusion result,

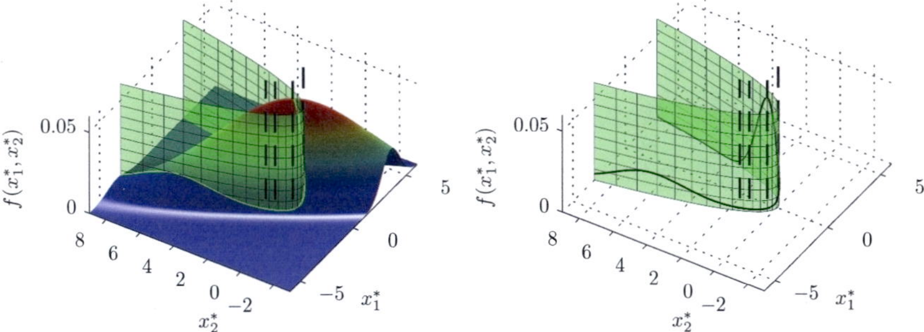

(a) Pseudo Gaussian density of sensor node 1. The original state space corresponds to the green manifold.

(b) True estimated density lies on the green manifold, which corresponds to Figure 4.5.

Figure 4.4: The transformation (4.13) yields the green manifold. Sensor 1 computes the density in Fig. (a), where estimates are fused by means of CI. In Fig. (b), the estimated density on the manifold is plotted, which corresponds to the green density in Fig. 4.5. The sensor's positions in the transformed state space are depicted by dashed black lines.

where all measurements are fused in a central instance with the prior. It captures the true state well, in contrast to the red density, where dependencies have been ignored for fusion. Evidently, the CI algorithm still provides a good fusion result.

The example elucidates how CI applied in the transformed space $\mathcal{S}^*$ flattens the modes of the density in the original state space $\mathcal{S}$, where the parameter ω for the fusion formulas (3.61) and (3.62) is chosen to minimize $\det(\mathbf{C}^*_{\mathrm{CI}})$. In terms of the underlying densities $f(\underline{x}\,|\,\mathcal{Z}_A) = \mathcal{N}(\underline{T}(\underline{x}); \hat{\underline{x}}^*_A, \mathbf{C}^*_A)$ and $f(\underline{x}\,|\,\mathcal{Z}_B) = \mathcal{N}(\underline{T}(\underline{x}); \hat{\underline{x}}^*_B, \mathbf{C}^*_B)$, the CI algorithm in $\mathcal{S}^*$ can be expressed as

$$
\begin{aligned}
f_{\mathrm{CI}}(\underline{x}\,|\,\mathcal{Z}_A \cup \mathcal{Z}_B) &= c \cdot \mathcal{N}(T(\underline{x}); \hat{\underline{x}}^*_A, \tfrac{1}{\omega}\mathbf{C}^*_A) \cdot \mathcal{N}(T(\underline{x}); \hat{\underline{x}}^*_B, \tfrac{1}{(1-\omega)}\mathbf{C}^*_B) \\
&= \hat{c} \cdot \underbrace{\left(\mathcal{N}(T(\underline{x}); \hat{\underline{x}}^*_A, \mathbf{C}^*_A)\right)^{\omega}}_{=c_A^{-1} f(\underline{x}\,|\,\mathcal{Z}_A)} \cdot \underbrace{\left(\mathcal{N}(T(\underline{x}); \hat{\underline{x}}^*_B, \mathbf{C}^*_B)\right)^{(1-\omega)}}_{=c_B^{-1} f(\underline{x}\,|\,\mathcal{Z}_B)} \\
&= \tilde{c} \cdot f^{\omega}(\underline{x}\,|\,\mathcal{Z}_A) \cdot f^{(1-\omega)}(\underline{x}\,|\,\mathcal{Z}_B) \,,
\end{aligned}
$$

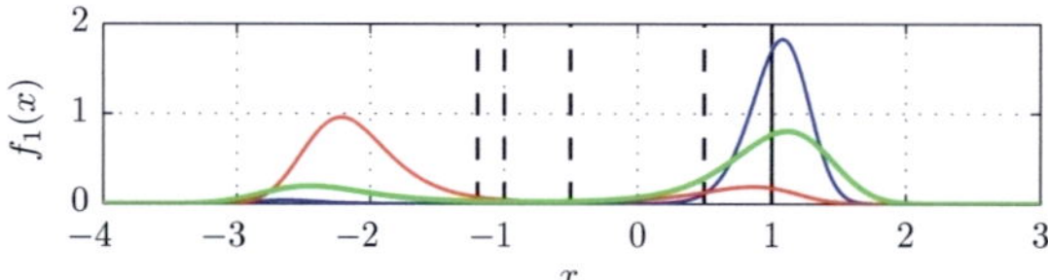

Figure 4.5: The positions of the four sensor nodes are depicted by the dashed black lines. The true state corresponds to the solid black line. By ignoring dependencies and naively fusing estimates at each time step, Sensor 1 computed the red estimated density. It preserves less probability mass around the true state. The optimal fusion result is drawn blue, and the green density is computed by employing CI in the transformed state space, which is also shown in Figure 4.4.

which is exactly the EMD update rule (4.7) being directly applied to the densities to be fused. In this regard, EMDs actually provide consistent estimates in the sense of covariance consistency in $\mathcal{S}^*$, i.e., relation (3.63), which confirms the considerations in [6, 93]. Hence, the concept of state-space transformations that entirely linearize the state estimation process also points to a direction to define consistency in nonlinear estimation problems. Another valuable feature of the space $\mathcal{S}^*$ is that it enables us to parameterize the set of possible dependencies between two estimates $(\hat{\underline{x}}_A^*, \mathbf{C}_A^*)$ and $(\hat{\underline{x}}_B^*, \mathbf{C}_B^*)$ by means of the cross-covariance matrix $\mathbf{C}_{AB}^*$. Possible dependencies are illustrated in the following example, which shall conclude this paragraph.

Example 4.3: Parameterization of possible fusion results

The same transformation $\underline{x}_k^* = [x_1^*, x_2^*]^T = [x_k, x_k^2]^T$ as in the previous Example 4.2 is considered. In the transformed state space $\mathcal{S}^*$, the local estimates are

$$(\hat{\underline{x}}_A^*, \mathbf{C}_A^*) = \left(\begin{bmatrix} 3 \\ 15 \end{bmatrix}, 3 \cdot \begin{bmatrix} 8 & -1 \\ -1 & 3 \end{bmatrix} \right) \tag{4.16}$$

and

$$(\hat{\underline{x}}_B^*, \mathbf{C}_B^*) = \left(\begin{bmatrix} -3 \\ 5 \end{bmatrix}, 2 \cdot \begin{bmatrix} 10 & 7 \\ 7 & 8 \end{bmatrix} \right), \tag{4.17}$$

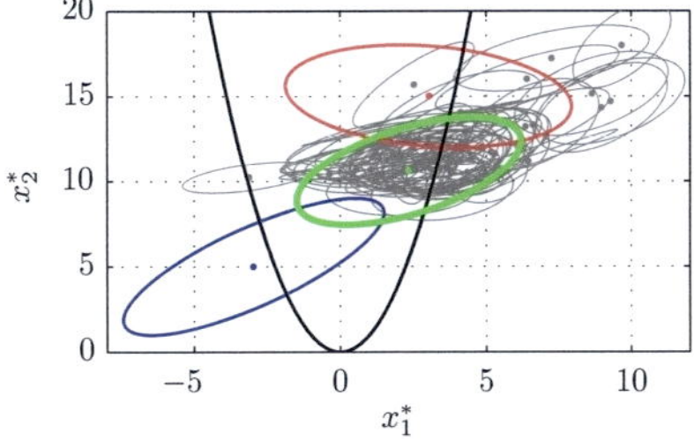 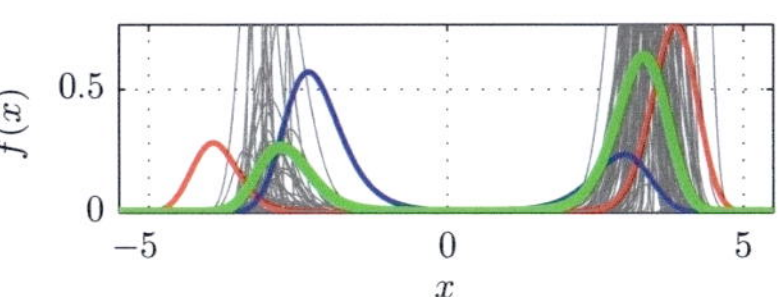

(a) Covariance ellipsoids in transformed state space $\mathcal{S}^*$. The green ellipsoid is the CI result for the red and blue ellipsoid.

(b) The original densities lie on the black manifold in Fig. (a). The same colors are used.

Figure 4.6: The blue and red estimates are fused. Fig. (a) shows the transformed state space, where several possible Bar-Shalom/Campo fusion results (gray) and the CI fusion result (green) are plotted. The black parabola represents the same manifold as in Figure 4.4, seen from above. Fig. (b) shows the corresponding densities of the original state space.

which are plotted in Figure 4.6 as the red and the blue covariance ellipsoids, respectively. 400 possible cross-covariance matrices $\mathbf{C}_{AB}^*$ have been determined randomly. For these, the Bar-Shalom/Campo rule, i.e., (3.48) and (3.49), has been applied to fuse (4.16) and (4.17). In Figure 4.6, the according results are drawn in gray. With CI, the green ellipsoid in Figure 4.6(a) has been computed, which correspond to the green EMD in Figure 4.6(b). It can be recognized that it preserves probability mass at the modes of the gray densities.

C   Prediction of Pseudo Gaussian Densities

For static systems, we have shown in Paragraph 4.3.1-A that a proper transformation $\underline{T}$ can easily be obtained by (4.11). Simultaneously to the sensor models, many deterministic systems can also be linearized through a state space transformation. Unfortunately, as mentioned before, a simultaneous linearization of stochastic system dynamics and stochastic sensor models is not possible in general, and even worse, linear system dynamics may become nonlinear through $\underline{T}$ while the sensor models turn into linear mappings. However, in many applications, a local prediction step for the state estimate is required at each node in a sensor network.

In order to obtain an approximate prediction of the transformed state $\underline{x}_k^*$, either the density (4.15) or the system (4.14) can be considered. In the former case, [165, 166] have shown how to predict exponentials of polynomials, which imply that (4.9) and (4.14) are polynomial mappings or polynomial approximations of them. In the latter case, feedback and Carleman linearizations [62, 64, 110] can be applied. Carleman linearization techniques turn systems into bilinear systems, where the product of two Gaussian random vectors can easily be reapproximated by the first two moments.

In this work, we propose to employ a naive sample-based approximation of the system dynamics:

1. Compute W samples $\{\underline{\xi}_j^*\}_{j=1,\dots,W}$ of the current estimate $(\hat{\underline{x}}_k^*, \mathbf{C}_k^*)$, which can be drawn randomly or deterministically.

2. Project each sample back onto manifold. This can be done by applying the inverse T_i^{-1} to each component of each sample $\underline{\xi}_j^*$. If T_i is only locally invertible on the subsets $\mathcal{S}_1^*, \mathcal{S}_2^*, \dots$, then T_i^{-1} is applied on each of these subsets. E.g., for a transformation like $T_i(x) = x^2 = \underline{x}_i^*$, we obtain the two samples $T_i^{-1}(\underline{\xi}_i^*) = \pm\sqrt{\underline{\xi}_i^*}$.

3. The projected samples can the be predicted through the nonlinear system model (4.14).

4. The predicted samples are transformed to $\mathcal{S}^*$ by means of $\underline{T}$. Sample mean and covariance matrix yield the predicted parameters $(\hat{\underline{x}}_{k+1}^{\mathrm{p},*}, \mathbf{C}_{k+1}^{\mathrm{p},*})$.

A second possibility is to express the system mapping in terms of the transformed state $\underline{x}_k^*$. The transformed system model can then be linearized by any standard technique for the purpose of computing the first two moments of the predicted state estimate.

Of course, an approximation of the system model entails the risk that nonlinear dependencies between estimates are neglected that lead to a situation as depicted in Figure 4.1(b). However, in order to counteract this issue and at the expense of higher computational demands, one can choose transformations to higher-order systems where nonlinearities become almost negligible.

4.3.3 Simulations

In order to evaluate the proposed concept, we will consider two scenarios. In the first scenario, a complicated measurement model is employed to transform a one-dimensional state space. In the second simulation, an object is tracked by means of three nodes equipped with quadratic distance sensors. The considered scenarios are extended versions of similar examples in [197].

A Complicated Measurement Model

At first, we confine ourselves to a static state $\underline{x} = 3$. It is observed by five nodes each of which uses the same sensor with the nonlinear measurement model

$$\underline{z} = \begin{bmatrix} 10 \cdot \cos(\underline{x} + 1) + \underline{x} - 5 \\ \underline{x}^2 \end{bmatrix} + \underline{v}$$

where $\underline{v}$ is a zero-mean noise term with covariance matrix

$$\mathbf{C}^z = 3 \cdot \begin{bmatrix} 3 & 1 \\ 1 & 4 \end{bmatrix} .$$

The state is transformed according to

$$\underline{T}(x) = \begin{bmatrix} 10 \cdot \cos(x + 1) + x - 5 \\ x^2 \end{bmatrix} .$$

Figure 4.7(a) shows the estimated pseudo Gaussian density after four measurement steps in each sensor node. The local estimates are interchanged after each local measurement and are fused by utilizing the CI algorithm from Section 3.5.3. The density that lies on the green manifold is depicted separately in Figure 4.7(b). In particular, it corresponds to the green EMD

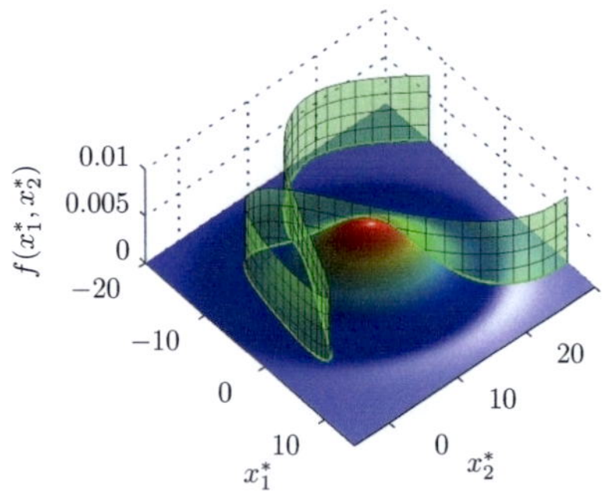

(a) CI fusion result after four time steps.

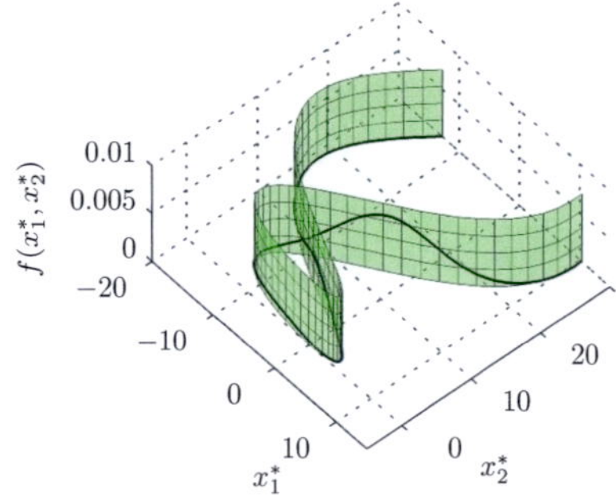

(b) Manifold cuts "true" density out of pseudo Gaussian density.

Figure 4.7: Fig. (a) shows the Pseudo Gaussian density that is obtain by employing the CI algorithm to fuse the estimates of the five sensors. The green density in Fig. 4.8 corresponds to the transformed density in Fig. (b).

fusion result in Figure 4.8. The blue density represents the optimal achievable result, where all measurements are processed centrally. Employing an extended Kalman filter (EKF) leads to the red density that is strongly biased, although a centralized architecture is used. The green result from the transformed state space flattens the modes of the blue density and captures the true state well.

The root mean squared error (RMSE) over 200 Monte Carlo runs, where 10 fusion steps have been performed in each run, is depicted in Figure 4.9(a). The results are more significant than in the similar setup of [197]: The blue line represents the optimum achieved by a centralized architecture. The solid green line is close to the blue one and is obtained by applying CI to the pseudo Gaussian estimates. The dashed green line also stems from a pseudo Gaussian density, but a naive fusion takes places, i.e., independence has erroneously been asserted. In Figure 4.9(b), the RMSE is computed when also a prediction step takes place at each time step. The dynamic evolution is modeled by means of

$$\boldsymbol{x}_{k+1} = 1.1 \cdot \boldsymbol{x}_k + \boldsymbol{w}_k$$

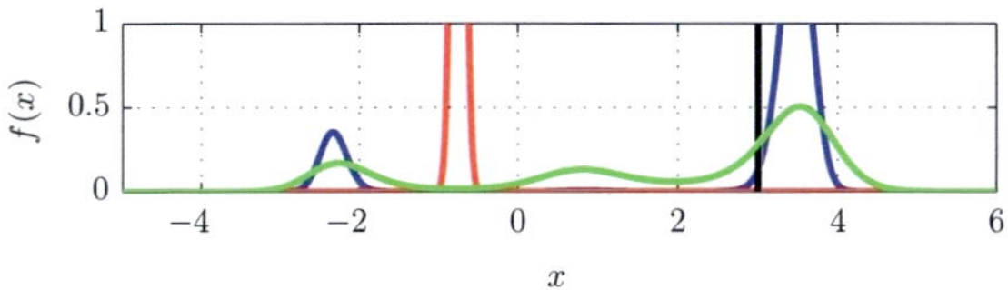

Figure 4.8: Corresponding densities in the original state space. Black line indicates true state. Green: CI fusion result of Figure 4.7. Blue: Optimal fusion result (centralized). Red: EKF (centralized).

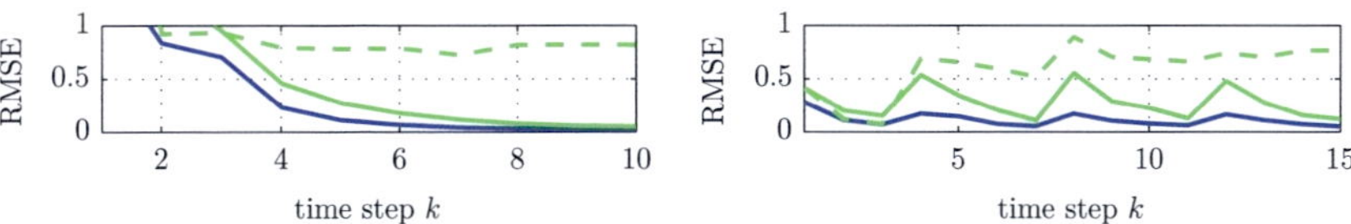

(a) RMSE in case of static system.　　　　**(b)** RMSE in case of dynamic system.

Figure 4.9: Blue line corresponds to the RMSE of the optimal solution. The solid green line is obtained by employing CI in the transformed state space. The dashed green line is obtained when ignoring dependencies.

where $\boldsymbol{w}_k$ has zero mean and variance 2. The prediction is sample-based as explained in Section 4.3.2-C. Again, the green line is closer to the blue one than the dashed line. Here, a horizon over 15 time steps is considered, and the plot slightly suggests that the naive fusion method may even lead to diverging fusion results.

B　Localization of Mobile Object

In the second scenario, a moving target is tracked by means of three sensor nodes located at $P^1 = [1,8]^\mathrm{T}$, $P^2 = [0,5]^\mathrm{T}$, and $P^3 = [4,6]^\mathrm{T}$. The sensor models for the distance measurements are

$$\underline{\boldsymbol{z}}_k^i = (\boldsymbol{x}_k - P_x^i)^2 + (\boldsymbol{y}_k - P_y^i)^2 + \boldsymbol{v}_k$$
$$= \mathbf{H}^* \underline{\boldsymbol{x}}_k^* + (P_x^i)^2 + (P_y^i)^2 + \boldsymbol{v}_k$$

with the transformed measurement mapping

$$\mathbf{H}^* = [-2P_x^i, 1, -2P_y^i, 1]$$

and the transformed state

$$\underline{x}_k^* = \underline{T}(\underline{x}) = \begin{bmatrix} x_k \\ x_k^2 \\ y_k \\ y_k^2 \end{bmatrix} . \tag{4.18}$$

The zero-mean measurement noise $\boldsymbol{v}_k$ has variance 16. The target moves linearly with constant velocity, i.e.,

$$\begin{bmatrix} \boldsymbol{x}_{k+1} \\ \boldsymbol{y}_{k+1} \end{bmatrix} = \begin{bmatrix} \boldsymbol{x}_k \\ \boldsymbol{y}_k \end{bmatrix} + \left(\begin{bmatrix} -1 \\ -0.5 \end{bmatrix} + \underline{\boldsymbol{w}}_k \right) ,$$

where the input noise $\underline{\boldsymbol{w}}_k$ has the covariance matrix $\mathbf{C}_k^w = \mathrm{diag}([0.8, 0.8])$. In this simulation, the prediction is performed by means of the second possibility explained in Section 4.3.2-C. For instance, the second component of (4.18), i.e.,

$$x_{k+1,2}^* = x_{k+1}^2 = \left(a \cdot \boldsymbol{x}_k + b \cdot (\hat{u}_k + \boldsymbol{w}_k) \right)^2 ,$$

is the square of a Gaussian random variable such that the first two moments can be computed analytically. Figure 4.10 shows the centralized solution, where only a single estimate is computed that is fused with all measurements at each time step. For the decentralized implementations in Figure 4.11 and Figure 4.12, only the local estimates at node P^1 are shown, which fuses its own estimate with every other estimate at each time step. In Figure 4.11, each node computes a local estimate, and these estimates are fused by ignoring dependencies. Sensor node P^1 apparently looses the track since the uncertainties are considerably underestimated. Figure 4.12 also shows the locally estimated densities of node P^1, but CI has been applied in order to interchange local estimates, and hence the estimation quality is significantly improved.

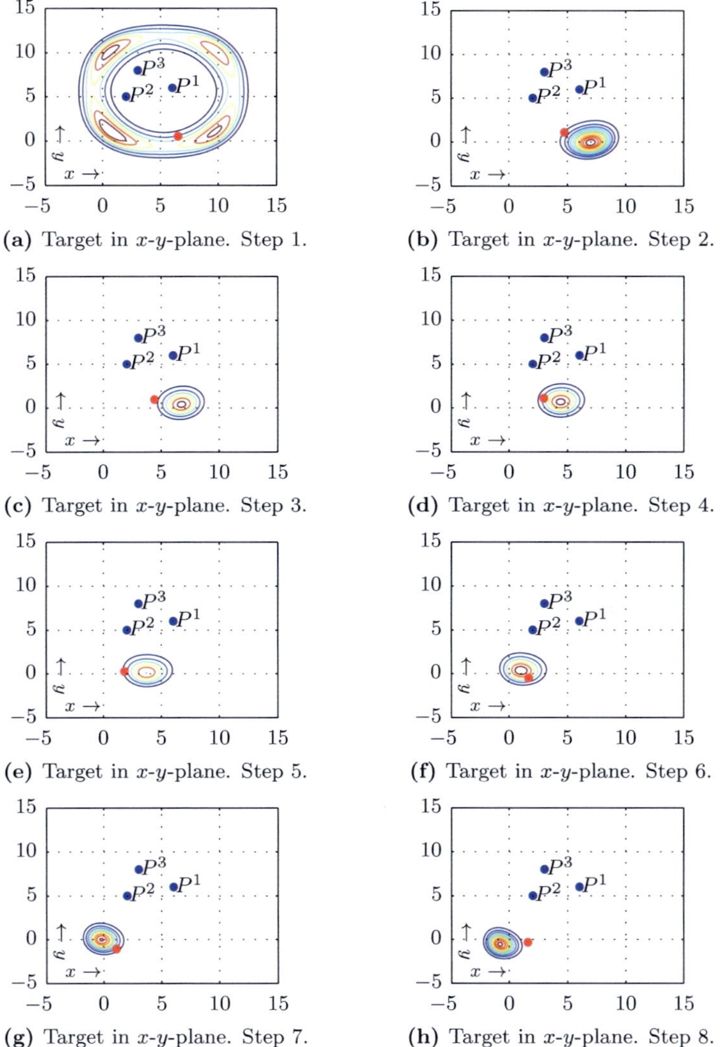

(a) Target in x-y-plane. Step 1.

(b) Target in x-y-plane. Step 2.

(c) Target in x-y-plane. Step 3.

(d) Target in x-y-plane. Step 4.

(e) Target in x-y-plane. Step 5.

(f) Target in x-y-plane. Step 6.

(g) Target in x-y-plane. Step 7.

(h) Target in x-y-plane. Step 8.

Figure 4.10: Probability density contours at different time instants. All measurements are fused in a center station, and the optimal solution is computed. Red dot marks the target's position.

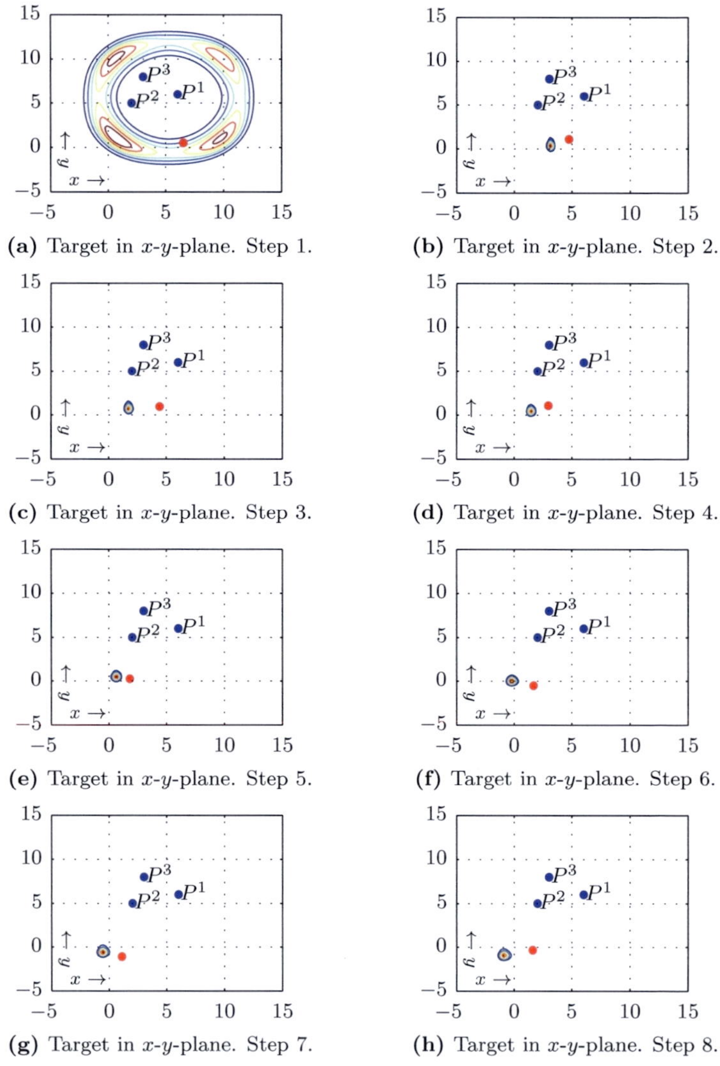

(a) Target in x-y-plane. Step 1.

(b) Target in x-y-plane. Step 2.

(c) Target in x-y-plane. Step 3.

(d) Target in x-y-plane. Step 4.

(e) Target in x-y-plane. Step 5.

(f) Target in x-y-plane. Step 6.

(g) Target in x-y-plane. Step 7.

(h) Target in x-y-plane. Step 8.

Figure 4.11: Probability density contours at different time instants calculated at sensor node P^1. A naive fusion takes place in $\mathcal{S}^*$. Red dot marks the target's position.

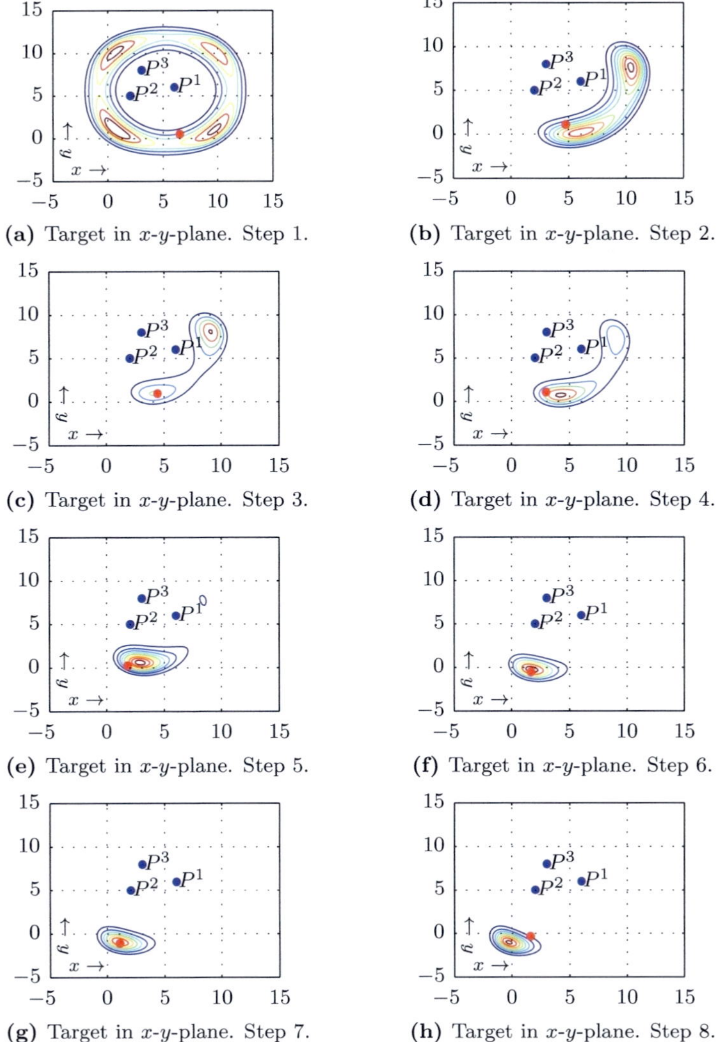

(a) Target in x-y-plane. Step 1.

(b) Target in x-y-plane. Step 2.

(c) Target in x-y-plane. Step 3.

(d) Target in x-y-plane. Step 4.

(e) Target in x-y-plane. Step 5.

(f) Target in x-y-plane. Step 6.

(g) Target in x-y-plane. Step 7.

(h) Target in x-y-plane. Step 8.

Figure 4.12: Probability density contours at different time instants calculated at sensor node P^1. The CI algorithm is applied in $\mathcal{S}^*$ in order to fuse the local estimates at each time step. Red dot marks the target's position.

4.4   Generalized Nonlinear Information Filtering

By employing pseudo Gaussian representations, the nonlinear estimation problem is reduced to a linear one. In the situation that state space transformations like (4.8) do not decrease the demand for communication and computing power, direct approximations of the underlying densities may prove to be more efficient. By this section, a general definition of an information space for arbitrary probability densities is provided. In particular, the filtering step can then be performed in a distributed fashion. The basic concepts behind this section have initially been studied in [182], and the results with respect to multisensor estimation problems that lie in the focus of this section have been published in [203]. In particular, we extend these results by an information-space representation of the EMD fusion rule.

4.4.1   Nonlinear Information Space

In [133], it has been shown that the information filter from Section 3.3.2 can be viewed as a log-likelihood formulation of Bayesian state estimation. More specifically, the Bayesian filtering step

$$
\begin{aligned}
f^{\mathrm{e}}(\underline{x}_k) &= \frac{f^{\mathrm{P}}(\underline{x}_k) \cdot f(\mathcal{Z}_k \mid \underline{x}_k)}{f(\mathcal{Z}_k)} \\
&= \frac{f^{\mathrm{P}}(\underline{x}_k) \cdot \prod_{i=1}^{N} f(\underline{\hat{z}}_k^i \mid \underline{x}_k)}{f(\mathcal{Z}_k)} ,
\end{aligned}
\tag{4.19}
$$

where $\mathcal{Z}_k = \{\underline{\hat{z}}_k^1, \ldots, \underline{\hat{z}}_k^N\}$ denotes the set of multiple observations at time instant k and the conditional independence (2.7) has been exploited, has the very useful representation as the sum

$$
\ln f^{\mathrm{e}}(\underline{x}_k) = \ln f^{\mathrm{P}}(\underline{x}_k) + \sum_{i=1}^{N} \ln f(\underline{\hat{z}}_k^i \mid \underline{x}_k) - \ln f(\mathcal{Z}_k)
\tag{4.20}
$$

of *log-densities*. In the following, we will refer to (4.20) as the *information space* formulation of the nonlinear estimation problem, while (4.19) is referred to as the *state space* formulation. This section aims at solving nonlinear multisensor problems by means of the simple information space representation (4.20).

The logarithm of $f(\mathcal{Z}_k \mid \underline{x}_k)$ turns into a sum due to the conditional independence of the measurements, but the joint probability density $f(\mathcal{Z}_k)$ of the received measurements cannot be split up in the same way. For the same reason, the Kalman filtering step cannot not easily be distributed, which can particularly be observed from inequality (3.4). More specifically, the Kalman filtering step implicitly computes the normalization constant $f(\mathcal{Z}_k)$. In contrast, the additive constant $\ln f(\mathcal{Z}_k)$ is, simply speaking, decoupled and excluded from the filtering process in the information space. Instead, it is implicitly calculated whenever the estimate is transformed back into the state space. For Gaussian densities, the sum $\sum_{i=1}^{N} \ln f(\hat{\underline{z}}_k^i \mid \underline{x}_k)$ in (4.20) directly corresponds to the sums (3.11) and (3.12) of information vectors and matrices. The information matrix is, more precisely, related to the Fisher information, and the Fisher information is related to derivatives of the log-likelihoods. Therefore, constant terms in (4.20) simply vanish. In the following, the reformulation (4.20) of the filtering problem will provide the basis for *nonlinear information filtering* with arbitrary densities and nonlinear models. An important goal is that the nonlinear information filter also remains unaffected of the constant term $\ln f(\mathcal{Z}_k)$ in order to be distributable.

4.4.2 Information Filtering with Logarithms of Densities

For the purpose of employing the simple multisensor filtering step (4.20) in the information space, an approximation technique and parameterization of the participating log-densities has to be derived that render a simple manipulation, i.e., addition and subtraction, possible. More precisely, we aspire the same simple data processing that is provided by the linear information filter. Apparently, only densities that are non-zero almost everywhere can be fused this way. This severe condition can be worked around by confining ourselves to that subset Ω of the probability space, over which all densities to be fused are non-zero. Excluding parts containing "no information" does not impair the estimation quality since the fusion result will also have zero probability mass there.

Filtering of Multiple Measurements   For the purpose of approximating the conditional log-densities in (4.20) and enabling an efficient fusion methodology, the participating functions will be represented by means of an orthonormal basis $\{\varphi_j\}_{j \geq 0}$ in the Hilbert space of square-integrable functions $L^2(\Omega)$ over the domain Ω. So, by confining ourselves to an M-dimensional subspace of $L^2(\Omega)$ spanned by the subset $\{\varphi_j\}_{j=1}^{M}$ of basis functions, an approximation of $\ln f_k^{\mathrm{P}}$ is given by

$$\ln f^{\mathrm{P}}(\underline{x}_k) \approx \sum_{j=1}^{M} \alpha_j^{\mathrm{P}} \cdot \varphi_j(\underline{x}_k) \tag{4.21}$$

with the coefficients

$$\begin{aligned} \alpha_j^{\mathrm{P}} &= \left\langle \ln f^{\mathrm{P}}, \varphi_j \right\rangle_{L^2} \\ &= \int_{\Omega} \ln f^{\mathrm{P}}(\underline{x}_k) \cdot \varphi_j(\underline{x}_k) \, d\underline{x}_k \ , \end{aligned} \tag{4.22}$$

where $\left\langle \, \cdot \, , \, \cdot \, \right\rangle_{L^2}$ denotes the inner product in $L^2(\Omega)$. Analogously, the log-likelihoods $\ln f(\hat{\underline{z}}^i \mid \, \cdot \,)$ of the individual sensor nodes are approximated by

$$\ln f(\hat{\underline{z}}_k^i \mid \underline{x}_k) \approx \sum_{j=1}^{M} \gamma_j^i \cdot \varphi_j(\underline{x}_k) \ , \tag{4.23}$$

where the coefficients

$$\begin{aligned} \gamma_j^i &= \left\langle \ln f(\hat{\underline{z}}_k^i \mid \, \cdot \,), \varphi_j \right\rangle_{L^2} \\ &= \int_{\Omega} \ln f(\hat{\underline{z}}_k^i \mid \underline{x}_k) \varphi_j(\underline{x}_k) \, d\underline{x}_k \end{aligned} \tag{4.24}$$

are also calculated by means of the inner product. An essential prerequisite for representing densities this way is that their logarithms need to be square-integrable, i.e., $\ln f \in L^2(\Omega)$. In the following, we therefore restrict our discussion to probability densities lying in the set

$$\mathcal{P}(\Omega) := \left\{ f \in L^1(\Omega) \mid \mathrm{supp}(f) = \Omega, \ln f \in L^2(\Omega) \right\} \ .$$

This idea of representing probability densities in the information space imposes the following condition on the domain Ω: Since a probability density f is a positive and L^1-integrable function, its logarithm can only be square-integrable if Ω is bounded. At first glance, requiring boundedness of Ω appears to be a harsh restriction, but the domain can, of course, be chosen large enough, so that all "interesting" parts of the participating densities are captured. For instance, we will particularly apply the presented approach to Gaussian noise terms in the following example. All told, every density function in the following discussions is considered to be non-zero over Ω and the domain Ω to be bounded.

The online applicability of the presented idea strongly depends on an efficient evaluation of the sensor log-likelihoods $\ln f(\hat{\underline{z}}_k^i \mid \cdot)$ for given measurements $\hat{\underline{z}}_k^i$, i.e., on the calculation of the coefficient vectors $[\gamma_1^i, \ldots, \gamma_M^i]^{\mathrm{T}}$ from the inner product (4.24). We commence the discussion on efficient log-likelihood computations with an example calculation for a nonlinear sensor model with a Gaussian perturbation. This example is also discussed in [203].

Example 4.4: Gaussian measurement noise
For the sake of simplicity, we consider a one-dimensional state and a scalar-valued measurement function

$$\hat{z}_k = h(\boldsymbol{x}_k) + \boldsymbol{v} \ ,$$

where $\boldsymbol{v}$ is a zero-mean normally distributed random variable with probability density

$$f^v(x) = \tfrac{1}{\sqrt{2\pi}\sigma} \exp\left\{ \tfrac{-(x)^2}{2\sigma^2} \right\}$$

and standard deviation σ. The log-likelihood can then simplified to

$$\ln f(\hat{z}_k \mid x_k) \overset{(2.11)}{=} \ln f^v(\hat{z}_k - h(x_k))$$
$$= -\tfrac{1}{2\sigma^2}(\hat{z}_k - h(x_k))^2 - C \ ,$$

where C is a constant. The coefficients (4.24) are hence given by

$$
\begin{aligned}
\gamma_j &= \langle \ln f^v(\hat{z}_k - h(\,\cdot\,)), \varphi_j \rangle_{L^2} \\
&= \int_\Omega \ln f^v(\hat{z}_k - h(x_k))\varphi_j(x_k)dx_k \\
&= -\tfrac{1}{2\sigma^2} \int_\Omega (\hat{z}_k - h(x_k))^2 \varphi_j(x_k)dx_k - \widehat{C} \\
&= \tfrac{1}{\sigma^2}\hat{z}_k \int_\Omega h(x_k)\varphi_j(x_k)dx_k - \tfrac{1}{2\sigma^2} \int_\Omega h^2(x_k)\varphi_j(x_k)dx_k - \widetilde{C} \ ,
\end{aligned}
$$

where all integrals can be computed offline in advance. Constant terms are subsumed in $\widehat{C}$ and $\widetilde{C}$. Finally, the coefficients are linearly dependent on specific measurement values $\hat{z}_k$ and can therefore easily be evaluated.

In order to deal with other more complex sensor models, the log-likelihoods $\ln f(\underline{z}_k^i \,|\, \underline{x}_k)$ can be interpreted as $n_z \times n_x$-dimensional functions in $L^2(\mathcal{Z} \times \Omega)$, where $\mathcal{Z}$ denotes the measurement space. Let $\{\psi_l\}_{l \geq 0}$ be an orthonormal basis in $L^2(\mathcal{Z})$. By virtue of the tensor product basis $\{\psi_l \otimes \varphi_j\}_{(l,j)} = \{\psi_l \cdot \varphi_j\}_{(l,j)}$, the log-likelihood $\ln f(\,\cdot\,|\,\cdot\,)$ can then be approximated by

$$
\ln f(\underline{z}_k^i \,|\, \underline{x}_k) \approx \sum_{l=1}^{L} \sum_{j=1}^{M} \beta_{l,j}^i \cdot \psi_l(\underline{z}_k^i) \cdot \varphi_j(\underline{x}_k) \tag{4.25}
$$

with

$$
\begin{aligned}
\beta_{l,j}^i &= \big\langle \ln f(\,\cdot\,|\,\cdot\,), \psi_l \cdot \varphi_j \big\rangle_{L^2} \\
&= \int_{\mathcal{Z}} \int_\Omega \ln f(\underline{z}_k^i \,|\, \underline{x}_k)\psi_l(\underline{z}_k^i)\varphi_j(\underline{x}_k)\,d\underline{x}_k d\underline{z}_k^i \ .
\end{aligned} \tag{4.26}
$$

The coefficients $\beta_{l,j}^i$ can be computed numerically in advance. For an obtained observation $\hat{\underline{z}}_k^i$, the corresponding function values $\psi_l(\hat{\underline{z}}_k^i)$ have to be inserted into (4.25). Thus, the coefficients γ_j^i for the calculation (4.23) of $\ln f(\hat{\underline{z}}_k^i \,|\,\cdot\,)$ are obtained by the sum

$$
\gamma_j^i = \sum_{l=1}^{L} \beta_{l,j}^i \cdot \psi_l(\hat{\underline{z}}_k^i) \ ,
$$

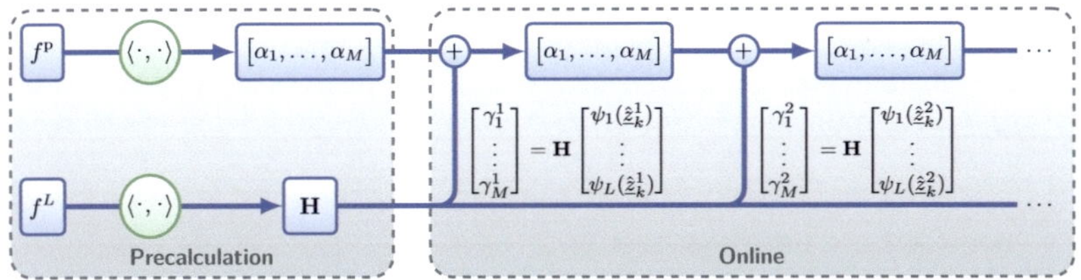

Figure 4.13: Filtering step [182] in terms of coefficient vectors. f^L denotes the likelihood $f(\,\cdot\mid\cdot\,)$ and $\mathbf{H}$ is a matrix with entries (4.26).

which can also be expressed in terms of a matrix-vector multiplication. As aforementioned at the beginning of this section, the normalizing constant $\ln f(\mathcal{Z}_k)$ in the calculation (4.20) is decoupled from the information filtering step, which provides the foundation for efficient distributed fusion algorithms. Accordingly, we have to demand that two log-densities $\ln f$ and $\ln f + C$ with constant offset C have the same coefficients (4.24), which implies

$$\left\langle \ln f, \varphi_j \right\rangle_{L^2} = \left\langle \ln f + C, \varphi_j \right\rangle_{L^2}$$
$$= \left\langle \ln f, \varphi_j \right\rangle_{L^2} + \left\langle C, \varphi_j \right\rangle_{L^2}$$

for every $j = 1, \ldots, m$. This means that the subspace spanned by $\{\varphi_j\}_{j=1}^m$ is orthogonal to the set of constant functions, i.e., $\left\langle C, \varphi_j \right\rangle_{L^2} = 0$. In particular, the complete basis $\{\varphi_j\}_{j\geq 0}$ of $L^2(\Omega)$ needs to contain the constant basis element $\varphi_0 = \left(\sqrt{\mathrm{vol}(\Omega)}\right)^{-1}$, so that all other basis elements are orthogonal, i.e., $0 = \langle \varphi_0, \varphi_i \rangle_{L^2}$ for $i \neq 0$. Hence, a wide variety of truncated basis expansions, e.g., Fourier, wavelet, and Legendre basis expansions, can be employed to approximate the log-densities.

With this spadework, the generalized filtering step (4.20) for N sensors can now be written in terms of the corresponding coefficient vectors (4.22) and (4.24), i.e.,

$$[\alpha_1^{\mathrm{e}}, \ldots, \alpha_M^{\mathrm{e}}]^{\mathrm{T}} = [\alpha_1^{\mathrm{P}}, \ldots, \alpha_M^{\mathrm{P}}]^{\mathrm{T}} + \sum_{i=1}^{N} [\gamma_1^i, \ldots, \gamma_M^i]^{\mathrm{T}} \,,$$

which yields the parameter vector $\underline{\alpha}_k^{\mathrm{e}} := [\alpha_1^{\mathrm{e}}, \ldots, \alpha_M^{\mathrm{e}}]^{\mathrm{T}}$ of the estimated log-density $\ln f^{\mathrm{e}}$. Apparently, this complies with the same simple fusion structure as in the linear case, i.e., (3.5) and (3.12), as illustrated in Figure 4.13. The two main prerequisites for the presented approach can be summarized as follows:

1. The bounded domain Ω is chosen such that the considered probability densities are non-zero over Ω.

2. The orthonormal basis $\{\varphi_j\}_{j=1}^M$ is orthogonal to constant functions.

EMD Fusion Rule in Information Space In linear estimation theory, the covariance intersection algorithm from Section 3.5.3 is related to a convex combination of information vectors and information matrices. Analogously, the EMD fusion rule (4.7) correspond to the convex combination

$$\ln f_{\mathrm{EMD}}(\underline{x} \,|\, \mathcal{Z}_A \cup \mathcal{Z}_B) = \ln f^{\omega}(\underline{x} \,|\, \mathcal{Z}_A) + \ln f^{1-\omega}(\underline{x} \,|\, \mathcal{Z}_A) - C$$
$$= \omega \ln f(\underline{x} \,|\, \mathcal{Z}_A) + (1 - \omega) \ln f(\underline{x} \,|\, \mathcal{Z}_A) - C$$

of log-densities. Hence, the coefficient vector of $\ln f_{\mathrm{EMD}}(\underline{x} \,|\, \mathcal{Z}_A \cup \mathcal{Z}_B)$ is also related to the convex combination

$$[\alpha_1, \ldots, \alpha_M]_{\mathrm{EMD}}^{\mathrm{T}} = \omega[\alpha_1, \ldots, \alpha_M]_A^{\mathrm{T}} + (1 - \omega)[\alpha_1, \ldots, \alpha_M]_B^{\mathrm{T}}$$

of the corresponding coefficient vectors of $\ln f(\underline{x} \,|\, \mathcal{Z}_A)$ and $\ln f(\underline{x} \,|\, \mathcal{Z}_B)$.

Prediction of Log-densities As for the linear information filter in Section 3.3.2, the benefits concerning distributed data fusion come at the expense of a more complicated prediction step. For nonlinear systems, the Chapman-Kolmogorov integral for predicting the conditional densities can in general not be solved in closed form and, in the information space, prediction becomes even more elaborate. So, either the log-densities are transformed back to state space at each prediction step or the prediction is expressed in information space. For deterministic or static systems, the second solution may be the best choice, but for stochastic systems, a transformation to state space can be inevitable, i.e., the Chapman-Kolmogorov integral becomes

$$\ln f^p(\underline{x}_{k+1}) = \ln \int_{\Omega} f(\underline{x}_{k+1} \,|\, \underline{x}_k) \exp\{\ln f^e(\underline{x}_k)\} \, d\underline{x}_k$$

for the log-densities. Each operation, i.e., the inverse transformation exp and the transformation ln of the integral, have, in general, to be approximated. In order to calculate the first operation, the coefficients of $\ln f^e$ can approximately be mapped to an L^2-basis in state space. In terms of the obtained coefficients, the integral can then be evaluated. Finally, the coefficients are mapped back to the information space basis. Simulations of the proposed concept can be found in [203]. Altogether, the implementation of the prediction step depends on the actual system model and a general solution cannot be stated.

While this subsection has focused on log-densities and their approximations, the following subsection will consider the implications on the probability densities themselves, i.e., the implications on the state space formulation.

4.4.3   Hilbert Space Structure on Probability Densities

For efficient nonlinear state estimation in state space, truncated Fourier [12, 29, 30] or wavelet [79] series expansions have also been successfully applied. In contrast to (4.21), densities are then directly approximated by

$$f^{\mathrm{P}}(\underline{x}_k) \approx \sum_{j=1}^{M} a_j^{\mathrm{P}} \cdot \phi_j(\underline{x}_k) \tag{4.27}$$

with $a_j^{\mathrm{P}} = \langle f^{\mathrm{P}}, \phi_j \rangle_{L^2}$. As expected, we face the same situation as in the linear case: While filtering is easy in information space and prediction is elaborate, the opposite holds for the state space approximation (4.27). Here, a reapproximation is required for every filtering step, which is a second issue besides the the problem of efficient multisensor data processing in state space. Another problem of the truncated series (4.27) is that, in general, it does not represent a valid probability density, i.e., it possibly does not integrate to one and it can even take negative function values. At this point, an essential advantage of the information space representation becomes apparent. Transforming (4.21) back to state space always yields a valid probability density. Even a single basis function φ_j corresponds to the probability density function

$$f_j(\underline{x}_k) = \frac{\exp\{\varphi_j(\underline{x}_k)\}}{\int_\Omega \exp\{\varphi_j(\underline{x}_k)\}\, d\underline{x}_k}$$

in state space and for the series expansion (4.21), we obtain

$$f(\underline{x}_k) \approx \frac{\prod_{j=1}^{M} \left(\exp\{\varphi_j(\underline{x}_k)\}\right)^{\gamma_j}}{\int_\Omega \prod_{j=1}^{M} \left(\exp\{\varphi_j(\underline{x}_k)\}\right)^{\gamma_j} d\underline{x}_k} \; .$$

This implies that the addition ”+“ and the multiplication ” $\cdot$ “ with a scalar $\alpha \in \mathbb{R}$ in information space correspond to the operations

$$f \oplus g := \frac{f(\,\cdot\,) \cdot g(\,\cdot\,)}{\int_\Omega f(\underline{x}) \cdot g(\underline{x})\, d\underline{x}} \tag{4.28}$$

and

$$a \odot f := \frac{f^a(\,\cdot\,)}{\int_\Omega f^a(\underline{x})\, d\underline{x}} \tag{4.29}$$

in state space, respectively. As a generalization of the Aitchison geometry [1], a Hilbert space $A(\Omega)$ on probability densities has been developed by means of these operations [54], where

$$\langle f, g \rangle_{A(\Omega)} := \frac{1}{2\mathrm{vol}(\Omega)} \int_\Omega \int_\Omega \ln \frac{f(\underline{x})}{f(\underline{y})} \ln \frac{g(\underline{x})}{g(\underline{y})}\, d\underline{x}\, d\underline{y}$$

$$= \int_\Omega \ln f(\underline{x}) \ln g(\underline{x})\, d\underline{x} - \frac{1}{\mathrm{vol}(\Omega)} \int_\Omega \ln f(\underline{x})\, d\underline{x} \int_\Omega \ln g(\underline{x})\, d\underline{x}$$

is the inner product in $A(\Omega)$. This product induces the vector space norm

$$\|f\|_{A(\Omega)} = \left[\int_\Omega \left(\ln f(\underline{x})\right)^2 d\underline{x} - \frac{1}{\mathrm{vol}(\Omega)}\left(\int_\Omega \ln f(\underline{x})\, d\underline{x}\right)^2 \right]^{\frac{1}{2}}$$

on $A(\Omega)$, which has, for example, been applied as an information measure for sensor management in [195]. Simply speaking, the spaces $A(\Omega)$ and $L^2(\Omega)$ can be related by the isometry $\ln : A(\Omega) \to L^2(\Omega)$, with which it can be proven that $A(\Omega)$ is a Hilbert space. For the construction of bases for the product space $A(\Omega) \times A(\Psi)$, the tensor product employed in (4.25) becomes

$$f \otimes_{A(\Omega)} g = \frac{\exp\{\ln f \cdot \ln g\}}{\int_\Omega \int_\Psi \exp\{\ln f(\underline{x}) \cdot \ln g(\underline{y})\} \, d\underline{y} d\underline{y}} \; .$$

Of course, likelihoods are not probability densities and therefore not elements of $A(\Omega)$, but this does not pose a problem due to the normalization factor in (4.28). In conclusion, the information space considered as the function space L^2 of square-integrable functions is strongly related to a Hilbert space structure in the state space. The vector space addition and multiplication correspond to the Bayesian update (4.28) and the power transformation (4.29), which is essentially a weighting of information. All approximation techniques in this space are compliant to operations on probabilities and therefore yield valid probability densities. In other words, complicated probability densities are approximated by sums of simpler probability densities, which makes the information space representation particularly attractive for stochastic state estimation.

The main downside of the proposed concept is that the information space involves logarithms of probability densities. In particular, function values that are close to zero can cause numerical instabilities when the logarithm is computed. In general, the filtering step can significantly be eased, but the prediction step may become disproportionately more complicated. However, the nonlinear information space constitutes a promising basis for further insights into efficient nonlinear multisensor filtering.

4.5 Nonlinear Federated Filtering

The same sources of dependent information that have been stated in Section 3.2 are encountered in nonlinear networked systems: Interchanging sensor data entails the risk of double-counting information, and sensor sites may utilize the same local process model. Common sensor information poses a minor problem if data is collected and fused in a data sink or central sensor node, and the network topology contains no cycles. Common process noise, on the contrary, has to be reckoned with whenever estimates are computed locally and are irregularly transferred at arbitrary instants of time. The federated Kalman filter, which has been introduced in Section 3.6.3, embodies an easy-to-implement solution and employs an inflated joint noise covariance matrix in order to keep the local tracks decorrelated. As a result, the local estimates are more conservative, but their fusion can be performed by means of a simple Kalman filtering step and yields an estimate with tighter second moment error statistics than CI would provide.

The idea behind federated Kalman filtering also shapes up to be a promising strategy for nonlinear estimation problems. Accordingly, we aspire to resolve the interdependencies among local prediction models into independence. Each node i in a sensor network computes an estimate of a "copy" $\underline{x}_k^i$ of the true state, which corresponds to the joint state vector

$$\begin{bmatrix} \underline{x}_k^1 \\ \vdots \\ \underline{x}_k^N \end{bmatrix} := \begin{bmatrix} \mathbf{I} \\ \vdots \\ \mathbf{I} \end{bmatrix} \underline{x}_k \; . \tag{4.30}$$

A decorrelation is therefore already required in the early beginning in order to initialize the local estimators and to distribute the prior density $f^{\mathrm{P}}(\underline{x}_0)$ for $\underline{x}_0$ among the sensor nodes. For this purpose, the prior density can be split into the product

$$f^{\mathrm{P}}(\underline{x}_0) = \left(f^{\mathrm{P}}(\underline{x}_0) \right)^{\omega_1} \cdot \ldots \cdot \left(f^{\mathrm{P}}(\underline{x}_0) \right)^{\omega_N} \tag{4.31}$$

with $\sum_{i=1}^{N} \omega_i = 1$ and $\omega_i > 0$. Each sensor node can be initialized by means of a factor of the product (4.31), which leads to the normalized local prior

$$f^{\mathrm{P},i}(\underline{x}_0) := \frac{\left(f^{\mathrm{P}}(\underline{x}_0) \right)^{\omega_i}}{\int_{\mathbb{R}^{n_x}} \left(f^{\mathrm{P}}(\underline{x}_0) \right)^{\omega_i} \, \mathrm{d}\underline{x}_0} \; .$$

The joint density for (4.30) then yields

$$f^{\mathrm{P}}\left(\underline{x}_0^1, \ldots, \underline{x}_0^N \right) = f^{\mathrm{P},1}(\underline{x}_0^1) \cdot \ldots \cdot f^{\mathrm{P},N}(\underline{x}_0^N) \tag{4.32}$$

The local prior estimates can now be assumed to be independent and therefore be fused according to (4.2), i.e.,

$$\begin{aligned} f^{\mathrm{P}}(\underline{x}_0) &= c \cdot f^{\mathrm{P}}\left(\underline{x}_0, \ldots, \underline{x}_0 \right) \\ &= c \cdot f^{\mathrm{P},1}(\underline{x}_0) \cdot \ldots \cdot f^{\mathrm{P},N}(\underline{x}_0) \; . \end{aligned} \tag{4.33}$$

with normalization constant c. The product form (4.31) shows a distinctive analogy to the EMD fusion rule (4.7), where, for a conservative fusion with (4.7), possibly dependent densities are each raised to the power ω_i. For the *nonlinear federated filter*, we pursue the opposite direction and split a single density into an exponential mixture of independent densities.

In compliance with the linear federated Kalman filter, the prediction models have to be relaxed in order to preserve the product from (4.32). It is easy to admit that a local measurement update by means of Bayes' rule (2.6) does not cause any dependencies. The system model $\tilde{\underline{a}}_k$ for the joint state (4.30), i.e.

$$
\begin{bmatrix} \underline{x}^1_{k+1} \\ \vdots \\ \underline{x}^N_{k+1} \end{bmatrix} = \tilde{\underline{a}}_k \left(\begin{bmatrix} \underline{x}^1_k \\ \vdots \\ \underline{x}^N_k \end{bmatrix}, \begin{bmatrix} \mathbf{I} \\ \vdots \\ \mathbf{I} \end{bmatrix} \underline{w}_k \right) := \begin{bmatrix} \underline{a}_k(\underline{x}^1_k, \underline{w}_k) \\ \vdots \\ \underline{a}_k(\underline{x}^N_k, \underline{w}_k) \end{bmatrix} , \tag{4.34}
$$

consists of multiple versions of the same model $\underline{a}_k$, and hence the system noise $\underline{w}_k$ effects dependencies among the predicted states. The prediction step is carried out by means of the Chapman-Kolmogorov integral (2.5), which yields

$$
f^{\mathrm{P}}(\underline{x}^1_{k+1}, \dots, \underline{x}^N_{k+1}) = \int_{\mathbb{R}^{n_x}} \cdots \int_{\mathbb{R}^{n_x}} f(\underline{x}^1_{k+1}, \dots, \underline{x}^N_{k+1} \mid \underline{x}^1_k, \dots, \underline{x}^N_k)
$$
$$
\cdot f^{\mathrm{e}}(\underline{x}^1_k, \dots, \underline{x}^N_k) \, \mathrm{d}\underline{x}^1_k \dots \mathrm{d}\underline{x}^N_k
$$

for the joint state space. This integral can only preserve the form (4.32) if the transition density can also be transformed into a product of densities. The subsequent subsections show how to maintain the product form after prediction steps.

4.5.1   Relaxed Prediction Model for Additive Gaussian Noise

At first, we confine ourselves to a Gaussian process noise $\underline{w}_k \sim \mathcal{N}(\underline{0}, \mathbf{C}^w_k)$ that additively affects the state transition according to

$$
\underline{x}_{k+1} = \underline{a}_k(\underline{x}_k) + \underline{w}_k \; . \tag{4.35}
$$

For the sake of simplicity, we do not take possible inputs into consideration. In line with (2.10), the transition density can then be written in terms of a Gaussian density as

$$f(\underline{x}_{k+1} \,|\, \underline{x}_k) = \mathcal{N}\left(\underline{x}_{k+1} - \underline{a}_k(\underline{x}_k); \underline{0}, \mathbf{C}_k^w\right) \,, \qquad (4.36)$$

and, with the system model (4.35), equation (4.34) for the predicted joint state can be simplified to

$$\begin{bmatrix} \boldsymbol{x}_{k+1}^1 \\ \vdots \\ \boldsymbol{x}_{k+1}^N \end{bmatrix} = \begin{bmatrix} \underline{a}_k(\boldsymbol{x}_k^1) \\ \vdots \\ \underline{a}_k(\boldsymbol{x}_k^N) \end{bmatrix} + \begin{bmatrix} \mathbf{I} \\ \vdots \\ \mathbf{I} \end{bmatrix} \underline{\boldsymbol{w}}_k \,. \qquad (4.37)$$

As a result, the transition density function (4.36) becomes the joint transition density

$$f(\underline{x}_{k+1}^1, \ldots, \underline{x}_{k+1}^N \,|\, \underline{x}_k^1, \ldots, \underline{x}_k^N)$$

$$= \mathcal{N}\left(\begin{bmatrix} \underline{x}_{k+1}^1 \\ \vdots \\ x_{k+1}^N \end{bmatrix} - \begin{bmatrix} \underline{a}_k(x_k^1) \\ \vdots \\ \underline{a}_k(x_k^N) \end{bmatrix} ; \begin{bmatrix} \underline{0} \\ \vdots \\ \underline{0} \end{bmatrix} , \begin{bmatrix} \mathbf{C}_k^w & \mathbf{C}_k^w & \cdots & \mathbf{C}_k^w \\ \mathbf{C}_k^w & \mathbf{C}_k^w & \ddots & \vdots \\ \vdots & \ddots & \ddots & \mathbf{C}_k^w \\ \mathbf{C}_k^w & \cdots & \mathbf{C}_k^w & \mathbf{C}_k^w \end{bmatrix} \right)$$

for the state transition model (4.37). In Chapter 3, we have become acquainted with several possibilities to conservatively replace correlated Gaussian estimates with independent ones. In particular, the linear federated Kalman filter in Section 3.6.3 employs the inflated process noise covariance matrix (3.66). In the same manner, we can decompose the joint

Gaussian transition density into the product

$$
\tilde{f}(\underline{x}^1_{k+1}, \ldots, \underline{x}^N_{k+1} \mid \underline{x}^1_k, \ldots, \underline{x}^N_k)
$$

$$
= \mathcal{N}\left(
\begin{bmatrix} \underline{x}^1_{k+1} \\ \vdots \\ \underline{x}^N_{k+1} \end{bmatrix}
-
\begin{bmatrix} \underline{a}_k(\underline{x}^1_k) \\ \vdots \\ \underline{a}_k(\underline{x}^N_k) \end{bmatrix}
;
\begin{bmatrix} \underline{0} \\ \vdots \\ \underline{0} \end{bmatrix}
,
\begin{bmatrix}
\frac{1}{\omega_1}\mathbf{C}^w_k & \mathbf{0} & \cdots & \mathbf{0} \\
\mathbf{0} & \frac{1}{\omega_2}\mathbf{C}^w_k & \ddots & \vdots \\
\vdots & \ddots & \ddots & \mathbf{0} \\
\mathbf{0} & \cdots & \mathbf{0} & \frac{1}{\omega_N}\mathbf{C}^w_k
\end{bmatrix}
\right)
$$

$$
= \underbrace{\mathcal{N}\left(\underline{x}^1_{k+1} - \underline{a}_k(\underline{x}^1_k); \underline{0}, \tfrac{1}{\omega_1}\mathbf{C}^w_k\right)}_{=:\tilde{f}(\underline{x}^1_{k+1} \mid \underline{x}^1_k)}
\cdots
\underbrace{\mathcal{N}\left(\underline{x}^N_{k+1} - \underline{a}_k(\underline{x}^N_k); \underline{0}, \tfrac{1}{\omega_N}\mathbf{C}^w_k\right)}_{=:\tilde{f}(\underline{x}^N_{k+1} \mid \underline{x}^N_k)}
$$

$$(4.38)$$

by employing the conservative bound (3.66) for the noise covariance matrix. If the density f^{e} to be predicted is given in a product form analogous to (4.32), also the prediction result can now be decomposed into the product

$$
\tilde{f}^{\mathrm{P}}(\underline{x}^1_{k+1}, \ldots, \underline{x}^N_{k+1})
$$

$$
= \int_{\mathbb{R}^{n_x}} \cdots \int_{\mathbb{R}^{n_x}} \tilde{f}(\underline{x}^1_{k+1} \mid \underline{x}^1_k) \ldots \tilde{f}(\underline{x}^N_{k+1} \mid \underline{x}^N_k) \cdot f^{\mathrm{e}}(\underline{x}^1_k) \ldots f^{\mathrm{e}}(\underline{x}^N_k) \, \mathrm{d}\underline{x}^1_k \ldots \mathrm{d}\underline{x}^N_k
$$

$$
= \underbrace{\left(\int_{\mathbb{R}^{n_x}} \tilde{f}(\underline{x}^1_{k+1} \mid \underline{x}^1_k) f^{\mathrm{e}}(\underline{x}^1_k) \, \mathrm{d}\underline{x}^1_k \right)}_{= f^{\mathrm{P}}(\underline{x}^1_{k+1})}
\cdots
\underbrace{\left(\int_{\mathbb{R}^{n_x}} \tilde{f}(\underline{x}^N_{k+1} \mid \underline{x}^N_k) f^{\mathrm{e}}(\underline{x}^N_k) \, \mathrm{d}\underline{x}^N_k \right)}_{= f^{\mathrm{P}}(\underline{x}^N_{k+1})} .
$$

The integrals can be computed on the individual sensor nodes and yield the locally predicted densities $f^{\mathrm{P}}(\underline{x}^i_{k+1})$ for the state estimate. In doing so, a fusion of local estimates by means of (4.33) provides a conservative result. This concept can be generalized to arbitrary noise densities and models, as discussed in the following subsection.

4.5.2 Arbitrary System Models

The nonlinear federated filter is strongly related to exponentials of densities: The decomposition (4.38) yields the factors $\tilde{f}(\underline{x}^i_{k+1} \mid \underline{x}^i_k)$ that can each be rewritten as

$$\tilde{f}(\underline{x}_{k+1}^i \,|\, \underline{x}_k^i) \;=\; \mathcal{N}\!\left(\underline{x}_{k+1}^i - \underline{a}_k(\underline{x}_k^i); \underline{0}, \tfrac{1}{\omega_i}\mathbf{C}_k^w\right)$$

$$=\; c^i\left(\mathcal{N}\!\left(\underline{x}_{k+1}^i - \underline{a}_k(\underline{x}_k^i); \underline{0}, \mathbf{C}_k^w\right)\right)^{\omega_i}$$

$$\overset{(4.36)}{=}\; c^i f^{\omega_i}(\underline{x}_{k+1}^i \,|\, \underline{x}_k^i)\ .$$

By following the same train of thought that leads to the EMD fusion rule (4.7), any joint transition density can be conservatively represented by an exponential mixture density

$$f_{\mathrm{EMD}}(\underline{x}_{k+1}^1, \ldots, \underline{x}_{k+1}^N \,|\, \underline{x}_k^1, \ldots, \underline{x}_k^N)$$
$$= c^1 f^{\omega_1}(\underline{x}_{k+1}^1 \,|\, \underline{x}_k^1) \cdot\ \ldots\ \cdot c^N f^{\omega_N}(\underline{x}_{k+1}^N \,|\, \underline{x}_k^N)$$

with $\sum_{i=1}^N \omega_i = 1$ and $\omega_i \geq 0$. Here, the EMD representation is used to decompose one density into several densities and not for fusing several densities into one density. Each factor $f^{\omega_i}(\underline{x}_{k+1}^i \,|\, \underline{x}_k^i)$ can locally be computed for any system model

$$\underline{x}_{k+1} = \underline{a}_k(\underline{x}_k, \underline{w}_k)$$

by means of (2.4) and exploited in the local prediction step

$$f^{\mathrm{p}}(\underline{x}_{k+1}^i) = \int_{\mathbb{R}^{n_x}} c^i f^{\omega_i}(\underline{x}_{k+1}^i \,|\, \underline{x}_k^i) \cdot f^{\mathrm{e}}(\underline{x}_k^i)\ \mathrm{d}\underline{x}_k^i$$

of sensor node i. Apparently, we employ (4.7) and the notion of consistency discussed in Section 4.2 in order to conservatively represent the transition density by a product of independent transition densities.

4.5.3 Simulations

The proposed nonlinear federated filter is evaluated with the help of a target tracking scenario, which is similar to the setup considered in Section 4.3.3-B. Five sensor nodes are positioned at $P^1 = [1,8]^{\mathrm{T}}$, $P^2 = [0,5]^{\mathrm{T}}$, $P^3 = [4,6]^{\mathrm{T}}$, $P^4 = [7,9]^{\mathrm{T}}$, and $P^5 = [3,12]^{\mathrm{T}}$. The two-dimensional state vector $[\boldsymbol{x}_k, \boldsymbol{y}_k]$ is observed by each sensor according to the nonlinear measurement models

$$\boldsymbol{z}_k^i = (\boldsymbol{x}_k - P_x^i)^2 + (\boldsymbol{y}_k - P_y^i)^2 + \boldsymbol{v}_k$$

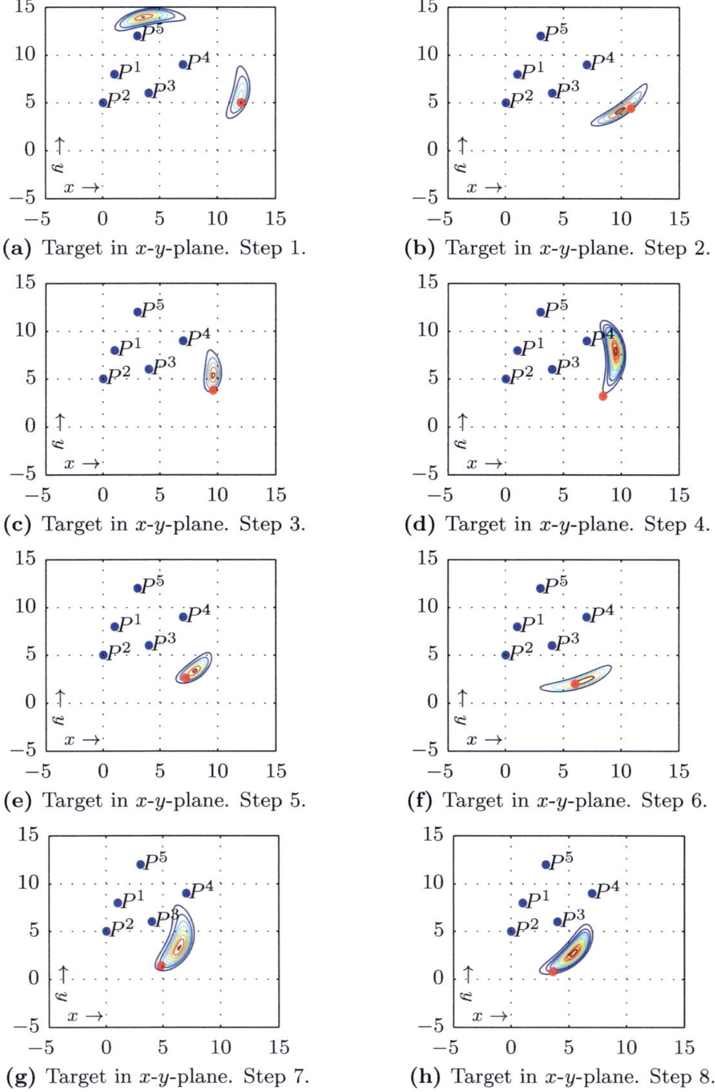

(a) Target in x-y-plane. Step 1.

(b) Target in x-y-plane. Step 2.

(c) Target in x-y-plane. Step 3.

(d) Target in x-y-plane. Step 4.

(e) Target in x-y-plane. Step 5.

(f) Target in x-y-plane. Step 6.

(g) Target in x-y-plane. Step 7.

(h) Target in x-y-plane. Step 8.

Figure 4.14: Probability density contours at different time instants. Centralized optimal solution. Red dot marks the target's position.

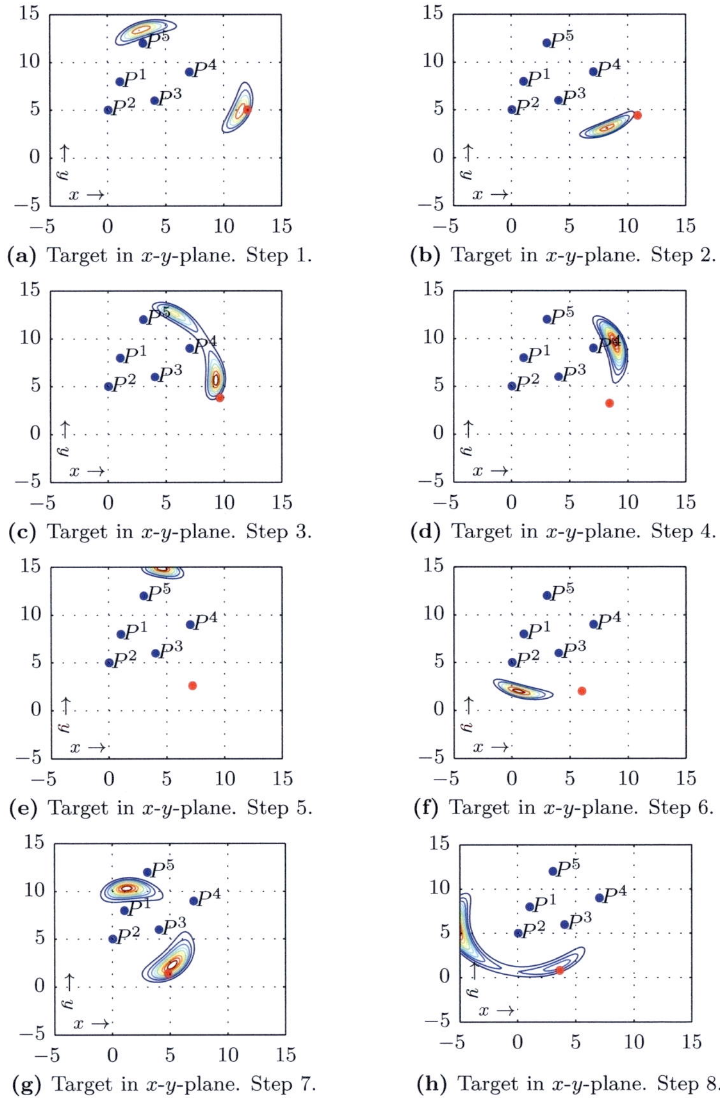

(a) Target in x-y-plane. Step 1.

(b) Target in x-y-plane. Step 2.

(c) Target in x-y-plane. Step 3.

(d) Target in x-y-plane. Step 4.

(e) Target in x-y-plane. Step 5.

(f) Target in x-y-plane. Step 6.

(g) Target in x-y-plane. Step 7.

(h) Target in x-y-plane. Step 8.

Figure 4.15: Probability density contours at different time instants show naive fusion results of local estimates; dependencies are ignored. Red dot marks the target's position.

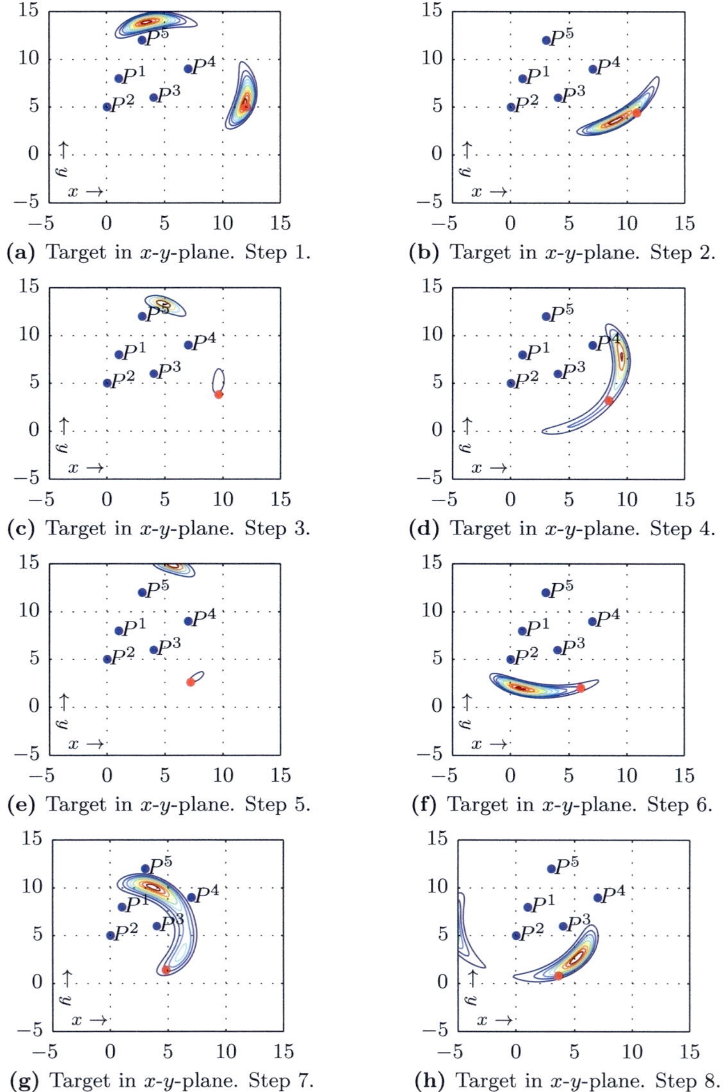

(a) Target in x-y-plane. Step 1.

(b) Target in x-y-plane. Step 2.

(c) Target in x-y-plane. Step 3.

(d) Target in x-y-plane. Step 4.

(e) Target in x-y-plane. Step 5.

(f) Target in x-y-plane. Step 6.

(g) Target in x-y-plane. Step 7.

(h) Target in x-y-plane. Step 8.

Figure 4.16: Probability density contours at different time instants. Nonlinear federated filtering with relaxed prediction step; local results are fused at each time step. Red dot marks the target's position.

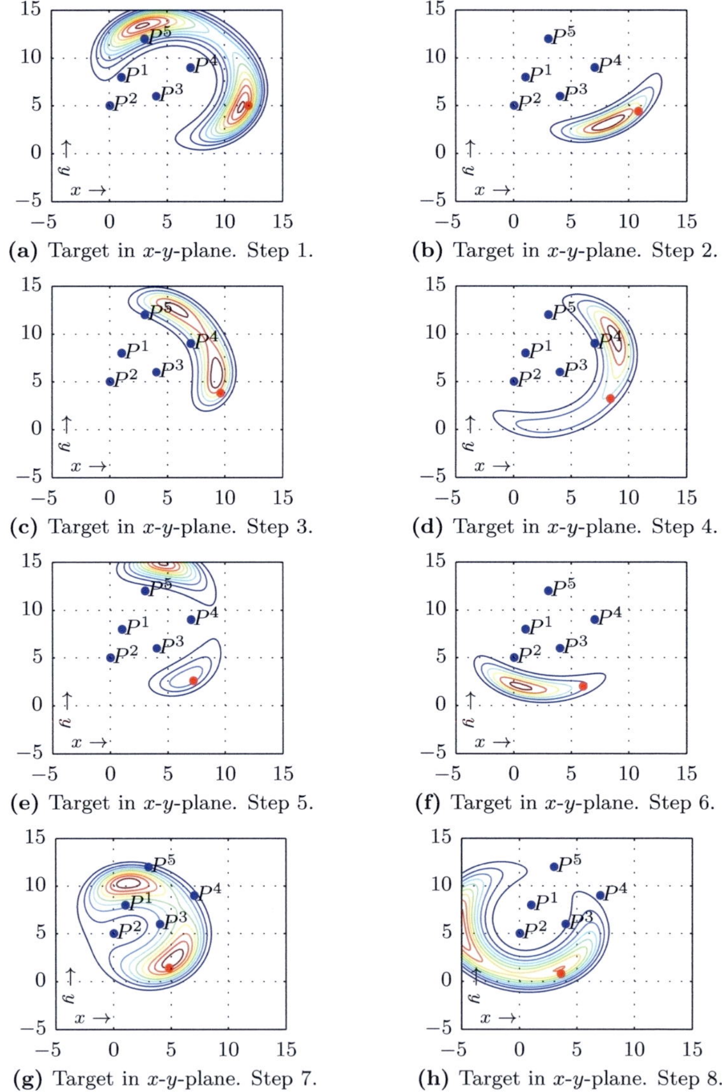

(a) Target in x-y-plane. Step 1.

(b) Target in x-y-plane. Step 2.

(c) Target in x-y-plane. Step 3.

(d) Target in x-y-plane. Step 4.

(e) Target in x-y-plane. Step 5.

(f) Target in x-y-plane. Step 6.

(g) Target in x-y-plane. Step 7.

(h) Target in x-y-plane. Step 8.

Figure 4.17: Probability density contours at different time instants. Local estimates are fused by means of the EMD fusion rule. Red dot marks the target's position.

with $\boldsymbol{v}_k \sim \mathcal{N}(0, 8^2)$. The prior density is a Gaussian density with high variance being distributed to the sensor nodes. Each calculated density is represented on a spatial grid of 1000×1000 points and the trapezoidal rule is utilized to compute the integrals. The state transition model is

$$\begin{bmatrix} \boldsymbol{x}_{k+1} \\ \boldsymbol{y}_{k+1} \end{bmatrix} = \begin{bmatrix} \boldsymbol{x}_k \\ \boldsymbol{y}_k \end{bmatrix} + \left(\begin{bmatrix} -1 \\ -0.5 \end{bmatrix} + \underline{\boldsymbol{w}}_k \right) ,$$

with $\underline{\boldsymbol{w}}_k$ having zero mean and covariance matrix $\mathbf{C}_k^w = \mathrm{diag}([4, 4])$. At every time step, the local estimation results are send to a data sink where they are fused, which is required to plot the estimated density. For Figure 4.14, a full-rate communication is inevitable since all measurements have to be send to the data sink, where the estimate is computed. This is the centralized solution, which serves as a ground truth. In Figure 4.15, each node employs the standard prediction step and possible dependencies are ignored. Figures 4.15(d)–(f) show that the track is lost. The nonlinear federated filter has been employed for the results in Figure 4.16 and preserves probability mass around the target's position. Using the standard prediction step and employing EMDs for fusion lead to the tracking result depicted in Figure 4.17, which is very conservative compared to the solution provided by the nonlinear federated filter. Nonlinear federated filtering hence provides more informative fusion results since it exploits additional knowledge about the network, i.e., the number of nodes and the absence of cycles. It is worth pointing out that Figure 4.17 also corresponds to the approach based on pseudo Gaussian densities studied in Section 4.3.

4.6   Conclusions from Chapter 4

Nonlinear state estimation is often in itself an arduous task, but in a networked system it becomes even more challenging. In distributed or decentralized systems, it is, in first place, difficult to get a clear picture of how dependencies are shaped. Without any further knowledge, possible fusion results of two locally estimated densities can be as arbitrary as their possible dependencies, which can even appear for Gaussian estimates to be haphazard. Hence, unknown dependencies among estimates have the unfavorable characteristic of possibly causing very unexpected fusion results. Accordingly, we are unable to constitute a conservative bound on the possible fusion results, unless more information about the networked estimation system can be utilized. As explained in Chapter 3, dependencies arise from cycles in the network topology and local state transition models that share a common process noise term. Hence, the same error model is incorporated multiple times when estimates are processed locally. This particular source of dependent information probably limits the arbitrariness of possible inderdependencies. Against this background, a notion of conservative fusion can then be defined that avoids being overconfident with regard to common sensor information and common process noise.

The exponential mixture density (EMD) fusion rule preserves the defined notion of conservativeness and constitutes a direct generalization of the CI algorithm to arbitrary probability densities. In many estimation problems, EMDs are also related to a higher-dimensional state space, where models and dependencies become linear. Estimates can there be represented by pseudo Gaussian densities, and dependencies among them can again be parameterized in terms of correlation coefficients, i.e., cross-covariance matrices. Applying CI to these pseudo Gaussian estimates in the transformed state space directly corresponds to the EMD fusion method being applied to the underlying densities in the original state space.

Like CI, many linear estimation concepts for networked systems are based on the information form of the Kalman filter. The information form is related to a log-likelihood formulation of the filtering step and can therewith also be used in nonlinear state estimation. The log-densities can be approximated by orthonormal bases representations such that the filtering step can be carried out as a sum of coefficient vectors. Furthermore, an orthonormal basis in the information space induces a new vector space structure in the original space of densities, where fusion and weighting of information are the corresponding vector space operations.

Similar to the linear case, dependency structures are often too conservatively bounded by the EMD fusion rule if no further knowledge can be exploited. In the absence of cycles in the network topology, only process noise causes dependent local estimates. As a generalization of the federated

Kalman filter, the dependent local transition densities can be "decorrelated" by means of an exponential mixture transition density. More specifically, each local transition density is raised to a power in order to prevent process noise to be incorporated multiple times.

CHAPTER
5

Conclusions and Future Research

Overview of Chapter 5

This thesis has laid its focus on state estimation problems in the simultaneous presence of stochastic and set-membership uncertainties. Centralized as well as distributed and decentralized estimation architectures have been studied, each of which imposes different requirements and constraints on the design of estimation algorithms. The result of an estimator is intended to be accompanied by a proper description of the involved uncertainties—for the presence of errors affecting the state estimation process prevents precise estimation results, which are less reliable if no information about involved uncertainties is available. Stochastic and set-membership uncertainties can be treated by the proposed Kalman-filter-like concept, which can also be implemented in networked systems. Unknown dependencies between locally processed data require conservative estimation methods, for which linear and nonlinear solutions have been proposed in this thesis.

5.1 Contributions in a Nutshell

The contributions of this thesis can be divided into three categories. First, an easy-to-implement estimation framework, together with a combined stochastic and set-membership uncertainty model, is proposed. Second, solutions to linear distributed and decentralized estimation problems are studied. Third, the treatment of unknown dependencies in nonlinear estimation problems is elucidated. In the following, we discuss the contributions in more detail.

5.1.1 Stochastic and Set-membership Uncertainties

The Kalman filter provides estimates with minimum mean squared error (MMSE) among all linear filters and is hence optimal when models are linear and the involved noise is Gaussian. Besides randomness, a common way to characterize uncertain quantities is the membership to a bounded set. Unknown but bounded parameters and perturbations represent a frequent problem in numerous applications. This thesis revealed that a combination of random and set-membership quantities can be achieved in a number of different ways. In particular, the addition of a Gaussian random vector and an ellipsoidal error bound can easily be expressed in terms of a set of translated versions of a Gaussian density. With this combination, a first generalization of the Kalman filtering scheme has been proposed, where the estimate is given by an ellipsoidal set of means and a covariance matrix. The shape matrix of the ellipsoid accounts for the set-membership uncertainty, and the covariance matrix represents the stochastic uncertainty. However, this set-valued Kalman filter still favors the minimization of the error covariance matrix; the shape matrix is merely a byproduct. Therefore, an alternative generalization of the Kalman filter has been studied that derives a Kalman gain minimizing the total MSE. The total MSE matrix is bounded by the sum of the error covariance matrix and the shape matrix. In contrast to a set of Gaussian densities, both matrices are considered as separate error characteristics of the point estimate. The resulting Kalman gain reveals to be a natural extension of the standard Kalman gain. In order to leave it free to the user to decide whether the stochastic or set-membership uncertainty shall primarily be minimized, an additional weighting parameter is introduced. Interestingly, for certain choices of the weighting parameter, the aforementioned set-valued Kalman filter or a purely set-membership estimator is attained, where in the latter case the covariance matrix can be regarded as a byproduct.

In general, the need for generalizations of probabilistic methods is controversially discussed. Some Bayesians are firmly convinced that uncertainties can be characterized probabilistically in any case. Even if this assertion complies with reality, it is the lack of knowledge about the underlying probabilities that clearly justifies the usage of set-membership methods. Instead of unnecessarily inflating linear estimation problems to complicated possibly nonlinear ones, the combination with set-membership methods simplifies the incorporation of any unknown error behavior, be it a systematic error such as a bias or a complicated unpredictable error. An important example for the employment of sets in purely stochastic setups has been studied against the treatment of nonlinearities in approximate Kalman filtering. Purely stochastic methods like extended or linear regression Kalman filters are complemented by bounds on linearization errors that significantly enhance the reliability of the estimation results.

Other applications prove the advantages of simultaneous stochastic and set-membership state estimation. For estimating the heart surface displacement, a reliable safety region has been computed that embraces any potential errors from the sensor devices. In model predictive control, different sources of uncertainty can efficiently be incorporated into the objective function to be optimized. In networks consisting of cheap and miniaturized sensor nodes, set-membership methods can contribute to reducing the required communication data rate and volume.

5.1.2 Distributed and Decentralized Estimation

In a network of sensors, data can either be collected and processed by a central unit or locally on each node in a distributed or decentralized fashion. Estimation algorithms need to be adapted to the considered network topology in order to allow an efficient and consistent data processing. For centralized architectures, the information form of the Kalman filter, which is an inverse covariance formulation, appears to be most appropriate since sensor data can efficiently be preprocessed along any communication path and never exceeds the dimension of the state vector. As the information filter displays a purely stochastic formulation, the incorporation of set-membership uncertainties has been elucidated in detail. A set of translated versions of a Gaussian density can be parameterized by a set of information vectors and a single information matrix. The generalized information filter then processes ellipsoids of information vectors that can be transformed back at the central node to an ellipsoidal error bound of the state estimate. The evaluation of Minkowski sums of multiple information ellipsoids can efficiently be distributed when a certain communication strategy is employed, which requires that the information matrix is computed first and send back to the fusing nodes.

The presence of a central instance that acquires and manages all the sensor data renders a thorough examination of possible dependencies unnecessary. In distributed and decentralized networks, where local estimates for the same state are processed autonomously on different nodes, the estimation quality of a centralized architecture cannot be reached in general: As the name suggests, local estimates only incorporate locally available data; the fusion of local estimates allow a considerable improvement of the estimation quality. However, although fusion of all local estimates implies that all available information is incorporated, the result is often worse than a centrally computed estimate since the conditional independence of measurements cannot be exploited in an optimal manner anymore. Local estimates are, in general, dependent of each other due to common process noise affecting each local state transition model and common sensor data that is double-counted. The fusion of estimates may yield overconfident and unreliable results if dependencies are not properly taken into account. If they are known, the Bar-Shalom/Campo formulas can be employed; if they are unknown, a suboptimal fusion result is obtained by means of the covariance intersection (CI) algorithm. An interesting fact to be emphasized is that set-membership methods are utilized to bound the effects of possible dependencies on the fusion result.

Since CI may provide very conservative fusion results, techniques have been proposed that allow to benefit from additional knowledge about the underlying dependency structure and to obtain less suboptimal fusion results. The conditional independence of current measurements can even be exploited automatically to tighten the bounds on unknown dependencies. Several further possibilities have been discussed to explicitly exploit knowledge about independence, such as conditional independence over a certain horizon of time steps or the absence of cycles in the network topology such that data is not double-counted.

5.1.3 Nonlinear Dependencies

In nonlinear estimation problems, dependencies cannot simply be represented by cross-covariance matrices, as it is the case for linear estimation problems. The first difficulty is to approximate and parameterize local estimates that are represented by conditional densities. The second difficulty is to determine and parameterize possible dependencies between these estimates. Dependencies are characterized by joint probability density functions. Even if the marginals, i.e., local estimates, are given, the spectrum of possible joint probability densities is not definable. As a result, a wide range of fusion results is possible, for which a conservative bound is to be derived if the dependencies are unknown. Fortunately, in the considered network topologies, this range can be cut down to dependencies that arise from common process noise and common sensor data. With exponential mixture densities (EMD), which embody a generalization of the CI algorithm, a conservative fusion rule can be defined that accounts for the aforementioned sources of dependencies.

The EMD fusion rule is also related to state-space transformations to higher-dimensional spaces, where the state transition and, in particular, the sensor models can be represented by linear models. Under certain conditions, the estimated conditional densities become pseudo Gaussian densities in the transformed state space. Applying the CI algorithm in the transformed state space directly coincides with the EMD fusion rule applied to the original densities. This relation reveals that fusion by means of EMDs guarantees covariance consistent results in any transformed state space, where the estimation problem becomes linear.

A second direction consists of transformations being directly applied to probability densities. By considering logarithms of densities, a nonlinear information space can be defined that induces a new vector space structure on the set of probability density function. When densities are approximated by truncated orthogonal series expansions in the information space, the filtering and fusion of information is carried out by means of simple additions of coefficient vectors. The EMD fusion rule becomes a simple convex combination in this information space. The same applies for the linear information space and the CI algorithm.

In many networks, common process noise is the only cause of interdependence. Analogously to the federated Kalman filter, the joint transition density can be approximated by a product of local transition densities. Each node employs one factor transition density for local prediction steps. The nonlinear federated Kalman filter ensures that fusion results are not overconfident but less conservative than EMD fusion results.

5.2   Directions for Future Research

"Science is always wrong. It never solves a problem without creating ten more." (George Bernard Shaw) We consider only three of them in this section. The results of this thesis reveal the following, particularly promising directions for prospective research.

5.2.1   Further Development of Stochastic and Set-membership State Estimation

The adjustable Kalman gains from Section 2.5 offer the degree of freedom to give priority either to the minimization of the stochastic uncertainty or to the minimization of the set-membership uncertainty. The adjustment can be made in each filtering step but cannot be revised at a later time step. An idea to warrant adjustability over the entire time horizon consists of employing two estimators in parallel, where the first one optimizes the covariance matrix and the other one optimizes the shape matrix. With bookkeeping of the correlations between both estimates, they can be fused at any time step by employing the Bar-Shalom/Campo formulas from Section 3.5.1-B. For the fusion, again an adjustable gain allows to balance between minimum stochastic and minimum set-membership error characteristics, and hence an adjustment is possible at any later time step.

A further direction is the stochastic and set-membership information filter, where in Section 3.4 only the covariance matrix is minimized, i.e., the information filter is a reformulation of the Kalman filter in Section 2.4.1, where only the standard Kalman gain has been considered. An information form of the advanced scheme from Section 2.5 is also desirable. First, a reformulation in terms of inverse covariance matrices as well as inverse shape matrices is to be found. Second, the Minkowski sum of multiple set-membership error bounds has to be computed efficiently in a distributed fashion, which is the more complicated part.

5.2.2 Simplification of Nonlinear Estimation Problems with Sets of Densities

Nonlinear state estimation comprises the problem of computing conditional densities of the state to be estimated. Unfortunately, closed-form solutions are a rarity and only available for very specific problems. Often, the complexity of the density representation increases continually with an advancing number of prediction and filtering steps. So, for almost any application, repeated reapproximations are inevitable, which can significantly reduce the reliability of the estimation results. An interesting direction to reduce the complexity has been pointed out in [106]. Densities difficult to represent and parameterize are bounded from below and above by means of two simple densities. Similarly, the estimation concept based on sets of probability densities from Section 2.3 can be employed. If, in particular, the true density lies in the convex hull of a set of Gaussian densities as in Section 2.4.1, a simple Kalman filtering scheme can be employed to process these densities. This idea is a generalization of the treatment of nonlinearities in Section 2.6, where the effects of nonlinearities are bounded by sets.

A similar approach is to characterize a density only by its first m moments and to consider the set of all densities that are compatible with these moments [190]. For instance, the Kalman filter can be applied if only the first two moments are known and represents the best linear estimator. However, the optimal strategy is to apply Bayes' rule elementwise to all compatible densities with the same first two moments and to derive a simple parameterization of the set of posterior densities. In many cases, this is the simpler approach, compared to an approximation of the true density.

5.2.3   Copula-based Dependencies and a Notion of Conservativeness

Chapter 4 demonstrates that the treatment of dependencies among estimated probability densities is a tremendous problem. With copulas [136], dependencies between two n_x-dimensional random vectors can be represented by a function defined over a unit hypercube $[0, 1]^{2n_x}$. Hence, an approximation and finite-dimensional parameterization of a copula is easier to achieve than for the corresponding joint density function, which is defined over the entire, possibly unbounded domain. It appears to be particularly promising to study how prediction and filtering steps affect the underlying copula in order to be capable of keeping track of the dependency structure. The copula can then be exploited to fuse local estimates, i.e., the marginals, by considering the corresponding copula densities. If the underlying copula is unknown, a conservative fusion result can be attained with the aid of maximum entropy copulas [146]. More precisely, the "most uncertain" copula with respect to an appropriate information measure is chosen for given marginals.

The minimum volume confidence set for a given probability level corresponds to a level set of the estimated probability density [137]. This relation can be useful to define conservativeness. For a fixed probability level, the according level set of a conservative estimate should include the level set of the true density. Also, the fusion of estimates might be definable through level sets, for instance, by computing intersections of level sets. In consequence, such a fusion method would link nonlinear stochastic state estimation with set-membership estimation methods, which resembles the covariance intersection algorithm, where covariance ellipsoids are intersected.

Lists of Figures and Examples

List of Figures

List of Examples

Bibliography

[1] J. Aitchison. *The Statistical Analysis of Compositional Data.* Monographs on Statistics and Applied Probability. Chapman Hall Ltd., London (UK), 1986.

[2] J. Ajgl, M. Šimandl, and J. Duník. Millman's Formula in Data Fusion. In *Proceedings of the 10th International PhD Workshop on Systems and Control*, pages 1–6, Hluboká nad Vltavou, Czech Republic, Sept. 2009. Institute of Information Theory and Automation, Academy of Sciences of the Czech Republic.

[3] D. L. Alspach and H. W. Sorenson. Nonlinear Bayesian Estimation Using Gaussian Sum Approximations. *IEEE Transactions on Automatic Control*, 17(4):439–448, Aug. 1972.

[4] M. Althoff, C. L. Guernic, and B. H. Krogh. Reachable Set Computation for Uncertain Time-varying Linear Systems. In *Proceedings of the 14th International Conference on Hybrid Systems: Computation and Control*, Chicago, Illinois, USA, Apr. 2011.

[5] T. Bailey, M. Bryson, H. Mu, J. Vial, L. McCalman, and H. Durrant-Whyte. Decentralised Cooperative Localisation for Heterogeneous Teams of Mobile Robots. In *Proceedings of the 2011 IEEE International Conference on Robotics and Automation (ICRA 2011)*, Shanghai, China, May 2011.

[6] T. Bailey, S. J. Julier, and G. Agamennoni. On Conservative Fusion of Information with Unknown Non-Gaussian Dependence. In *Proceedings of the 15th International Conference on Information Fusion (Fusion 2012)*, Singapore, July 2012.

[7] Y. Bar-Shalom and L. Campo. On the Track-to-Track Correlation Problem. *IEEE Transactions on Automatic Control*, 26(2):571–572, Apr. 1981.

[8] Y. Bar-Shalom and L. Campo. The Effect of the Common Process Noise on the Two-Sensor Fused-Track Covariance. *IEEE Transactions on Aerospace and Electronic Systems*, 22(6):803–805, Nov. 1986.

[9] Y. Bar-Shalom and X.-R. Li. *Multitarget Multisensor Tracking: Principles and Techniques*. YBS Publishing, 1995.

[10] S. Basu. Variations of Posterior Expectations for Symmetric Unimodal Priors in a Distribution Band. Technical report, Department of Statistics, Purdue University, 1991.

[11] S. Basu. Ranges of Posterior Probability over a Distribution Band. *Journal of Statistical Planning and Inference*, 44:149–166, 1995.

[12] R. Beard, J. Gunther, J. Lawton, and W. Stirling. Nonlinear Projection Filter Based on Galerkin Approximation. *AIAA Journal of Guidance Control and Dynamics*, 22(2):258–266, 1999.

[13] E. F. Beckenbach and R. E. Bellman. *Inequalities*. Springer-Verlag, 1965.

[14] Y. Ben-Haim. *Engineering Design Reliability Handbook*, chapter Info-gap Decision Theory For Engineering Design. Or: Why 'Good' is Preferable to 'Best', pages 11.1–11.30. CRC Press, Boca Raton, 2005.

[15] Y. Ben-Haim. *Info-Gap Decision Theory: Decisions Under Severe Uncertainty*. Academic Press, 2. edition, 2006.

[16] A. R. Benaskeur. Consistent Fusion of Correlated Data Sources. In *Proceedings of the 2002 28th Annual Conference of the IEEE Industrial Electronics Society (IECON2002)*, volume 4, pages 2652–2656, Nov. 2002.

[17] A. Benavoli. *Modelling and Efficient Fusion of Uncertain Information: Beyond the Classical Probability Approach*. Ph.D. thesis, University of Florence, Dec. 2007.

[18] A. Benavoli, M. Zaffalon, and E. Miranda. Reliable Hidden Markov Model Filtering Through Coherent Lower Previsions. In *Proceedings of the 12th International Conference on Information Fusion (Fusion 2009)*, pages 1743–1750, July 2009.

[19] A. Benavoli, M. Zaffalon, and E. Miranda. Robust Filtering Through Coherent Lower Previsions. *IEEE Transactions on Automatic Control*, 56(7):1567–1581, July 2011.

[20] J. O. Berger. *Statistical Decision Theory and Bayesian Analysis*. Springer Series in Statistics, 1985.

[21] J. O. Berger. An Overview of Robust Bayesian Analysis. *Test*, 3:5–124, 1994.

[22] J. O. Berger and L. M. Berliner. Robust Bayes and Empirical Bayes Analysis with ϵ-Contaminated Priors. *The Annals of Statistics*, 14(2):461–486, 1986.

[23] A. Bernardini and F. Tonon. *Bounding Uncertainty in Civil Engineering*. Springer, 2010.

[24] D. P. Bertsekas and I. B. Rhodes. Recursive State Estimation for a Set-membership Description of Uncertainty. *IEEE Transactions on Automatic Control*, 16:117–128, 1971.

[25] F. Beutler, M. F. Huber, and U. D. Hanebeck. Gaussian Filtering using State Decomposition Methods. In *Proceedings of the 12th International Conference on Information Fusion (Fusion 2009)*, Seattle, Washington, USA, July 2009.

[26] D. F. Bizup and D. E. Brown. The Over-Extended Kalman Filter – Don't Use It! In *Proceedings of the Sixth International Conference on Information Fusion (Fusion 2003)*, pages 40–46, Cairns, Australia, July 2003.

[27] F. Bourgault and H. F. Durrant-Whyte. Communication in General Decentralized Filters and the Coordinated Search Strategy. In *Proceeedings of the 7th International Conference on Information Fusion (Fusion 2004)*, Stockholm, Sweden, June 2004.

[28] S. Boyd, L. E. Ghaoui, E. Feron, and V. Balakrishnan. *Linear Matrix Inequalities in System and Control Theory*, volume 15 of *Studies in Applied Mathematics*. Society for Industrial and Applied Mathematics (SIAM), Philadelphia, Pennsylvania, USA, June 1994.

[29] D. Brunn, F. Sawo, and U. D. Hanebeck. Efficient Nonlinear Bayesian Estimation based on Fourier Densities. In *Proceedings of the 2006 IEEE International Conference on Multisensor Fusion and Integration for Intelligent Systems (MFI 2006)*, pages 312–322, Heidelberg, Germany, Sept. 2006.

[30] D. Brunn, F. Sawo, and U. D. Hanebeck. Nonlinear Multidimensional Bayesian Estimation with Fourier Densities. In *Proceedings of the 2006 IEEE Conference on Decision and Control (CDC 2006)*, pages 1303–1308, San Diego, California, USA, Dec. 2006.

[31] J. Capitán, L. Merino, F. Caballero, and A. Ollero. Delayed-state Information Filter for Cooperative Decentralized Tracking. In *Proceedings of the 2009 IEEE International Conference on Robotics and Automation (ICRA 2009)*, Kobe, Japan, May 2009.

[32] J. Capitán, L. Merino, F. Caballero, and A. Ollero. Decentralized Delayed-State Information Filter (DDSIF): A New Approach for Cooperative Decentralized Tracking. *Robotics and Autonomous Systems*, 59(6):376–388, June 2011.

[33] N. A. Carlson. Federated Filter for Fault-tolerant Integrated Navigation Systems. In *Proceedings of the IEEE Position Location and Navigation Symposium (PLANS'88)*, pages 110–119, Orlando, Florida, USA, 1988. Record 'Navigation into the 21st Century' (IEEE Cat. No.88CH2675-7).

[34] N. A. Carlson. Information-sharing Approach to Federated Kalman Filtering. In *Proceedings of the IEEE Aerospace and Electronics Conference 1988 (NAECON 1988)*, Dayton, Ohio, USA, May 1988.

[35] N. A. Carlson. Federated Square Root Filter for Decentralized Parallel Processors. *IEEE Transactions on Aerospace and Electronic Systems*, 26:517–525, 1990.

[36] K. Casey, A. Lim, and G. Dozier. A Sensor Network Architecture for Tsunami Detection and Response. *International Journal of Distributed Sensor Networks*, 4(1):27–42, Jan. 2008. Selected Papers in Innovations and Real-Time Applications of Distributed Sensor Networks.

[37] K.-C. Chang, R. K. Saha, and Y. Bar-Shalom. On Optimal Track-to-Track Fusion. *IEEE Transactions on Aerospace and Electronic Systems*, 33(4):1271–1276, Oct. 1997.

[38] L. Chen, P. O. Arambel, and R. K. Mehra. Fusion under Unknown Correlation – Covariance Intersection as a Special Case. In *Proceedings of the Fifth International Conference on Information Fusion (Fusion 2002)*, Annapolis, Maryland, USA, July 2002.

[39] F. L. Chernousko. Guaranteed Estimation of Reachable Sets for Controlled Systems. In V. Arkin, A. Shiraev, and R. Wets, editors, *Stochastic Optimization*, volume 81 of *Lecture Notes in Control and Information Sciences*, pages 619–631. Springer Berlin / Heidelberg, 1986.

[40] F. L. Chernousko. *State Estimation for Dynamic Systems*. CRC Press, 1994.

[41] C.-Y. Chong and S. Mori. Convex Combination and Covariance Intersection Algorithms in Distributed Fusion. In *Proceedings of the 4th International Conference on Information Fusion (Fusion 2001)*, Montréal, Québec, Canada, Aug. 2001.

[42] C.-Y. Chong and S. Mori. Graphical Models for Nonlinear Distributed Estimation. In *Proceedings of the 7th International Conference on Information Fusion (Fusion 2004)*, Stockholm, Sweden, June 2004.

[43] D. Clark, S. J. Julier, R. P. S. Mahler, and B. Ristic. Robust Multi-object Sensor Fusion with Unknown Correlations. In *Sensor Signal Processing for Defence (SSPD 2010)*, London, United Kingdom, Sept. 2010.

[44] P. L. Combettes. The Foundations of Set-theoretic Estimation. *Proceedings of the IEEE*, 81(2):182–208, Feb. 1993.

[45] F. G. Cozman. Calculation of Posterior Bounds Given Convex Sets of Prior Probability Measures and Likelihood Functions. *Journal of Computational and Graphical Statistics*, 8(4):824–838, Dec. 1999.

[46] R. E. Curry, W. E. vander Velde, and J. E. Potter. Nonlinear Estimation With Quantized Measurements–PCM, Predictive Quantization, and Data Compression. *IEEE Transactions on Information Theory*, 16(2):152–161, Mar. 1970.

[47] L. DeRobertis and J. A. Hartigan. Bayesian Inference Using Intervals of Measures. *The Annals of Statistics*, 9(2):235–244, Mar. 1981.

[48] J. L. Doob. *Stochastic Processes*. John Wiley & Sons, 1995.

[49] A. Doucet, N. de Freitas, and N. Gordon, editors. *Sequential Monte Carlo Methods in Practice*, chapter 1, pages 6–12. Statistics for Engineering and Information Science. Springer-Verlag, 2001.

[50] J.-P. Drécourt, H. Madsen, and D. Rosbjerg. Bias Aware Kalman Filters: Comparison and Improvements. *Advances in Water Resources*, 29(5):707–718, May 2006.

[51] Z. Duan. *State Estimation with Unconventional and Networked Measurements*. Ph.D. thesis, University of New Orleans, Apr. 2010.

[52] Z. Duan, V. P. Jilkov, and X.-R. Li. State Estimation with Quantized Measurements: Approximate MMSE Approach. In *Proceedings of the 11th International Conference on Information Fusion (Fusion 2008)*, Cologne, Germany, July 2008.

[53] Z. Duan, X.-R. Li, and V. P. Jilkov. State Estimation with Point and Set Measurements. In *Proceedings of the 13th International Conference on Information Fusion (Fusion 2010)*, Edinburgh, United Kingdom, July 2010.

[54] J. J. Egozcue, J. L. Díaz-Barrero, and V. Pawlosky-Glahn. Hilbert Space of Probability Density Functions Based on Aitchison Geometry. *Acta Mathematica Sinica, English Series*, 22(4):1175–1182, July 2006.

[55] D. Ellsberg. *Risk, ambiguity, and the Savage Axioms*. The RAND Corporation, Santa Monica, California, USA, Aug. 1961.

[56] M. Emmelmann, B. Bochow, and C. Kellum, editors. *Vehicular Networking: Automotive Applications and Beyond (Intelligent Transportation Systems)*. John Wiley & Sons, 2010.

[57] S. Ferson, V. Kreinovich, L. Ginzburg, D. S. Myers, and K. Sentz. Constructing Probability Boxes and Dempster-Shafer Structures. Technical Report SAND2002-4015, Sandia National Laboratories, 2003.

[58] S. Ferson, V. Kreinovich, J. Hajagos, W. Oberkampf, and L. Ginzburg. Experimental Uncertainty Estimation and Statistics for Data Having Interval Uncertainty. Technical Report SAND2007-0939, Sandia National Laboratories, 2007.

[59] D. Fränken and A. Hüpper. Improved Fast Covariance Intersection for Distributed Data Fusion. In *Proceedings of the 8th International Conference on Information Fusion (Fusion 2005)*, Philadelphia, Pennsylvania, USA, July 2005.

[60] B. Friedland. Treatment of Bias in Recursive Filtering. *IEEE Transactions on Automatic Control*, 14(4):359–367, Aug. 1969.

[61] M. Fu, C. E. de Souza, and Z.-Q. Luo. Finite-Horizon Robust Kalman Filter Design. *IEEE Transactions on Signal Processing*, 49(9):2103–2112, Sept. 2001.

[62] A. Germani and C. Manes. The Polynomial Extended Kalman Filter as an Exponential Observer for Nonlinear Discrete-Time Systems. In *Proceedings of the 47th IEEE Conference on Decision and Control (CDC 2008)*, Cancún, Mexico, Dec. 2008.

[63] A. Germani, C. Manes, and P. Palumbo. Polynomial Extended Kalman Filtering for Discrete-Time Nonlinear Stochastic Systems. In *Proceedings of the 42nd IEEE Conference on Decision and Control (CDC 2003)*, Maui, Hawaii, USA, Dec. 2003.

[64] A. Germani, C. Manes, and P. Palumbo. Polynomial Extended Kalman Filter. *IEEE Transactions on Automatic Control*, 50(12):2059–2064, Dec. 2005.

[65] R. Gielen, S. Olaru, and M. Lazar. On Polytopic Approximations as a Modelling Framework for Networked Control Systems. In *Proceedings of the 2008 International Workshop on Assessment and Future Directions on Nonlinear Model Predictive Control*, Pavia, Italy, Sept. 2008.

[66] A. Gning, B. Ristic, and L. Mihaylova. Bernoulli Particle/Box-Particle Filters for Detection and Tracking in the Presence of Triple Measurement Uncertainty. *IEEE Transactions on Signal Processing*, 60(5):2138–2151, May 2012.

[67] F. Govaers, A. Charlish, and W. Koch. On the Decorrelated Distributed Kalman Filter under Measurement Origin Uncertainty. In *Proceedings of the 15th International Conference on Information Fusion (Fusion 2012)*, Singapore, July 2012.

[68] F. Govaers and W. Koch. Distributed Kalman Filter Fusion at Arbitrary Instants of Time. In *Proceedings of the 13th International Conference on Information Fusion (Fusion 2010)*, Edinburgh, United Kingdom, July 2010. IEEE.

[69] F. Govaers and W. Koch. On the Globalized Likelihood Function for Exact Track-To-Track Fusion at Arbitrary Instants of Time. In *Proceedings of the 14th International Conference on Information Fusion (Fusion 2011)*, Chicago, Illinois, USA, July 2011.

[70] M. Grabe. *Measurement Uncertainties in Science and Technology*. Springer, 2005.

[71] S. Grime and H. F. Durrant-Whyte. Data Fusion in Decentralized Sensor Networks. *Control Engineering Practice*, 2(5):849–863, Oct. 1994.

[72] U. D. Hanebeck. Recursive Nonlinear Set–Theoretic Estimation Based on Pseudo–Ellipsoids. In *Proceedings of the 2001 IEEE International Conference on Multisensor Fusion and Integration for Intelligent Systems (MFI 2001)*, pages 159–164, Baden–Baden, Germany, Aug. 2001.

[73] U. D. Hanebeck. Optimal Filtering of Nonlinear Systems Based on Pseudo Gaussian Densities. In *Proceedings of the 13th IFAC Symposium on System Identification (SYSID 2003)*, pages 331–336, Rotterdam, Netherlands, Aug. 2003.

[74] U. D. Hanebeck and K. Briechle. New Results for Stochastic Prediction and Filtering with Unknown Correlations. In *Proceedings of the 2001 IEEE International Conference on Multisensor Fusion and Integration for Intelligent Systems (MFI 2001)*, pages 147–152, Baden–Baden, Germany, Aug. 2001.

[75] U. D. Hanebeck, K. Briechle, and J. Horn. A Tight Bound for the Joint Covariance of Two Random Vectors with Unknown but Constrained Cross–Correlation. In *Proceedings of the 2001 IEEE International Conference on Multisensor Fusion and Integration for Intelligent Systems (MFI 2001)*, pages 85–90, Baden–Baden, Germany, Aug. 2001.

[76] U. D. Hanebeck and J. Horn. New Estimators for Mixed Stochastic and Set Theoretic Uncertainty Models: The Scalar Measurement Case. In *Proceedings of the 1999 IEEE Conference on Decision and Control (CDC 1999)*, pages 1934–1939, Phoenix, Arizona, USA, Dec. 1999.

[77] U. D. Hanebeck and J. Horn. New Estimators for Mixed Stochastic and Set Theoretic Uncertainty Models: The Vector Case. In *Proceedings of the 5th European Control Conference (ECC 1999)*, Karlsruhe, Germany, Sept. 1999.

[78] U. D. Hanebeck, J. Horn, and G. Schmidt. On Combining Statistical and Set Theoretic Estimation. *Automatica*, 35(6):1101–1109, June 1999.

[79] A. Hekler, M. Kiefel, and U. D. Hanebeck. Nonlinear Bayesian Estimation with Compactly Supported Wavelets. In *Proceedings of the 2010 IEEE Conference on Decision and Control (CDC 2010)*, Atlanta, Georgia, USA, Dec. 2010.

[80] T. Henningsson. Recursive State Estimation for Linear Systems with Mixed Stochastic and Set-Bounded Disturbances. In *Proceedings of the 47th IEEE Conference on Decision and Control (CDC 2008)*, Cancún, Mexico, Dec. 2008.

[81] M. Hua, T. Bailey, P. Thompson, and H. Durrant-Whyte. Decentralised Solutions to the Cooperative Multi-Platform Navigation Problem. In *IEEE Transactions on Aerospace and Electronic Systems*, volume 47, pages 1433–1449, Apr. 2011.

[82] M. F. Huber and U. D. Hanebeck. Gaussian Filter based on Deterministic Sampling for High Quality Nonlinear Estimation. In *Proceedings of the 17th IFAC World Congress (IFAC 2008)*, volume 17, Seoul, Republic of Korea, July 2008.

[83] P. J. Huber. *Robust Statistics*. Wiley, New York, 1981.

[84] M. B. Hurley. An Information Theoretic Justification for Covariance Intersection and its Generalization. In *Proceedings of the 5th International Conference on Information Fusion (Fusion 2002)*, Annapolis, Maryland, USA, July 2002.

[85] L. Jaulin, M. Kieffer, O. Didrit, and É. Walter. *Applied Interval Analysis: With Examples in Parameter and State Estimation, Robust Control and Robotics*. Springer, London, 2001.

[86] E. T. Jaynes. Confidence Intervals vs Bayesian Intervals. *Foundations of Probability Theory, Statistical Inference, and Statistical Theories of Science*, 2:175–257, 1976.

[87] A. H. Jazwinski. *Stochastic Processes and Filtering Theory*. Dover Publications Inc., 2007.

[88] A. Jøsang. Artificial Reasoning with Subjective Logic. In *Proceedings of the Second Australian Workshop on Commonsense Reasoning*, Perth, Australia, Dec. 1997.

[89] A. Jøsang. A Logic for Uncertain Probabilities. *International Journal of Uncertainty, Fuzziness and Knowledge-Based Systems*, 9(3):279–311, 2001.

[90] S. J. Julier. An Empirical Study into the Use of Chernoff Information for Robust, Distributed Fusion of Gaussian Mixture Models. In *Proceedings of the 9th International Conference on Information Fusion (Fusion 2006)*, Florence, Italy, July 2006.

[91] S. J. Julier. Fusion without Independence (keynote abstract). In *Proceedings of the IET Seminar on Tracking and Data Fusion: Algorithms and Applications*, pages 1–5, Birmingham, United Kingdom, Apr. 2008.

[92] S. J. Julier. Estimating and Exploiting the Degree of Independent Information in Distributed Data Fusion. In *Proceedings of the 12th International Conference on Information Fusion (Fusion 2009)*, Seattle, Washington, USA, July 2009.

[93] S. J. Julier, T. Bailey, and J. K. Uhlmann. Using Exponential Mixture Models for Suboptimal Distributed Data Fusion. In *IEEE Nonlinear Statistical Signal Processing Workshop (NSSPW 2006)*, pages 160–163, Cambridge, United Kingdom, Sept. 2006.

[94] S. J. Julier and J. J. LaViola Jr. An Empirical Study into the Robustness of Split Covariance Addition (SCA) for Human Motion Tracking. In *Proceedings of the 2004 American Control Conference (ACC 2004)*, Boston, Massachusetts, USA, June 2004.

[95] S. J. Julier and J. K. Uhlmann. A New Extension of the Kalman Filter to Nonlinear Systems. In *Proceedings the 11th International Symposium on Aerospace/Defense Sensing, Simulation and Controls (AeroSense 1997)*, Orlando, Florida, USA, 1997.

[96] S. J. Julier and J. K. Uhlmann. A Non-divergent Estimation Algorithm in the Presence of Unknown Correlations. In *Proceedings of the IEEE American Control Conference (ACC 1997)*, volume 4, pages 2369–2373, Albuquerque, New Mexico, USA, June 1997.

[97] S. J. Julier and J. K. Uhlmann. Unscented Filtering and Nonlinear Estimation. *Proceedings of the IEEE*, 92(3):401–422, 2004.

[98] S. J. Julier and J. K. Uhlmann. Using Covariance Intersection for SLAM. *Robotics and Autonomous Systems*, 55(1):3–20, 2007.

[99] S. J. Julier and J. K. Uhlmann. *Handbook of Multisensor Data Fusion: Theory and Practice*, chapter General Decentralized Data Fusion with Covariance Intersection, pages 319–343. CRC Press, 2nd ed. edition, 2009.

[100] S. J. Julier and J. K. Uhlmann. *Handbook of Multisensor Data Fusion: Theory and Practice*, chapter Data Fusion in Nonlinear Systems, pages 345–368. CRC Press, 2nd ed. edition, 2009.

[101] T. Kailath, A. H. Sayed, and B. Hassibi. *Linear Estimation*. Prentice Hall, 2 edition, 2000.

[102] A. Kallapur, I. R. Petersen, and S. Anavatti. A Discrete-time Robust Extended Kalman Filter. In *Proceedings of the 2009 American Control Conference (ACC 2009)*, St. Louis, Missouri, USA, June 2009.

[103] R. E. Kalman. A New Approach to Linear Filtering and Prediction Problems. *Transactions of the ASME - Journal of Basic Engineering*, pages 35–45, 1960.

[104] J. D. Kenney, D. R. Poole, G. C. Willden, B. A. Abbott, A. P. Morris, R. N. McGinnis, and D. A. Ferrill. Precise Positioning with Wireless Sensor Nodes: Monitoring Natural Hazards in All Terrains. In *Proceedings of the 2009 IEEE International Conference on Systems, Man, and Cybernetics*, San Antonio, Texas, USA, Oct. 2009.

[105] J. M. Keynes. *A Treatise on Probability*. Macmillan, London, 1921.

[106] V. Klumpp, D. Brunn, and U. D. Hanebeck. Approximate Nonlinear Bayesian Estimation Based on Lower and Upper Densities. In *Proceedings of the 9th International Conference on Information Fusion (Fusion 2006)*, Florence, Italy, July 2006.

[107] F. H. Knight. *Risk, Uncertainty, and Profit*. Boston, New York, Houghton Mifflin Company, 1921.

[108] W. Koch. On Optimal Distributed Kalman Filtering and Retrodiction at Arbitrary Communication Rates for Maneuvering Targets. In *Proceedings of the 2008 IEEE International Conference on Multisensor Fusion and Integration for Intelligent Systems (MFI 2008)*, Seoul, Republic of Korea, Aug. 2008.

[109] W. Koch. Exact Update Formulae for Distributed Kalman Filtering and Retrodiction at Arbitrary Communication Rates. In *Proceedings of the 12th International Conference on Information Fusion (Fusion 2009)*, pages 2209–2216, Seattle, Washington, USA, July 2009.

[110] K. Kowalski and W.-H. Steeb. *Nonlinear Dynamical Systems and Carleman Linearization*. World Scientific Publishing Company, Singapore, 1991.

[111] A. B. Kurzhanski and I. Vályi. *Ellipsoidal Calculus for Estimation and Control.* Birkhäuser, 1997.

[112] M. Lavine. Sensitivity in Bayesian Statistics: The Prior and the Likelihood. *Journal of the American Statistical Association*, 86(414):396–399, June 1991.

[113] C. L. Lawson and R. J. Hanson. *Solving Least Squares Problems.* Society for Industrial and Applied Mathematics (SIAM), 1995.

[114] T. Lefebvre, H. Bruyninckx, and J. de Schutter. Kalman Filters for Non-linear Systems: A Comparison of Performance. *International Journal of Control*, 77(7):639–653, May 2004.

[115] T. Lefebvre, H. Bruyninckx, and J. de Schutter. *Nonlinear Kalman Filtering for Force-Controlled Robot Tasks*, volume 19 of *Springer Tracts in Advanced Robotics*. Springer, 2005.

[116] E. L. Lehmann and G. Casella. *Theory of Point Estimation.* Springer, 2. edition, 1998.

[117] I. Levi. *The Enterprise of Knowledge: An Essay on Knowledge, Credal Probobility, and Chance.* MIT Press, 1983.

[118] M. E. Liggins, C.-Y. Chong, I. Kadar, M. G. Alford, V. Vannicola, and S. Thomopoulis. Distributed Fusion Architectures and Algorithms for Target Tracking. In *Proceedings of the IEEE*, volume 85, pages 95–107, 1997.

[119] M. E. Liggins II, D. L. Hall, and J. Llinas, editors. *Handbook of Multisensor Data Fusion: Theory and Practice.* The Electrical Engineering and Applied Signal Processing Series. CRC Press, 2nd ed. edition, 2009.

[120] R. P. S. Mahler. Optimal/Robust Distributed Data Fusion: A Unified Approach. In *Proceedings of SPIE*, volume 4052, 2000.

[121] R. P. S. Mahler. *Statistical Multisource-Multitarget Information Fusion.* Artech House, 2007.

[122] R. P. S. Mahler. General Bayes Filtering of Quantized Measurements. In *Proceedings of the 14th International Conference on Information Fusion (Fusion 2011)*, Chicago, Illinois, USA, July 2011.

[123] A. Makarenko and H. F. Durrant-Whyte. Decentralized Bayesian Algorithms for Active Sensor Networks. *Information Fusion*, 7(4):418–433, Dec. 2006. Special Issue on the Seventh International Conference on Information Fusion – Part I.

[124] E. S. Manolakos, E. Logaras, and F. Paschos. Wireless Sensor Network Application for Fire Hazard Detection and Monitoring. In N. Komninos, O. Akan, P. Bellavista, J. Cao, F. Dressler, D. Ferrari, M. Gerla, H. Kobayashi, S. Palazzo, S. Sahni, X. S. Shen, M. Stan, J. Xiaohua, A. Zomaya, and G. Coulson, editors, *Sensor Applications, Experimentation, and Logistics*, Lecture Notes of the Institute for Computer Sciences, Social Informatics and Telecommunications Engineering. Springer Berlin Heidelberg, 2010.

[125] M. Milanese and A. Vicino. Optimal Estimation Theory for Dynamic Systems with Set-membership Uncertainty: An Overview. *Automatica*, 27(6):997–1009, Nov. 1991.

[126] E. Miranda. A Survey of the Theory of Coherent Lower Previsions. *International Journal of Approximate Reasoning*, 48(2):628–658, 2008.

[127] E. Miranda, I. Couso, and P. Gil. Random Sets as Imprecise Random Variables. *Journal of Mathematical Analysis and Applications*, 307(1):32–47, July 2005.

[128] E. Miranda, I. Couso, and P. Gil. Approximations of Upper and Lower Probabilities by Measurable Selections. *Information Sciences*, 180(8):1407–1417, Apr. 2010.

[129] E. Miranda and G. de Cooman. Marginal Extension in the Theory of Coherent Lower Previsions. *International Journal of Approximate Reasoning*, 46(1):188–225, Sept. 2007. Special Section: Random Sets and Imprecise Probabilities (Issues in Imprecise Probability).

[130] D. R. Morrell and W. C. Stirling. Convex Bayes Decision Theory. *IEEE Transactions on Systems, Man, and Cybernetics*, 21(1):173–183, Jan. 1991.

[131] D. R. Morrell and W. C. Stirling. Set-Valued Filtering and Smoothing. *IEEE Transactions on Systems, Man, and Cybernetics*, 21(1):184–193, Jan. 1991.

[132] D. R. Morrell and W. C. Stirling. An Extended Set-Valued Kalman Filter. In *Proceedings of the Third International Symposium on Imprecise Probabilities and Their Applications (ISIPTA 2003)*, pages 395–405, Lugano, Switzerland, July 2003.

[133] A. G. O. Mutambara. *Decentralized Estimation and Control for Multisensor Systems*. CRC Press, Inc., Boca Raton, Florida, USA, 1998.

[134] R. B. Nelsen. *An Introduction to Copulas*. Springer Series in Statistics. Springer, 2nd edition, 2006.

[135] H. T. Nguyen. *An Introduction to Random Sets*. Chapman and Hall, 2006.

[136] W. Niehsen. Information Fusion based on Fast Covariance Intersection Filtering. In *Proceedings of the 5th International Conference on Information Fusion (Fusion 2002)*, Annapolis, Maryland, USA, July 2002.

[137] J. Nuñez Garcia, Z. Kutalik, K.-H. Cho, and O. Wolkenhauer. Level Sets and Minimum Volume Sets of Probability Density Functions. *International Journal of Approximate Reasoning*, 34(1):25–47, Sept. 2003.

[138] R. Olfati-Saber. Distributed Kalman Filtering for Sensor Networks. In *Proceedings of the 46th IEEE Conference on Decision and Control (CDC 2007)*, pages 5492–5498, New Orleans, Louisiana, USA, Dec. 2007.

[139] L.-L. Ong, M. Ridley, B. Upcroft, S. Kumar, T. Bailey, S. Sukkarieh, and H. Durrant-Whyte. A Comparison of Probabilistic Representations for Decentralised Data Fusion. In *Proceedings of the 2005 International Conference on Intelligent Sensors, Sensor Networks and Information Processing Conference (ISSNIP 2005)*, pages 187–192, Dec. 2005.

[140] L.-L. Ong, B. Upcroft, M. Ridley, T. Bailey, S. Sukkarieh, and H. Durrant-Whyte. Consistent Methods for Decentralised Data Fusion using Particle Filters. In *Proceedings of the 2006 IEEE International Conference on Multisensor Fusion and Integration for Intelligent Systems (MFI 2006)*, pages 85–91, Heidelberg, Germany, Sept. 2006.

[141] A. Papoulis and S. U. Pillai. *Probability, Random Variables, and Stochastic Processes*. Mcgraw-Hill Publ.Comp., 4 edition, 2002.

[142] I. R. Petersen, M. R. James, and P. Dupuis. Minimax Optimal Control of Stochastic Uncertain Systems with Relative Entropy Constraints. *IEEE Transactions on Automatic Control*, 45:398–412, 2000.

[143] K. B. Petersen and M. S. Pedersen. The Matrix Cookbook. Online, Technical University of Denmark, Oct. 2008.

[144] K. S. J. Pister, J. M. Kahn, and B. E. Boser. Smart Dust: Wireless Networks of Millimeter-Scale Sensor Nodes. Highlight Article in 1999 Electronics Research Laboratory Research Summary, University of California, Berkely, USA, 1999.

[145] D.-B. Pougaza and A. Mohammad-Djafari. Maximum Entropies Copulas. In *American Institute of Physics Conference Proceedings*, volume 1305 of *Mathematical and Statistical Physics*, pages 329–336, July 2010.

[146] D.-B. Pougaza, A. Mohammad-Djafari, and J.-F. Bercher. Link Between Copula and Tomography. *Pattern Recognition Letters*, 31(12):2258–2264, Oct. 2010.

[147] A. Rauh and U. D. Hanebeck. Moment–Based Prediction Step for Nonlinear Discrete-Time Dynamic Systems Using Exponential Densities. In *Proceedings of the 44th IEEE Conference on Decision and Control and European Control Conference (CDC-ECC 2005)*, Sevilla, Spain, Dec. 2005.

[148] S. Reece and S. Roberts. Robust, Low-Bandwidth, Multi-Vehicle Mapping. In *Proceedings of the 8th International Conference on Information Fusion (Fusion 2005)*, volume 2, Philadelphia, Pennsylvania, USA, July 2005.

[149] B. Ristic. Bayesian Estimation With Imprecise Likelihoods: Random Set Approach. *IEEE Signal Processing Letters*, 18(7):395–398, July 2011.

[150] J. A. Roecker and C. D. McGillem. Comparison of Two-sensor Tracking Methods Based on State Vector Fusion and Measurement Fusion. *IEEE Transactions on Aerospace and Electronic Systems*, 24(4):447–449, July 1988.

[151] L. Ros, A. Sabater, and F. Thomas. An Ellipsoidal Calculus Based on Propagation and Fusion. *IEEE Transactions on Systems, Man, and Cybernetics*, 32(4):430–442, Aug. 2002.

[152] F. Salmon. Recipe for Disaster: The Formula That Killed Wall Street. *Wired Magazine*, 17.03, Mar. 2009.

[153] F. Sawo, D. Brunn, and U. D. Hanebeck. Parameterized Joint Densities with Gaussian and Gaussian Mixture Marginals. In *Proceedings of the 9th International Conference on Information Fusion (Fusion 2006)*, Florence, Italy, July 2006.

[154] F. C. Schweppe. *Uncertain Dynamic Systems*. Prentice-Hall, 1973.

[155] G. Shafer. *A Mathematical Theory of Evidence*. Princeton University Press, 1976.

[156] U. Shaked and C. E. de Souza. Robust Minimum Variance Filtering. *IEEE Transactions on Signal Processing*, 43(11):2474–2483, Nov. 1995.

[157] V. Shin, Y. Lee, and T.-S. Choi. Generalized Millman's Formula and Its Application for Estimation Problems. *Signal Processing*, 86(2):257–266, 2006.

[158] J. Sijs. *State Estimation in Networked Systems*. Ph.D. thesis, Technische Universiteit Eindhoven, Apr. 2012.

[159] J. Sijs and M. Lazar. Empirical Case-studies of State Fusion via Ellipsoidal Intersection. In *Proceedings of the 14th International Conference on Information Fusion (Fusion 2011)*, Chicago, Illinois, USA, July 2011.

[160] J. Sijs, M. Lazar, and P. P. J. van den Bosch. State-Fusion with Unknown Correlation: Ellipsoidal Intersection. In *Proceedings of the 2010 American Control Conference (ACC 2010)*, Baltimore, Maryland, USA, June 2010.

[161] D. Simon. *Optimal State Estimation: Kalman, H Infinity, and Nonlinear Approaches*. John Wiley & Sons, 2006.

[162] P. Smets. Imperfect Information: Imprecision and Uncertainty. In A. Motro and P. Smets, editors, *Uncertainty Management in Information Systems*. Springer, 1996.

[163] H. W. Sorenson. Least-Squares Estimation: From Gauss to Kalman. *IEEE Spectrum*, 7(7):63–68, July 1970.

[164] J. R. Taylor. *Introduction to Error Analysis: The Study of Uncertainties in Physical Measurements*. University Science Books, Sausalito, California, USA, 1997.

[165] B. Tonkes and A. D. Blair. Decentralised Data Fusion with Exponentials of Polynomials. In *Proceedings of the 2007 IEEE/RSJ International Conference on Intelligent Robots and Systems (IROS 2007)*, San Diego, California, USA, Oct. 2007.

[166] B. Tonkes and A. D. Blair. Deriving Sensor Models and Non-linear Filtering for Exponentials of Polynomials. In *Proceedings of the 2007 Australasian Conference on Robotics and Automation (ARAA 2007)*, Brisbane, Australia, Dec. 2007.

[167] J. K. Uhlmann. General Data Fusion for Estimates with Unknown Cross Covariances. In *Proceedings of SPIE Aerosense Conference*, volume 2755, pages 536–547, Orlando, Florida, USA, Apr. 1996.

[168] M. Üney, D. E. Clark, and S. J. Julier. Information Measures in Distributed Multitarget Tracking. In *Proceedings of the 14th International Conference on Information Fusion (Fusion 2011)*, Chicago, Illinois, USA, July 2011.

[169] P. Walley. *Statistical Reasoning with Imprecise Probabilities*, volume 42 of *Monographs on Statistics and Applied Probability*. Chapman and Hall, 1991.

[170] Y. Wang and X.-R. Li. A Fast and Fault-Tolerant Convex Combination Fusion Algorithm under Unknown Cross-Correlation. In *Proceedings of 12th International Conference on Information Fusion (Fusion 2009)*, Seattle, Washington, USA, July 2009.

[171] Y. Wang and X.-R. Li. Distributed Estimation Fusion under Unknown Cross-correlation: An Analytic Center Approach. In *Proceedings of the 13th International Conference on Information Fusion (Fusion 2010)*, Edinburgh, United Kingdom, July 2010.

[172] B. A. Warneke, M. D. Scott, B. S. Leibowitz, L. Zhou, C. L. Bellew, J. A. Chediakt, J. M. Kahn, B. E. Boser, and K. S. Pister. An Autonomous 16mm^3 Solar-Powered Node for Distributed Wireless Sensor Networks. In *Proceedings of IEEE Sensors 2002*, pages 1510–1515, Orlando, Florida, USA, June 2002.

[173] K. Weichselberger. The Theory of Interval-probability as a Unifying Concept for Uncertainty. *International Journal of Approximate Reasoning*, 24:149–170(22), May 2000.

[174] K. Weichselberger. *Elementare Grundbegriffe einer allgemeineren Wahrscheinlichkeitsrechnung I*. Heidelberg: Physica-Verlag, 2001.

[175] K. Weichselberger and T. Augustin. On the Symbiosis of Two Concepts of Conditional Interval Probability. In *Proceedings of the Third International Symposium on Imprecise Probabilities and Their Applications (ISIPTA 2003)*, pages 608–629, 2003.

[176] G. Werner-Allen, J. Johnson, M. Ruiz, J. Lees, and M. Welsh. Monitoring Volcanic Eruptions with a Wireless Sensor Network. In *Proceeedings of the Second European Workshop on Wireless Sensor Networks (EWSN 2005)*, pages 108–120, Istanbul, Turkey, Feb. 2005.

[177] M. A. Woodbury. *Inverting Modified Matrices*. Number 42 in Statistical Research Group, Memorandum Report. Princeton University, Princeton, New Jersey, USA, 1950.

[178] L. A. Zadeh. Fuzzy Sets. *Information and Control*, 8(3):338–353, 1965.

[179] L. A. Zadeh. Book Review: A Mathematical Theory of Evidence. *AI Magazine*, pages 81–83, 1984.

Supervised Student Theses

[180] M. Berg. Bewertung kartenbasierter Steigungsprofilinformation für prädiktive Applikationen im Fahrzeug. Diploma thesis, Intelligent Sensor-Actuator-Systems Laboratory, Karlsruhe Institute of Technology (KIT), 2010.

[181] D. Itte. Assoziationsfreie Methoden zum Tracking von mehreren Objekten. Diploma thesis, Intelligent Sensor-Actuator-Systems Laboratory, Karlsruhe Institute of Technology (KIT), 2011.

[182] M. Nagel. Approximation von Wahrscheinlichkeitsdichten mithilfe einer Verallgemeinerung der Aitchison Geometrie. Minor thesis, Intelligent Sensor-Actuator-Systems Laboratory, Karlsruhe Institute of Technology (KIT), 2010.

[183] N. Petkov. Schätzen mit Mengen von Wahrscheinlichkeitsdichten unter Berücksichtigung von Linearisierungsfehlern. Minor thesis, Intelligent Sensor-Actuator-Systems Laboratory, Karlsruhe Institute of Technology (KIT), 2010.

[184] F. Pfaff. Weiterentwicklung mengenbasierter Kalman-Filtermethoden für Multi-Sensorsysteme. Diploma thesis, Intelligent Sensor-Actuator-Systems Laboratory, Karlsruhe Institute of Technology (KIT), 2011.

[185] M. Reinhardt. Fusion unsicherer Information unter unbekannten Korrelationen. Minor thesis, Intelligent Sensor-Actuator-Systems Laboratory, Karlsruhe Institute of Technology (KIT), 2010.

[186] M. Reinhardt. Strategien zur Identifikation von Abhängigkeiten und zur dezentralen Informationsfusion in verteilten Sensornetzen. Diploma thesis, Intelligent Sensor-Actuator-Systems Laboratory, Karlsruhe Institute of Technology (KIT), 2011.

Own Publications

[187] M. Baum, B. Noack, F. Beutler, D. Itte, and U. D. Hanebeck. Optimal Gaussian Filtering for Polynomial Systems Applied to Association-free Multi-Target Tracking. In *Proceedings of the 14th International Conference on Information Fusion (Fusion 2011)*, Chicago, Illinois, USA, July 2011.

[188] M. Baum, B. Noack, and U. D. Hanebeck. Extended Object and Group Tracking with Elliptic Random Hypersurface Models. In *Proceedings of the 13th International Conference on Information Fusion (Fusion 2010)*, Edinburgh, United Kingdom, July 2010.

[189] M. Baum, B. Noack, and U. D. Hanebeck. Random Hypersurface Mixture Models for Tracking Multiple Extended Objects. In *Proceedings of the 50th IEEE Conference on Decision and Control (CDC 2011)*, Orlando, Florida, USA, Dec. 2011.

[190] A. Benavoli and B. Noack. Pushing Kalman's Idea to the Extremes. In *Proceedings of the 15th International Conference on Information Fusion (Fusion 2012)*, Singapore, July 2012.

[191] E. Bogatyrenko, B. Noack, and U. D. Hanebeck. Reliable Estimation of Heart Surface Motion under Stochastic and Unknown but Bounded Systematic Uncertainties. In *Proceedings of the 2010 IEEE/RSJ International Conference on Intelligent Robots and Systems (IROS 2010)*, Taipei, Taiwan, Oct. 2010.

[192] A. Hekler, D. Lyons, B. Noack, and U. D. Hanebeck. Nonlinear Model Predictive Control Considering Stochastic and Systematic Uncertainties with Sets of Densities. In *Proceedings of the IEEE*

Multi-Conference on Systems and Control (MSC 2010), Yokohama, Japan, Sept. 2010.

[193] V. Klumpp, B. Noack, M. Baum, and U. D. Hanebeck. Combined Set-Theoretic and Stochastic Estimation: A Comparison of the SSI and the CS Filter. In *Proceedings of the 13th International Conference on Information Fusion (Fusion 2010)*, Edinburgh, United Kingdom, July 2010.

[194] D. Lyons, A. Hekler, B. Noack, and U. D. Hanebeck. Maße für Wahrscheinlichkeitsdichten in der informationstheoretischen Sensoreinsatzplanung. In F. Puente León, K.-D. Sommer, and M. Heizmann, editors, *Verteilte Messsysteme*, pages 121–132. KIT Scientific Publishing, Mar. 2010.

[195] D. Lyons, B. Noack, and U. D. Hanebeck. A Log-Ratio Information Measure for Stochastic Sensor Management. In *Proceedings of the IEEE International Conference on Sensor Networks, Ubiquitous, and Trustworthy Computing (SUTC 2010)*, Newport Beach, California, USA, June 2010.

[196] B. Noack, M. Baum, and U. D. Hanebeck. Automatic Exploitation of Independencies for Covariance Bounding in Fully Decentralized Estimation. In *Proceedings of the 18th IFAC World Congress (IFAC 2011)*, Milan, Italy, Aug. 2011.

[197] B. Noack, M. Baum, and U. D. Hanebeck. Covariance Intersection in Nonlinear Estimation Based on Pseudo Gaussian Densities. In *Proceedings of the 14th International Conference on Information Fusion (Fusion 2011)*, Chicago, Illinois, USA, July 2011.

[198] B. Noack, V. Klumpp, D. Brunn, and U. D. Hanebeck. Nonlinear Bayesian Estimation with Convex Sets of Probability Densities. In *Proceedings of the 11th International Conference on Information Fusion (Fusion 2008)*, pages 1–8, Cologne, Germany, July 2008.

[199] B. Noack, V. Klumpp, and U. D. Hanebeck. State Estimation with Sets of Densities considering Stochastic and Systematic Errors. In *Proceedings of the 12th International Conference on Information Fusion (Fusion 2009)*, Seattle, Washington, USA, July 2009.

[200] B. Noack, V. Klumpp, D. Lyons, and U. D. Hanebeck. Modellierung von Unsicherheiten und Zustandsschätzung mit Mengen von Wahrscheinlichkeitsdichten. *tm - Technisches Messen, Oldenbourg Verlag*, 77(10):544–550, Oct. 2010.

[201] B. Noack, V. Klumpp, D. Lyons, and U. D. Hanebeck. Systematische Beschreibung von Unsicherheiten in der Informationsfusion mit Mengen von Wahrscheinlichkeitsdichten. In F. Puente León, K.-D. Sommer, and M. Heizmann, editors, *Verteilte Messsysteme*, pages 167–178. KIT Scientific Publishing, Mar. 2010.

[202] B. Noack, V. Klumpp, N. Petkov, and U. D. Hanebeck. Bounding Linearization Errors with Sets of Densities in Approximate Kalman Filtering. In *Proceedings of the 13th International Conference on Information Fusion (Fusion 2010)*, Edinburgh, United Kingdom, July 2010.

[203] B. Noack, D. Lyons, M. Nagel, and U. D. Hanebeck. Nonlinear Information Filtering for Distributed Multisensor Data Fusion. In *Proceedings of the 2011 American Control Conference (ACC 2011)*, San Francisco, California, USA, June 2011.

[204] B. Noack, F. Pfaff, and U. D. Hanebeck. Combined Stochastic and Set-membership Information Filtering in Multisensor Systems. In *Proceedings of the 15th International Conference on Information Fusion (Fusion 2012)*, Singapore, July 2012.

[205] B. Noack, F. Pfaff, and U. D. Hanebeck. Optimal Kalman Gains for Combined Stochastic and Set-Membership State Estimation. In *Proceedings of the 51st IEEE Conference on Decision and Control (CDC 2012)*, Maui, Hawaii, USA, Dec. 2012.

[206] M. Reinhardt, B. Noack, M. Baum, and U. D. Hanebeck. Analysis of Set-theoretic and Stochastic Models for Fusion under Unknown Correlations. In *Proceedings of the 14th International Conference on Information Fusion (Fusion 2011)*, Chicago, Illinois, USA, July 2011.

[207] M. Reinhardt, B. Noack, and U. D. Hanebeck. Closed-form Optimization of Covariance Intersection for Low-dimensional Matrices. In *Proceedings of the 15th International Conference on Information Fusion (Fusion 2012)*, Singapore, July 2012.

[208] M. Reinhardt, B. Noack, and U. D. Hanebeck. Decentralized Control Based on Globally Optimal Estimation. In *Proceedings of the 51st IEEE Conference on Decision and Control (CDC 2012)*, Maui, Hawaii, USA, Dec. 2012.

[209] M. Reinhardt, B. Noack, and U. D. Hanebeck. On Optimal Distributed Kalman Filtering in Non-ideal Situations. In *Proceedings of the 15th International Conference on Information Fusion (Fusion 2012)*, Singapore, July 2012.

[210] M. Reinhardt, B. Noack, and U. D. Hanebeck. The Hypothesizing Distributed Kalman Filter. In *Proceedings of the 2012 IEEE International Conference on Multisensor Fusion and Integration for Intelligent Systems (MFI 2012)*, Hamburg, Germany, Sept. 2012.

[211] J. Schmid, F. Beutler, B. Noack, U. D. Hanebeck, and K. D. Müller-Glaser. An Experimental Evaluation of Position Estimation Methods for Person Localization in Wireless Sensor Networks. In P. J. Marrón and K. Whitehouse, editors, *Proceedings of the 8th European Conference on Wireless Sensor Networks (EWSN 2011)*, volume 6567, pages 147–162, Bonn, Germany, Feb. 2011. Springer.

Karlsruhe Series on
Intelligent Sensor-Actuator-Systems

Edited by Prof. Dr.-Ing. Uwe D. Hanebeck // ISSN 1867-3813

Band 1   **Oliver Schrempf**
Stochastische Behandlung von Unsicherheiten in
kaskadierten dynamischen Systemen. 2008
ISBN 978-3-86644-287-0

Band 2   **Florian Weißel**
Stochastische modell-prädiktive Regelung nichtlinearer
Systeme. 2009
ISBN 978-3-86644-348-8

Band 3   **Patrick Rößler**
Telepräsente Bewegung und haptische Interaktion in
ausgedehnten entfernten Umgebungen. 2009
ISBN 978-3-86644-346-4

Band 4   **Kathrin Roberts**
Modellbasierte Herzbewegungsschätzung für
robotergestützte Interventionen. 2009
ISBN 978-3-86644-353-2

Band 5   **Felix Sawo**
Nonlinear state and parameter estimation of spatially
distributed systems. 2009
ISBN 978-3-86644-370-9

Band 6   **Gregor F. Schwarzenberg**
Untersuchung der Abbildungseigenschaften eines
3D-Ultraschall-Computertomographen zur Berechnung der
3D-Abbildungsfunktion und Herleitung einer optimierten
Sensorgeometrie. 2009
ISBN 978-3-86644-393-8

Karlsruhe Series on Intelligent Sensor-Actuator-Systems

Edited by Prof. Dr.-Ing. Uwe D. Hanebeck // ISSN 1867-3813

Band 7
Marco Huber
Probabilistic Framework for Sensor Management. 2009
ISBN 978-3-86644-405-8

Band 8
Frederik Beutler
Probabilistische modellbasierte Signalverarbeitung zur
instantanen Lageschätzung. 2010
ISBN 978-3-86644-442-3

Band 9
Marc Peter Deisenroth
Efficient Reinforcement Learning using
Gaussian Processes. 2010
ISBN 978-3-86644-569-7

Band 10
Evgeniya Ballmann
Physics-Based Probabilistic Motion Compensation of
Elastically Deformable Objects. 2012
ISBN 978-3-86644-862-9

Band 11
Peter Krauthausen
Learning Dynamic Systems for Intention Recognition in
Human-Robot-Cooperation. 2013
ISBN 978-3-86644-952-7

Band 12
Antonia Pérez Arias
Haptic Guidance for Extended Range Telepresence. 2013
ISBN 978-3-7315-0035-3

Band 13
Marcus Baum
Simultaneous Tracking and Shape Estimation
of Extended Objects. 2013
ISBN 978-3-7315-0078-0

Die Bände sind unter www.ksp.kit.edu als PDF frei verfügbar oder als Druckausgabe bestellbar.

Karlsruhe Series on
Intelligent Sensor-Actuator-Systems

Edited by Prof. Dr.-Ing. Uwe D. Hanebeck // ISSN 1867-3813

Band 14   **Benjamin Noack**
State Estimation for Distributed Systems with Stochastic
and Set-membership Uncertainties. 2014
ISBN 978-3-7315-0124-4

Die Bände sind unter www.ksp.kit.edu als PDF frei verfügbar oder als Druckausgabe bestellbar.